Human Developmental Toxicants
Aspects of Toxicology and Chemistry

James L. Schardein & Orest T. Macina

To My Good Friend Gary with Respect Regards, Jim Schard

CRC Taylor & Francis
Taylor & Francis Group
Boca Raton   London   New York

CRC is an imprint of the Taylor & Francis Group,
an informa business

CRC Press
Taylor & Francis Group
6000 Broken Sound Parkway NW, Suite 300
Boca Raton, FL 33487-2742

International Standard Book Number-10: 0-8493-7229-1 (Hardcover)
International Standard Book Number-13: 978-0-8493-7229-2 (Hardcover)

Library of Congress Cataloging-in-Publication Data

Schardein, James L.
    Human developmental toxicants : aspects of toxicology and chemistry / James L. Schardein, Orest T. Macina.
        p. cm.
    Includes bibliographical references and index.
    ISBN 0-8493-7229-1 -- ISBN 1-4200-0675-4
    1. Pediatric toxicology. 2. Fetus--Effect of drugs on. 3. Fetus--Effect of chemicals on. I. Macina, Orest T. II. Title.

RG627.6.D79S35 2006
618.92'98--dc22   2006044602

Visit the Taylor & Francis Web site at
http://www.taylorandfrancis.com

and the CRC Press Web site at
http://www.crcpress.com

Foreword

Much attention has focused on the identification of drugs and chemicals that produce malformations following human exposure during *in utero* development. However, as noted by the authors of this monograph, that is only one of the four types of adverse effects that may occur following exposure (or treatment) during development. Over the past several decades, clinicians and developmental scientists have established that developmental toxicity includes not only structural malformations but also growth retardation and death, as well as functional (including behavioral) abnormalities. Research by these clinicians and developmental scientists has also pointed out that vulnerable periods for developmental toxicology may begin prior to conception and extend well beyond birth.

The work of Schardein and Macina in this monograph provides a unique resource that links chemistry with developmental toxicity profiles of the pharmaceuticals and industrial chemicals that represent the majority of presently known human developmental toxicants to which pregnant women may be exposed, either therapeutically or through the workplace or home environment. The use of *human* data as the initial source of comparison of toxicological and chemical properties is logical, because the target of toxicity of greatest priority is the human species. Human data are supplemented with available animal data for comparative purposes and to discern any "animal models" of the corresponding human effect. The chemistry component entails the chemical structure as well as a set of computationally calculated physicochemical and topological parameters that represent the steric, transport, and electronic properties of the selected molecules. The inclusion of chemical property data represents a new focus on attempts to understand chemically induced developmental toxicity.

As significant as this work is in assisting our understanding of developmental toxicology, it is also essential to note that we are just at the threshold. Much remains to be done to improve our ability to understand why and how a chemical may alter the many different steps occurring during development. The calculated properties presented within this monograph (and on the accompanying CD) can be utilized by interested investigators in deriving structure–activity relationship (SAR) models linking the chemical structure and properties with the observed human and animal developmental toxicity data. Successful SAR models for developmental toxicity would be an invaluable adjunct to the risk assessment process as well as in the investigation of the mechanistic basis of developmental events.

This critical work will improve both our ability to predict chemicals that may produce developmental toxicity as well as to provide insight into the chemical properties responsible for the observed effects on human development.

Donald R. Mattison, M.D.
Captain, U.S. Public Health Service
Senior Advisor to the Directors of the National Institute of Child Health and
Human Development (NICHD) and the Center for Research for Mothers and Children (CRMC)
Branch Chief, Obstetric and Pediatric Pharmacology Branch
National Institutes of Health, U.S. Department of Health and Human Services

Preface

The Human Developmental Toxicants Database (HumDevTox) is a chemical structure–chemical property–biological activity database for 50 known agents that adversely affect human development as a result of exposure prior to conception or during prenatal and postnatal development. The developmental effects elicited include growth retardation, death, structural abnormality, and functional deficits. These effects vary from single endpoints of marginal or even questionable validity, to severe, proven effects of teratogenesis or death. The database also includes available animal data for each of the human developmental toxicants identified and discussed in this book.

The electronic component of the database consists of three-dimensional structures and 49 calculated physicochemical and topological properties for each of the agents. The complete database is in the form of an SD file, and it includes the three-dimensional chemical structures, calculated physicochemical and topological properties, and the associated biological data in humans and animals. The construction of a database consisting of the chemical structures and properties of human developmental toxicants and the associated animal developmental data provides a valuable resource for the biomedical scientific community. To our knowledge, a detailed database such as this for human developmental toxicity does not exist in the public domain. This unique database will serve as a reference source for toxicologists, teratologists, chemists, and other scientists interested in mammalian development, and as a starting point for investigating the chemical requirements necessary for exhibiting human developmental toxicity as well as the differences in various species.

DEVELOPMENTAL TOXICOLOGY

With thousands of drugs already available and 300 new ones approved for marketing over the past decade alone (Lacy et al., 2004), together with >70,000 chemicals circulating in the environment (Fagin et al., 1996), there is increasing concern for the safety of pregnant women and their offspring. This is so because a high percentage of them are exposed to these agents, despite the rigorous testing of all chemical agents before they reach the marketplace.

It has been established for over 30 years that there are four classes of embryo/fetal toxicity, or more properly, developmental toxicity, in mammalian species, including humans (Wilson, 1973). In simplest terms, these are growth retardation, death, malformation or terata, and functional deficit. While it has been commonplace to term those agents that induce malformations as "teratogens," it is equally proper to term agents that affect one or more of these classes as "developmental toxicants." This term, to our knowledge, is attributable to scientists at the U.S. Environmental Protection Agency (EPA), formulated in 1980 and publicly defined in an EPA guideline document some 6 years later (U.S. EPA, 1980, 1986). It was coined to denote those agents that induce any one or more of the four classes of developmental toxicity, as defined in those documents. The term has since been used in regulatory documents and by investigators in other publications. Adverse effects comprising these classes are shown in Table 1.

The classes of developmental toxicity demonstrate a continuum, many times appearing together (e.g., growth retarded fetuses may have structural malformations, of which some may be lethal and some may be associated with functional deficiencies). While teratogens have been emphasized in importance in pregnancy studies, all classes are of equal importance in assessing developmental toxicity, whether it be in animals or in humans. The natural history of developmental parameters in humans is shown in Table 2.

TABLE 1
Adverse Endpoints Comprising Classes of Developmental Toxicity in Animals and Humans

Class	Endpoints	
	Animals	**Humans**
Growth retardation	Reduced fetal body weight	Intrauterine growth retardation (IUGR), low birth weight, prematurity, microcephaly
Death	Embryolethality, abortion, postnatal mortality	Spontaneous abortion, stillbirth, fetal wastage, perinatal mortality
Malformation	Minor/major congenital (structural) abnormalities, anatomical (developmental) variations	Minor/major congenital (structural) abnormalities
Functional deficit	Postnatal behavioral alterations, developmental delay	Mental retardation/deficiency, metabolic alteration, altered social behavior, neurological deficit, developmental delay

TABLE 2
Normal Incidence Patterns of Adverse Developmental Effects in Humans

Developmental Effect	Normal Incidence (%)	Ref.
Growth retardation		
Intrauterine growth retardation (IUGR)	3–10	Seeds, 1984
Low birth weight	7.9	Hamilton et al., 2004
Prematurity	6.4–9.2	Chez et al., 1976
Death		
Spontaneous abortion (<20 weeks)	20	Abortion statistics, 1995
Early embryonic/fetal	11–25	Hook, 1981
Late fetal	1	Hook, 1981
Stillbirth	2	Rosenberg, 1984
Neonatal	1	Hook, 1981
Infant	1.4	Hook, 1981
Pregnancy loss (total)	31	Wilcox et al., 1988
Malformation		
Minor	14	Hook, 1981
Major	2–4	Rosenberg, 1984; VanRegemorter et al., 1984
Defects at birth	2–3	Hook, 1981
Defects at 1 yr	6–7	Hook, 1981
Functional deficit		
Children in need of special education	10–15	Gaddes, 1980
Mild mental retardation	0.6	Hook, 1981
Severe mental retardation	0.3–0.4	Hook, 1981

The contribution of drug and chemical agent exposures to these statistics is not known with certainty. One respected clinician placed environmental agents as responsible for birth defects in humans on the order of <1% of the total (Brent, 2001). Unfortunately, similar estimates for other developmental toxicity parameters are not available. However, as stated above, concern is currently high, because approximately 75% of women consume one or more therapeutic drugs during their pregnancies (Rayburn et al., 1982), and most likely, an equally great number are exposed to chemicals in the home as well as in the environment during pregnancy.

A number of publications in the past and in the present decade have largely addressed the issue of drug and chemical induction of congenital malformations in humans (Folb and Dukes, 1990; Abrams, 1990; Persaud, 1990; Needleman and Bellinger, 1994; Scialli et al., 1995; Gilstrap and Little, 1998; Friedman and Polifka, 2000; Schardein, 2000; Yankowitz and Niebyl, 2001; Schaefer, 2001; Shepard and Lemire, 2004; Weiner and Buhimschi, 2004; Briggs et al., 2005). However, little emphasis has been placed on developmental toxicity in humans as a whole.

Because of this deficiency, it is the objective of this project to prepare brief, concise, thorough, up-to-date, and useful summaries of clinically important developmental toxicants in humans. It is our intention in this survey of representative developmental toxicants to emphasize growth, viability, and functional changes that have been recorded in the literature examined, in addition to the induction of congenital malformations. Laboratory animal studies have been included in this survey in comparison to the human clinical situations, as they have been predictive in many ways of the human potential for developmental toxicity. In this regard, of the approximately 44 recognized human teratogens, all have been corroborated in one or more species of laboratory animal (Schardein, 2000). Comparisons of effective doses and routes of administration, defect concordance, and definitions of animal "models" have been made in all instances where data are available.

Details of the developmental toxicology in animals and humans are provided on the CD that accompanies this book.

COMPUTATIONAL CHEMISTRY

It is accepted that the biological activity of a chemical is a function of its properties. These properties can be physicochemical or topological in nature and may arise from the chemical structure (i.e., the types and arrangement of atoms that constitute a molecular entity). The central paradigm within structure–activity relationship (SAR) studies is that the chemical structure dictates the properties, which, in turn, give rise to the observed biological activity.

Chemical structure is central to the language of chemistry. Structure is defined in two primary ways: the connectivity between atoms and the three-dimensional arrangement that the atoms adopt within a molecule. The structure of each compound within the database was obtained from the National Library of Medicine's Web site (http://sis.nlm.nih.gov/Chem/ChemMain.html). Each structure was subjected to conformational analysis about selected rotatable bonds (Lennard-Jones 6-12 potential; 10° rotational increment) and subsequent full geometry optimization (MM2 force field) utilizing Molecular Modeling Pro (MMP; http://www.ChemSW.com). The resulting low-energy three-dimensional chemical structures are stored in individual MOL files (MDL; http://www.mdli.com). Simplified Molecular Input Line Entry Specification (SMILES; http://www.daylight.com) codes were generated for each structure as an additional representation of the atom–bond connectivity within chemicals. Providing the individual chemical structures will also allow investigators to perform their own calculations utilizing their respective computational chemistry software. A traditional two-dimensional structure diagram is provided within the text for each of the respective chemicals discussed.

Chemicals were submitted to algorithms within MMP to calculate the following 20 physico-chemical properties: molecular weight, molecular volume, density, surface area, logP (octanol–water partition coefficient), HLB (hydrophilic–lipophilic balance), solubility parameter, dispersion, polarity, hydrogen bonding, H (hydrogen) bond acceptor, H (hydrogen) bond donor, percent hydrophilic surface, MR (molar refractivity), water solubility, hydrophilic surface area, polar surface area, HOMO (highest occupied molecular orbital), LUMO (lowest unoccupied molecular orbital), and dipole. These parameters characterize molecular size, transport, electronic properties, and the ability to engage in intermolecular interactions. The physicochemical parameters vary in accuracy and calculated values depending on the algorithms utilized.

SciQSAR-2D (SciVision, Inc.) was utilized to calculate 29 topological indices: simple connectivity indices (x0, x1, x2, xp3, xp4, xp5, xp6, xp7, xp8, xp9, xp10), valence connectivity indices (xv0, xv1, xv2, xvp3, xvp4, xvp5, xvp6, xvp7, xvp8, xvp9, xvp10), and kappa indices (k0, k1, k2, k3, ka1, ka2, ka3). Topological indices characterize the connectivity (of various orders; i.e., path one, path two, path three) between the atoms comprising a molecular entity, as well as size and degree of branching. One of the advantages of this type of parameter is that the values are invariant (there is one way to calculate them), unlike physicochemical parameters with which the calculated values may differ due to different algorithms or molecular conformations.

The original literature detailing the algorithms utilized to calculate the above physicochemical and topological properties (in order of database appearance) are provided under the Chemical section within the References.

The electronic database consisting of the individual three-dimensional chemical structures and physicochemical/topological properties together with the associated biological data is stored as an SD file (MDL; www.mdli.com), which is a standard file format for transferring linked chemical and biological data between computational chemistry software. The SD file has the advantage that, with the appropriate software, the molecular structure can be visualized together with the calculated properties and biological activities. In addition to the SD file and individual MOL files, an Excel file of the database listing the calculated parameters and associated biological data is also provided for investigators without access to chemical structure viewing software. All of the electronic files are provided on the accompanying CD.

A summary of the calculated 49 physicochemical and topological parameters is listed in Table 3 for the first database entry, Aminopterin.

Histograms plotting the distribution of compounds according to the calculated physicochemical and topological parameters are listed in Appendixes I and II. A discussion of the histograms can be found in the concluding chapter of this book.

CONCLUSION

The agents in this survey, numbering 50, were selected rather arbitrarily, but their selection was considered in light of (1) their importance in commerce, and, most importantly, their importance in public health considerations, (2) the availability of quality data in humans (and animals), and (3) their representation for affecting the various classes of developmental toxicity. Some affect a single class, others affect all four classes. There are approximately 70 developmental toxicants known. However, we are satisfied that the 50 agents selected for this project are representative of the group. We hope their inclusion here with up-to-date information pertinent to their adverse toxic properties when used in pregnancy should help allay concerns to public health. The agents excluded are shown in Table 4, together with the reasons for their exclusion. Inorganic agents that are metals or mixtures are not included, because detailed computational chemical analysis as applied here cannot be conducted on such agents.

TABLE 3
Calculated Parameters for Aminopterin

Parameter	Value	Units
		Physicochemical
Molecular weight	440.418	g/mol
Molecular volume	361.87	A^3
Density	1.493	g/cm^3 (with fragment corrections)
Surface area	441.97	A^2
LogP	−4.001	log ([oct]/[water])
HLB	21.158	(hydrophilic mw/total mw) × 20
Solubility parameter	32.668	$J^{(0.5)}$/cm$^{(1.5)}$
Dispersion	27.188	$J^{(0.5)}$/cm$^{(1.5)}$
Polarity	8.861	$J^{(0.5)}$/cm$^{(1.5)}$
Hydrogen bonding	15.793	$J^{(0.5)}$/cm$^{(1.5)}$
H bond acceptor	3.6	Sum of partial atomic charges < −0.15
H bond donor	2.13	Sum of partial atomic charges > 0.20
Percent hydrophilic surface	98.34	(hydrophilic surface area/total surface area) × 100
MR	117.696	Molar refractivity (unitless)
Water solubility	−1.817	log (mol/M^3)
Hydrophilic surface area	434.63	A^2
Polar surface area	228.81	A^2
HOMO	−8.821	eV (single point MOPAC/AM1 calculation)
LUMO	−1.551	eV (single point MOPAC/AM1 calculation)
Dipole	5.270	debye (single point MOPAC/AM1 calculation)
		Topological (unitless)
x0	23.250	Zero-order simple connectivity index
x1	15.223	First-order simple connectivity index
x2	14.203	Second-order simple connectivity index
xp3	10.778	Third-order path simple connectivity index
xp4	8.491	Fourth-order path simple connectivity index
xp5	6.953	Fifth-order path simple connectivity index
xp6	4.834	Sixth-order path simple connectivity index
xp7	3.129	Seventh-order path simple connectivity index
xp8	2.099	Eighth-order path simple connectivity index
xp9	1.617	Ninth-order path simple connectivity index
xp10	1.046	Tenth-order path simple connectivity index
xv0	16.648	Zero-order valence connectivity index
xv1	9.366	First-order valence connectivity index
xv2	6.728	Second-order valence connectivity index
xvp3	4.372	Third-order valence connectivity index
xvp4	2.766	Fourth-order valence connectivity index
xvp5	1.810	Fifth-order valence connectivity index
xvp6	0.973	Sixth-order valence connectivity index
xvp7	0.527	Seventh-order valence connectivity index
xvp8	0.303	Eighth-order valence connectivity index
xvp9	0.186	Ninth-order valence connectivity index
xvp10	0.099	Tenth-order valence connectivity index
k0	46.960	Zero-order kappa shape index
k1	26.602	First-order kappa shape index
k2	12.630	Second-order kappa shape index
k3	8.033	Third-order kappa shape index
ka1	23.081	First-order kappa–alpha shape index
ka2	10.145	Second-order kappa–alpha shape index
ka3	6.212	Third-order kappa–alpha shape index

TABLE 4
Known Developmental Toxicants Excluded from This Treatise

Agent(s)	Reason Excluded	See Chapter Number
ACE inhibitors: enalapril, lisinopril	Another representative member of group included	18
Aminoglycosides: kanamycin, dihydrostreptomycin	Another representative member of group included	20
Coumarins: acenoprocoumon, phenprocoumon	No longer marketed in the United States, better representative of group included	34
Goitrogens: carbimazole, others	Another more representative member of group included	21
Iodides	Metal (inorganic)	—
Lead	Metal (inorganic)	—
Lithium	Metal (inorganic)	—
Methandriol	No longer marketed in the United States, other representatives included	13, 37
Methyl mercury	Metal (inorganic)	—
Methylthiouracil	No longer marketed, another representative of group included	29
PCBs	Mixture	—
Progestins: hydroxyprogesterone	Other representatives of group included	30, 41, 45
Sartans: losartan, candesartan, telmisartan	Another representative member of group included	47
Tobacco smoke	Mixture	—
Trimethadione	Largely replaced by a similar agent (included) in the United States	14

REFERENCES

TOXICOLOGICAL

Abrams, R. S. (1990). *Will It Hurt the Baby? The Safe Use of Medications during Pregnancy and Breastfeeding*, Addison-Wesley, Reading, MA.

Brent, R. L. (2001). The cause and prevention of human birth defects: What have we learned in the past 50 years? *Cong. Anom.* 41: 3–21.

Briggs, G. G. et al. (2005). *Drugs in Pregnancy and Lactation. A Reference Guide to Fetal and Neonatal Risk*, Seventh ed., Lippincott Williams & Wilkins, Philadelphia.

Chez, R. A. et al. (1976). High risk pregnancies: Obstetrical and perinatal factors. In *Prevention of Embryonic, Fetal, and Perinatal Disease*, R. L. Brent and M. I. Harris, Eds., DHEW Publ. (NIH)76-853, pp. 67–95.

Fagin, D. et al. (1996). *Toxic Deception. How the Chemical Industry Manipulates Science, Bends the Law, and Endangers Your Health*, Carol Publishing Group, Secaucus, NJ.

Folb, P. I. and Dukes, M. N. (1990). *Drug Safety in Pregnancy*, Elsevier, Amsterdam.

Friedman, J. M. and Polifka, J. E. (2000). *Teratogenic Effects of Drugs. A Resource for Clinicians (TERIS)*, Second ed., Johns Hopkins University Press, Baltimore, MD.

Gaddes, W. H. (1980). *Learning Disabilities and Brain Function*, Springer-Verlag, New York.

Gilstrap, L. C. and Little, B. B. (1998). *Drugs and Pregnancy*, Second ed., Chapman & Hall, New York.

Hamilton, B. E. et al. (2004). Births: Preliminary data for 2003. *Nat. Vital Stat. Rep.* 53: 1–17.

Hook, E. B. (1981). Human teratogenic and mutagenic markers in monitoring about point sources of pollution. *Environ. Res.* 25: 178–203.

Lacy, C. F. et al. (2004). *Drug Information Handbook (Pocket), 2004-2005*, Lexi-Comp., Inc., Hudson, OH.

Needleman, H. L. and Bellinger, D., Eds. (1994). *Prenatal Exposure to Toxicants. Developmental Consequences*, Johns Hopkins University Press, Baltimore, MD.

Persaud, T. V. N. (1990). *Environmental Causes of Human Birth Defects*, Charles C Thomas, Springfield, IL.

Rayburn, W. F. et al. (1982). Counseling by telephone. A toll-free service to improve prenatal care. *J. Reprod. Med.* 27: 551–556.

Rosenberg, M. J. (1984). Practical aspects of reproductive surveillance. In *Reproduction: The New Frontier in Occupational and Environmental Health Research,* Proceedings of the 5th Annual RMCOEH Occupational and Environmental Health Conference, 1983, J. R. Lockey, G. K. Lemasters, and W. R. Keye, Eds., Alan R. Liss, New York, pp. 147–156.

Schaefer, C. (Ed.) (2001). *Drugs during Pregnancy and Lactation. Handbook of Prescription Drugs and Comparative Risk Assessment*, Elsevier, Amsterdam.

Schardein, J. L. (2000). *Chemically Induced Birth Defects*, Third ed., Marcel Dekker, New York.

Scialli, A. R. et al. (1995). *Reproductive Effects of Chemical, Physical, and Biologic Agents, Reprotox*, Johns Hopkins University Press, Baltimore, MD.

Seeds, J. W. (1984). Impaired fetal growth: Definition and clinical diagnosis. *Obstet. Gynecol.* 64: 303.

Shepard, T. H. and Lemire, R. J. (2004). *Catalog of Teratogenic Agents*, Eleventh ed., Johns Hopkins University Press, Baltimore, MD.

U.S. EPA. (1980). Assessment of risks to human reproduction and to development of the human conceptus from exposure to environmental substances. *NTIS DE82-007897*, pp. 99–116.

U.S. EPA. (1986). Guidelines for the Health Assessment of Suspect Developmental Toxicants. *Fed. Regist.* 51 (#185): 34028-34040, September 14.

VanRegemorter, N. et al. (1984). Congenital malformations in 10,000 consecutive births in a university hospital: Need for genetic counseling and prenatal diagnosis. *J. Pediatr.* 104: 386–390.

Weiner, C. P. and Buhimschi, C. (2004). *Drugs for Pregnant and Lactating Women*, Churchill Livingstone, Philadelphia.

Wilcox, A. J. et al. (1988). Incidence of early loss of pregnancy. *N. Engl. J. Med.* 319: 189.

Wilson, J. G. (1973). *Environment and Birth Defects*, Academic Press, New York.

Yankowitz, J. and Niebyl, J. R. (2001). *Drug Therapy in Pregnancy*, Third ed., Lippincott Williams & Wilkins, Philadelphia.

CHEMICAL

Physicochemical parameters (programmed by Norgwyn Montgomery Software Inc., www.norgwyn.com, and may vary from the values obtained with other programs):

Molecular mechanics (MM2 force field): Burkert, U. and Allinger, N. L. (1982). *Molecular Mechanics ACS Monograph 177*, American Chemical Society, Washington, D.C.

Molecular volume, HLB, surface area, hydrophilic surface area, percent hydrophilic surface area: decriptions of these proprietary methods can be downloaded from www.norgwyn.com.

Log P: Hansch, C. and Leo, A. (1979). *Substituent Constants for Correlation Analysis in Chemistry and Biology*, John Wiley & Sons, New York.

Solubility parameter, dispersion, polarity, hydrogen bonding: van Krevelen, D. W. (1990). *Properties of Polymers*, Elsevier, Amsterdam, pp. 200–225.

H bond acceptor/donor: Del Re, G. (1958). A simple MO-LCAO method for the calculation of charge distributions in saturated organic molecules. *J. Chem. Soc.* 4031–4040.

MR: Lyman, W. F. et al. (1982). *Handbook of Chemical Property Estimation Methods*, McGraw-Hill, New York, chap. 12.

Water solubility: Klopman G. et al. (1992). Estimation of aqueous solubility of organic molecules by the group contribution approach. Application to the study of biodegradation. *J. Chem. Inf. Comput. Sci.* 32: 474–482.

Polar surface area: Ertl, P. et al. (2000). Fast calculation of molecular polar surface area as a sum of fragment-based contributions and its application to the prediction of drug transport properties. *J. Med. Chem.* 43: 3714–3717.

MOPAC/AM1 (HOMO, LUMO, dipole): Dewar, M. J. S. et al. (1985). Development and use of quantum mechanical models. 76. AM1: A new general purpose quantum mechanical molecular model. *J. Am. Chem. Soc.* 107(13): 3902–3909.

Topological parameters (programmed by SciVision, Inc.):

Devillers, J. and Balaban, A. T. (Eds.) (1999). *Topological Indices and Related Descriptors in QSAR and QSPR*, Gordon and Breach Science Publishers, Amsterdam, chap. 7 (**simple and valence indices**), chap. 10 (**kappa indices**).

Acknowledgments

Financial support for the construction of the Human Developmental Toxicants database was provided by the National Institutes of Health under contract #263-MD-415075.

The authors appreciate the support and encouragement of Donald R. Mattison, M.D., senior advisor to the directors of the National Institute of Child Health and Human Development, Center for Research for Mothers and Children (NICHD/CRMC) and branch chief, Obstetric and Pediatric Pharmacology Branch of the National Institutes of Health.

One of the authors (JLS) would like to thank Mrs. Barbara Stoffer for her excellent work in the collection of pertinent publications to this work.

The Authors

James L. Schardein, M.S., a fellow of the Academy of Toxicological Sciences, is an internationally recognized expert and leader in developmental toxicology. His professional career of some 47 years was in the scientific area of reproductive and developmental toxicology conducted at a major pharmaceutical company and several principal contract research laboratories. His research interests have mainly focused on laboratory animal teratology, with associations to human clinical teratology. His research has involved original experimental animal studies, and his laboratory was one of the first in the industry to investigate the effects of candidate pharmaceuticals on the developing animal model with respect to the induction of congenital malformations. Management and research direction responsibilities followed. He has served as an officer for several national peer scientific societies, has served on the editorial boards of several international journals in the developmental toxicology field, and has published over 150 abstracts, manuscripts, book chapters, and two textbooks. He is certified in toxicology by the Academy of Toxicological Sciences, and he was recognized by a number of biographical dictionaries, including several editions of *Who's Who, 5,000 Personalities of the World,* and *Sterling's Who's Who.* He is currently an independent consultant to the pharmaceutical and chemical industries, governmental agencies, and the legal profession.

Orest T. Macina, Ph.D., has nearly two decades of professional experience in the application of the tools and techniques of computational chemistry to problems of biological interest. His research interests are in the derivation of quantitative structure activity relationships (QSAR) utilizing standard statistical approaches as well as advanced data mining algorithms. He has industrial experience in the pharmaceutical area and academic experience in the toxicological field. As a result of his industrial experience, he holds several patents related to the discovery of cardiovascular and antifungal agents. While a faculty member at the Graduate School of Public Health, University of Pittsburgh, Pennsylvania, he developed graduate-level courses, supervised the research of M.S. and Ph.D. students, and was instrumental in developing a graduate Ph.D. track in computational toxicology. He has contributed to a number of publications regarding the application of computational chemistry to the pharmaceutical and toxicological fields (including developmental toxicity). He is currently the principal of Macina Informatics, providing computational toxicology services to the chemical and pharmaceutical industries and to government agencies.

Contents

1 Aminopterin

Chemical name: 4-Aminopteroylglutamic acid

CAS #: 54-62-6

SMILES: c12c(nc(cn1)CNc3ccc(cc3)C(NC(CCC(O)=O)C(O)=O)=O)c(nc(n2)N)N

INTRODUCTION

Aminopterin, a formerly used antimetabolite (folic acid antagonist) antineoplastic agent for the treatment of acute leukemia in children, has now been largely replaced in this category by methotrexate, and is now used mainly as a rodenticide. The drug blocks important actions of folic acid conversion to folinic acid in nucleic acid metabolism and cytopoiesis. Because of this property, it has also been used as an abortifacient in women. It was effective in cancer therapy, because it incorporated readily into cells and was slowly excreted. Folic acid antagonists like aminopterin that inhibit dihydrofolate result in cell death during the S-phase of the cell cycle (Skipper and Schobel, 1973).

DEVELOPMENTAL TOXICOLOGY

ANIMALS

In animal studies, this agent had teratogenic potential by the oral route, inducing multiple defects, in rats (Sansone and Zunin, 1954), dogs (Earl et al., 1975), and swine (Earl et al., 1975). It did not induce malformations in cats (Khera, 1976) or in monkeys (Wilson, 1968) by the oral route, or in mice (Thiersch and Philips, 1950) by the intraperitoneal route, although abortion (in cats and monkeys) and embryocidal effects (in mice) were reported in these species at doses of 0.1 mg/kg and higher. Effective oral doses in the responsive species were in the range of 0.0125 to 0.05 mg/kg. In rabbits, intravenous doses of 15 mg/kg were teratogenic and embryolethal (Goeringer and DeSesso, 1990). Some sheep given 5 or 10 mg aminopterin by subcutaneous injection aborted, and some had ear and skeletal defects in their offspring, along with other developmental toxicity, including reduced fetal size and embryolethality (James and Keeler, 1968). The agent in animal studies then, has a variety of adverse developmental effects in a wide variety of species.

TABLE 1
Developmental Toxicity Profile of Aminopterin in Humans

Case Number	Malformations	Growth Retardation	Death	Functional Deficit	Ref.
1	Brain		✔		Thiersch, 1952
2	Lip/palate		✔		Thiersch, 1952
3	Brain		✔		Thiersch, 1952
4	Brain		✔		Thiersch, 1956
5	Skull, limbs				Meltzer, 1956
6	Multiple: skull, digits, ears, face, palate, axial skeleton, limbs	✔	✔		Warkany et al., 1959
7	Multiple: skull, brain, limbs, ears, face, palate		✔		Emerson, 1962
8	"Gross multiple severe anomalies incompatible with life"		✔		Goetsch, 1962
9	Brain		✔		deAlvarez, 1962
10	Brain		✔		deAlvarez, 1962
11[a]	Multiple: skull, digits, limbs		✔		Werthemann, 1963
12	Multiple: skull, face, ears	✔	✔		Shaw and Steinbach, 1968; Shaw, 1972; Shaw and Rees, 1980
13	Multiple: similar to case #6 plus limbs, eyes				Gautier, 1969; Brandner and Nussle, 1969 (Patrick case)
14[b]	Multiple: face, palate, ears, testes, skull, digits		✔		Hermann and Opitz, 1969
15	Multiple: limbs, skull, face, teeth, testes (at 4 yr)	✔		✔	Cited, Smith, 1970; Howard and Rudd, 1977 (Rudd case)
16	Multiple: brain, skull, ears, face, palate, axial skeleton, genital, limbs, digits			✔	Reich et al., 1978
17	Head and face abnormalities at birth (surgically corrected), limbs				Reich et al., 1978
18[b]	Multiple: skull, face, palate, testes, limbs, digits				Reich et al., 1978
19	Multiple: limbs, face, ears				Gellis and Finegold, 1979; Char, 1979
20	Multiple: facial dysmorphia similar to case #6, skull, digits, skin, palate, limbs				Hill and Tennyson, 1984

[a] Also treated with thalidomide.
[b] Drug intake uncertain, may be methotrexate treatment. Face abnormality components included jaw, eyes, nose, and hair.

HUMANS

Given to human subjects in the decades of the 1950s through the 1970s, aminopterin was associated with 20 cases of malformation and associated developmental toxicity due to unsuccessful abortion attempts. These cases are of great interest to teratologists and clinicians, for they represent one of the very few teratologic experiments performed in the human (Warkany, 1978). These cases are tabulated in Table 1. The malformations were characterized in full (Schardein, 2000). Prominent in a number of these cases were skull malformations. Wide fontanelles, synostosis of sutures, and partial or absent ossification of a number of bones, including the frontal, parietal, and occipital bones were observed. Micrognathia was also usually present, giving the head a peculiar globular "clover-leaf" shape. The head is large and brachycephalic, due to either hydrocephalus or craniosynostosis. The hair is oftentimes swept back, the eyes prominent, and the ears low set. Ocular hypertelorism and wide nasal bridge are also usual features. Several of the infants had associated limb deformities, including talipes equinovarus and mesomelic shortening of forearms, and most of the abortuses and infants who died shortly after birth had cerebral anomalies, notably, anencephaly, hydrocephaly, meningomyelocele, and hypoplasia. A few of the affected infants are of low birth weight, and survivors are generally shorter in height than normal. Mentality has been variable, ranging from normal to low IQ and poor speech development. Several of the patients survive. One (case #12), at 17 $\frac{1}{2}$ yr of age some 25 yr ago, was still improving developmentally and socially, and was considered normal for the teenager (Shaw and Rees, 1980). No further information on this individual has appeared in print. Prognosis for self-support of several of the surviving cases (about one-half of the reported cases) has been predicted. Oddly, the sheep is considered an animal model for defects appearing in humans.

   Where histories of the above cases are complete, dosages eliciting the developmental effects were in the range of 10 to 41 mg/day. When cited, treatment was limited to the first trimester (<12 weeks), with the sixth through eighth gestational weeks defined as the critical period (Feldkamp and Carey, 1993). The risk of malformation is suggested to be about 44%, as some 25 of the 45 exposed cases due to failed abortion were reported as normal in case reports (Thiersch, 1952, 1956; Harris, 1953; Cariati, 1955; Smith et al., 1958; Goetsch, 1962). Review articles on the subject of aminopterin developmental toxicity were published (Warkany, 1978; Lloyd et al., 1999).

CHEMISTRY

Aminopterin is a large heterocyclic structure that can participate in hydrogen bonding interactions, both as an acceptor and as a donor. It is hydrophilic and has a relatively large polar surface area in comparison to the other human developmental toxicants. The calculated physicochemical and topological properties are listed below.

PHYSICOCHEMICAL PROPERTIES

Parameter	Value
Molecular weight	440.418 g/mol
Molecular volume	361.87 A^3
Density	1.495 g/cm^3
Surface area	441.97 A^2
LogP	−4.001
HLB	21.158
Solubility parameter	32.668 $J^{(0.5)}/cm^{(1.5)}$
Dispersion	27.188 $J^{(0.5)}/cm^{(1.5)}$
Polarity	8.861 $J^{(0.5)}/cm^{(1.5)}$
Hydrogen bonding	15.793 $J^{(0.5)}/cm^{(1.5)}$
H bond acceptor	3.6
H bond donor	2.13
Percent hydrophilic surface	98.34
MR	117.696
Water solubility	−1.817 log (mol/M^3)
Hydrophilic surface area	434.63 A^2
Polar surface area	228.81 A^2
HOMO	−8.821 eV
LUMO	−1.551 eV
Dipole	5.270 debye

TOPOLOGICAL PROPERTIES (UNITLESS)

Parameter	Value
x0	23.250
x1	15.223
x2	14.203
xp3	10.778
xp4	8.491
xp5	6.953
xp6	4.834
xp7	3.129
xp8	2.099
xp9	1.617
xp10	1.046
xv0	16.648
xv1	9.366
xv2	6.728
xvp3	4.372
xvp4	2.766
xvp5	1.810
xvp6	0.973
xvp7	0.527
xvp8	0.303
xvp9	0.186
xvp10	0.099
k0	46.960
k1	26.602
k2	12.630
k3	8.033
ka1	23.081
ka2	10.145
ka3	6.212

REFERENCES

Brandner, M. and Nussle, D. (1969). Foetopathie du a l'aminopterine ovec stenose congenitale de l'espace medullaire des os tubulaires longs. *Amm. Radiol. (Paris)* 12: 703–712.

Cariati, A. (1955). [A case of acute hemocytoblastic leukemia and pregnancy]. *Riv. Ostet. Ginecol.* 10: 785–796.

Char, F. (1979). Denouement and discussion: Aminopterin embryopathy syndrome. *Am. J. Dis. Child.* 133: 1189–1190.

deAlvarez, R. R. (1962). Discussion to: An evaluation of aminopterin as an abortifacient. *Am. J. Obstet. Gynecol.* 83: 1476–1477.

Earl, F. L., Miller, E., and Van Loon, E. J. (1975). Beagle dog and miniature swine as a model for teratogenesis evaluation. *Teratology* 11:16A.

Emerson, D. J. (1962). Congenital malformation due to attempted abortion with aminopterin. *Am. J. Obstet. Gynecol.* 84: 356–357.

Feldkamp, M. and Carey, J. C. (1993). Clinical teratology counseling and consultation case report: Low dose methotrexate exposure in the early weeks of pregnancy. *Teratology* 47: 533–539.

Gautier, E. (1969). Demonstrations cliniques, Embryopathie de l'aminopterin, kwashiorkor, enfant maltraite, listeriose congenitale et saturnisme, maladie de Weil. *Schweiz. Med. Wochenschr.* 99: 33–42.

Gellis, S. S. and Feingold, M. (1979). Aminopterin embryopathy syndrome. *Am. J. Dis. Child.* 133: 1189–1190.

Goeringer, G. C. and DeSesso, J. M. (1990). Developmental toxicity in rabbits of the antifolate aminopterin and its amelioration by leucovorin. *Teratology* 41: 560–561.

Goetsch, C. (1962). An evaluation of aminopterin as an abortifacient. *Am. J. Obstet. Gynecol.* 83: 1474–1477.

Harris, L. J. (1953). Leukaemia and pregnancy. *Can. Med. Assoc. J.* 68: 234–236.

Hermann, J. and Opitz, J. M. (1969). An unusual form of acrocephalosyndactyly. *Birth Defects* 5: 39–42.

Hill, R. M. and Tennyson, L. M. (1984). Drug-induced malformations in humans. In *Drug Use in Pregnancy*, L. Stern, Ed., Adis Health Science Press, Balgowlah, Australia, pp. 99–133.

Howard, N. J. and Rudd, N. L. (1977). The natural history of aminopterin-induced embryopathy. *Birth Defects* 13: 85–93.

James, L. F. and Keeler, R. F. (1968). Teratogenic effects of aminopterin in sheep. *Teratology* 1: 407–412.

Khera, K. S. (1976). Teratogenicity studies with methotrexate, aminopterin, and acetylsalicylic acid in domestic cats. *Teratology* 14: 21–28.

Lloyd, M. E. et al. (1999). The effects of methotrexate on pregnancy, fertility and lactation. *Q. J. Med.* 92: 551–563.

Meltzer, H. J. (1956). Congenital anomalies due to attempted abortion with 4-aminopteroylglutamic acid. *JAMA* 161: 1253.

Reich, E. W. et al. (1978). Recognition of adult patients of malformation induced by folic acid antagonists. *Birth Defects* 14: 139–160.

Sansone, G. and Zunin, C. (1954). Embriopatie sperimentali da somministrazione di antifolici. *Acta Vitaminol. (Milano)* 8: 73–79.

Schardein, J. L. (2000). *Chemically Induced Birth Defects*, Third ed., Marcel Dekker, New York, pp. 585–587.

Shaw, E. B. (1972). Fetal damage due to maternal aminopterin ingestion. Follow-up at age 9 years. *Am. J. Dis. Child.* 124: 93–94.

Shaw, E. B. and Rees, E. L. (1980). Fetal damage due to aminopterin ingestion — followup at 17 $\frac{1}{2}$ years of age. *Am. J. Dis. Child.* 134: 1172–1173.

Shaw, E. B. and Steinbach, H. L. (1968). Aminopterin-induced fetal malformation. Survival of infant after attempted abortion. *Am. J. Dis. Child.* 115: 477–482.

Skipper, H. T. and Schobel, F. M. (1973). Quantitative and cytokinetic studies in experimental tumor models. In *Cancer Medicine*, J. F. Holland and E. Frei, Eds., Lea & Febiger, Philadelphia, pp. 629–650.

Smith, D. W. (1970). *Recognizable Patterns of Human Malformation*, W. B. Saunders, Philadelphia.

Smith, R. B. W., Sheehy, T. W., and Rothbert, H. (1958). Hodgkin's disease and pregnancy. *Arch. Intern. Med.* 102: 777–789.

Thiersch, J. B. (1952). Therapeutic abortions with a folic acid antagonist 4-aminopteroylglutamic acid administered by the oral route. *Am. J. Obstet. Gynecol.* 63: 1298–1304.

Thiersch, J. B. (1956). The control of reproduction in rats with the aid of antimetabolites and early experiences with antimetabolites as abortifacient agents in man. *Acta Endocrinol. Suppl. 1(Copenh.)* 28: 37–45.

Thiersch, J. B. and Philips, F. S. (1950). Effect of 4-amino-pteroylglutamic acid (aminopterin) on early pregnancy. *Proc. Soc. Exp. Biol. Med.* 74: 204–208.

Warkany, J. (1978). Aminopterin and methotrexate: Folic acid deficiency. *Teratology* 17: 353–357.

Warkany, J., Beaudry, P. H., and Hornstein, S. (1959). Attempted abortion with aminopterin (4-aminopteroylglutamic acid). Malformations of the child. *Am. J. Dis. Child.* 97: 274–281.

Werthemann, A. (1963). Allgemeine und spezielle Probleme bei der Analyse von Missbildungsursachen, in Sonderheit bei Thalidomid- und Aminopterin-schaden. *Schweiz. Med. Wochenschr.* 93: 223–227.

Wilson, J. G. (1968). Teratological and reproductive studies in non-human primates. In *Papers from Second International Workshop Teratology*, H. Nishimura (Ed.), University of Kyoto, April 1–5, pp. 176–201.

2 Busulfan

Chemical name: 1,4-Butanediol dimethanesulfonate

CAS #: 55-98-1

SMILES: O(S(C)(=O)=O)CCCCOS(C)(=O)=O

INTRODUCTION

Busulfan is an alkylating chemical used as an antineoplastic agent, with selective action confined to myelosuppression (e.g., myelogenous leukemia and other bone marrow disorders). The drug reacts with the N-7 position of guanosine and interferes with DNA replication and transcription of RNA (Lacy et al., 2004). Its activity against leukemia was reported as early as 1953 (*PDR*, 2002). It is available commercially as Busulfex® or Myleran®, and it has a pregnancy risk factor of D (drug labeling states that it "may cause fetal harm when administered to a pregnant woman").

DEVELOPMENTAL TOXICOLOGY

ANIMALS

Both mice (Pinto Machado, 1966) and rats (Weingarten et al., 1971) evidenced the full spectrum of developmental toxicity (malformation, fetal growth retardation, and embryolethality) when administered busulfan by the intraperitoneal route on various days of the organogenesis period to gravid animals. Effective doses were in the range of 10 to 50 mg/kg/day. Further, rats given subteratogenic doses of the drug showed postnatal behavioral disturbances (Malakhovsky, 1969). Ovarian dysgenesis, a feature also observed in the human, was reported in rats following a single dose of 10 mg/kg on gestation day 13 (Heller and Jones, 1964). Testicular degeneration was reported in rats on the same regimen (Vanhems and Bousquet, 1972) and resulted in sterile progeny when given at a dose of 10 mg/kg/day over a 3-day interval during organogenesis (Bollag, 1954). The drug is clearly a reproductive toxicant as well as a developmental one.

HUMANS

In the human, seven cases of malformation or other adverse developmental effects were recorded, as shown in Table 1. Doses effective in eliciting developmental toxicity were in the range of 2 to 6 mg/day, and treatment was limited, except for in cases #3 and #4, to the first trimester. The recommended human therapeutic dose for busulfan is 0.12 to 4 mg/kg/day po. It should be emphasized that no specific pattern of malformation is obvious among the recorded cases, although toxicity

TABLE 1
Developmental Toxicity Profile of Busulfan in Humans

Case Number	Malformations	Growth Retardation	Death	Functional Deficit	Ref.
1	Multiple: palate, eye, genitals, ovary	✔	✔		Diamond et al., 1960
2	Present, but unspecified		✔		deRezende et al., 1965
3	None	✔			Dugdale and Fort, 1967
4	None (unrelated renal defect)	✔			Boros and Reynolds, 1977
5	Brain	✔	✔		Abramovici et al., 1978
6	Multiple: unspecified				Szentcsiki et al., 1982
7	Multiple: brain, esophagus, heart, genitals, adrenal, umbilicus	✔	✔		Zuazu et al., 1991

is apparent with use of this drug, and it is generally considered potentially teratogenic. Intrauterine growth retardation (IUGR) was observed in most of the recorded cases; one investigator stated that 40% of infants resulting from maternal treatment with antineoplastic agents, including busulfan, were of low birth weight (Nicholson, 1968). Death was also a common occurrence. There have been no reported effects on postnatal function. Based on a fairly large number of normal pregnancies following first trimester treatment with busulfan (Moloney, 1964; Zuazu et al., 1991; see Schardein, 2000), the risk to developmental toxicity in the human would appear to be approximately 21%.

CHEMISTRY

Busulfan is a smaller, slightly hydrophilic compound consisting of two polar functional groups separated by a four-carbon scaffold. The calculated physicochemical and topological properties for busulfan are listed below.

PHYSICOCHEMICAL PROPERTIES

Parameter	Value
Molecular weight	246.305 g/mol
Molecular volume	197.63 A^3
Density	1.404 g/cm^3
Surface area	271.51 A^2
LogP	−0.592
HLB	18.932
Solubility parameter	14.493 $J^{(0.5)}/cm^{(1.5)}$
Dispersion	14.493 $J^{(0.5)}/cm^{(1.5)}$
Polarity	0.000 $J^{(0.5)}/cm^{(1.5)}$
Hydrogen bonding	0.000 $J^{(0.5)}/cm^{(1.5)}$
H bond acceptor	0.85
H bond donor	0.16
Percent hydrophilic surface	88.66
MR	64.790
Water solubility	1.564 log (mol/M^3)
Hydrophilic surface area	240.73 A^2
Polar surface area	99.38 A^2
HOMO	−11.880 eV
LUMO	−2.280 eV
Dipole	2.133 debye

TOPOLOGICAL PROPERTIES (UNITLESS)

Parameter	Value
x0	11.243
x1	6.207
x2	7.036
xp3	2.604
xp4	1.664
xp5	1.052
xp6	0.655
xp7	0.406
xp8	0.188
xp9	0.281
xp10	0.000
xv0	9.727
xv1	7.527
xv2	6.047
xvp3	2.323
xvp4	1.466
xvp5	0.934
xvp6	0.556
xvp7	0.248
xvp8	0.227
xvp9	0.206
xvp10	0.000
k0	10.627
k1	14.000
k2	5.778
k3	13.091
ka1	13.820
ka2	5.640
ka3	12.912

REFERENCES

Abramovici, A., Shaklai, M., and Pinkhas, J. (1978). Myeloschisis in a six weeks embryo of a leukemic woman treated by busulfan. *Teratology* 18: 241–246.

Bollag, W. (1954). [Cytostatica in pregnancy]. *Schweiz. Med. Wochenschr.* 84: 393–395.

Boros, S. J. and Reynolds, J. W. (1977). Intrauterine growth retardation following third-trimester exposure to busulfan. *Am. J. Obstet. Gynecol.* 129: 111–112.

deRezende, J., Coslovsky, S., and deAguiar, P. B. (1965). Leucemia e gravidez. *Rev. Ginecol. Obstet.* 117: 46–50.

Diamond, I., Anderson, M. M., and McCreadie, S. R. (1960). Transplacental transmission of busulfan (Myleran) in a mother with leukemia. Production of fetal malformation and cytomegaly. *Pediatrics* 25: 85–90.

Dugdale, M. and Fort, A. T. (1967). Busulfan treatment of leukemia during pregnancy. *JAMA* 199: 131–133.

Heller, R. H. and Jones, H. W. (1964). Production of ovarian dysgenesis in the rat and human by busulfan. *Am. J. Obstet. Gynecol.* 89: 414–420.

Lacy, C. F. et al. (2004). *Drug Information Handbook (Pocket), 2004–2005*, Lexi-Comp. Inc., Hudson, OH.

Malakhovsky, V. G. (1969). Behavioral disturbances in rats receiving teratogenic agents antenatally. *Biull. Eksp. Biol. Med.* 68: 1230–1232.

Moloney, W. C. (1964). Management of leukemia in pregnancy. *Ann. NY Acad. Sci.* 114: 857–867.

Nicholson, H. O. (1968). Cytotoxic drugs in pregnancy. Review of reported cases. *J. Obstet. Gynaecol. Br. Commonw.* 75: 307–312.

PDR® (Physicians' Desk Reference®). (2002). Medical Economics Co. Inc., Montvale, NJ.

Pinto Machado, J. (1966). [The embryotoxic and teratogenic action of busulfan (1,4-dimethanesulfonyl-oxybutane) in the mouse]. *Acta Obstet. Gynaecol. Hisp. Lusit.* 15: 201–212.

Schardein, J. L. (2000). *Chemically Induced Birth Defects,* Third ed., Marcel Dekker, New York, pp. 580–581.

Szentcsiki, M. et al. (1982). [Pregnancy during busulfan therapy of chronic granulocytic leukemia]. *Orv. Hetil.* 123: 1307–1308.

Vanhems, E. and Bousquet, J. (1972). Influence du busulphan sur le developpement du testcule du rat. *Annu. Endocrinol.* 33: 119–128.

Weingarten, P. L., Ream, J. R., and Pappas, A. M. (1971). Teratogenicity of Myleran against musculo-skeletal tissues in the rat. *Clin. Orthop.* 75: 236.

Zuazu, J. et al. (1991). Pregnancy outcome in hematologic malignancies. *Cancer* 67: 703–709.

3 Cyclophosphamide

Chemical name: *N,N-Bis* (2-chloroethyl)tetrahydro-2H-1,3,2-oxazaphosphorin-2-amine-2-oxide

CAS #: 6055-19-2

SMILES: P1(N(CCCl)CCCl)(NCCCO1)=O

INTRODUCTION

Cyclophosphamide is an alkylating antineoplastic agent that acts against a wide variety of oncologic and nononcologic conditions (e.g., transplantation prophylaxis, severe rheumatoid disorders) in various therapeutic categories. The drug prevents cell division by cross-linking DNA strands and decreasing DNA synthesis. It is cell-cycle-phase nonspecific (Lacy et al., 2004). The mechanism by which this occurs is apparently through its metabolites phosphoramide mustard and acrolein (Mirkes, 1985). It is available commercially as Cytoxan® and Neosar® among others, and has a pregnancy risk factor of D (inferring it "may cause fetal harm when administered to a pregnant woman").

DEVELOPMENTAL TOXICOLOGY

ANIMALS

Like many other antineoplastic agents, cyclophosphamide elicits a full spectrum of developmental toxicity (malformation, fetal weight inhibition, and embryolethality) in animals. In mice (Hackenberger and Kreybig, 1965), rats (Murphy, 1962), and rabbits (Gerlinger and Clavert, 1964), the drug given intraperitoneally during 2, 3, or 4 days during organogenesis induced multiple malformations (central nervous system, limb, digit, palate, and jaw) in a dosage range of 2 to 50 mg/kg/day. In rats, the drug was teratogenic when administered as early as day 4 of gestation, even prior to implantation (Brock and Kreybig, 1964). In a primate (rhesus species), two syndromes of congenital defects were observed, depending on when treatment occurred: cleft lip/palate and eye defects when administered 10 mg/kg/day on gestation days 27 and 29, and craniofacial dysmorphia on gestation days 32 to 40 by the intramuscular route (Wilk et al., 1978). In contrast to susceptible species, in sheep, cyclophosphamide given 25 mg/kg orally over a wide but late range of single days in gestation caused no developmental toxicity (Dolnick et al., 1970).

Cyclophosphamide is also a reproductive toxicant in animals and has additionally shown male-mediated effects on development. Treating male rats in fertility-type studies with cyclophosphamide resulted in behavioral deficits in F_1 offspring in one study (Adams et al., 1981). In another study, there was a twofold increase in preimplantation loss from treated males siring dams in rats (Hales et al., 1986). Finally, in still another male rat multigenerational reproduction study, developmental anomalies were produced in the third-generation offspring, and physical and behavioral changes were produced in three successive generations (Dulioust et al., 1989; Auroux et al., 1990). Studies suggest that the chemical's effect on female gametes and, subsequently, on future reproduction is influenced by the stage of oocyte maturation at the time of exposure (Meirow et al., 2001).

HUMANS

A distinct phenotype of developmental toxicity was established from at least 11 case reports in humans, as shown in Table 1. Doses required to elicit the phenotype in humans ranged from daily doses over short periods of up to 400 mg/day to a total of 2100 mg over the course of treatment. Therapeutic doses are on the order of 50 to 100 mg/m²/day orally. All treatments where so indicated covered at least the first trimester.

A specific syndrome of defects (embryopathy) was identified to include congenital malformations of digits, palate, ears, facies, and skin. Intrauterine growth retardation (IUGR) was recorded in some cases, and it has been stated that up to 40% of delivered infants from mothers treated with antineoplastic drugs, including cyclophosphamide, have low birth weight (Nicholson, 1968). Abortion or early postnatal death was also recorded in some of the cases, and fetal loss has been a noted characteristic in mothers treated with antineoplastic agents as a group, including cyclophosphamide (Selevan et al., 1985). Functional deficiency, including especially developmental delay, and neurologic deficits were also observed in some cases. With the recorded number of healthy infants born following first trimester cyclophosphamide treatment as being approximately 20 (Lergier et al., 1974; see Schardein, 2000), the risk of malformation is about 40%.

As was the experience in animal studies, congenital malformations (syndactyly, tetralogy of Fallot) were reported in offspring in which there was paternal treatment with cyclophosphamide (combined with other antineoplastic drug treatment (Russell et al., 1976). A child was also reported to have multiple anomalies due to treatment with cyclophosphamide of the father over several years

TABLE 1
Developmental Toxicity Profile of Cyclophosphamide in Humans

Case Number	Malformations	Growth Retardation	Death	Functional Deficit	Ref.
1	Multiple: digits, palate, nose, skin				Greenberg and Tanaka, 1964
2	Digits, heart		✔		Toledo et al., 1971
3	Unspecified "gross" defects		✔		Sosa Munoz et al.,1983
4	Minor unspecified defects		✔		Sosa Munoz et al.,1983
5	Urogenital				Murray et al., 1984
6	Multiple: digits, palate, face, ears, skin			✔	Kirshon et al., 1988
7	Multiple: brain, face, ear, skull, palate, limbs, digits	✔			Mutchinick et al., 1992
8	Multiple: cartilage, esophagus, vessels, renal, genital	✔		✔	Zemlickis et al., 1993
9	"Embryopathy": brain, skull, face, ears, palate, digits	✔		✔	Enns et al., 1999
10	Embryopathy (similar to previous cases)				Vaux et al., 2002
11	Empryopathy				Paladini et al., 2004

(Evenson et al., 1984). Several useful reviews were published on the use of cyclophosphamide during pregnancy (Mirkes, 1985; Gilchrist and Friedman, 1989; Matalon et al., 2004).

CHEMISTRY

Cyclophosphamide is average in size compared to the other human developmental toxicants. It is hydrophilic and capable of interacting primarily as a hydrogen bond acceptor. The parent structure ultimately yields electrophilic metabolites (phosphoramide mustard). The calculated physicochemical and topological properties for cyclophosphamide are as follows.

PHYSICOCHEMICAL PROPERTIES

Parameter	Value
Molecular weight	261.087 g/mol
Molecular volume	204.39 A^3
Density	1.317 g/cm^3
Surface area	270.48 A^2
LogP	−2.957
HLB	14.027
Solubility parameter	21.034 $J^{(0.5)}/cm^{(1.5)}$
Dispersion	17.966 $J^{(0.5)}/cm^{(1.5)}$
Polarity	6.258 $J^{(0.5)}/cm^{(1.5)}$
Hydrogen bonding	8.970 $J^{(0.5)}/cm^{(1.5)}$
H bond acceptor	1.86
H bond donor	0.22
Percent hydrophilic surface	67.34
MR	64.627
Water solubility	1.184 log (mol/M^3)
Hydrophilic surface area	182.12 A^2
Polar surface area	44.73 A^2
HOMO	−10.671 eV
LUMO	0.533 eV
Dipole	3.678 debye

TOPOLOGICAL PROPERTIES (UNITLESS)

Parameter	Value
x0	10.441
x1	6.726
x2	5.489
xp3	4.256
xp4	3.306
xp5	2.488
xp6	1.120
xp7	0.493
xp8	0.246
xp9	0.102
xp10	0.000
xv0	10.323
xv1	7.244

Continued.

Parameter	Value
xv2	5.884
xvp3	4.595
xvp4	3.920
xvp5	2.869
xvp6	1.291
xvp7	0.624
xvp8	0.305
xvp9	0.098
xvp10	0.000
k0	14.240
k1	12.071
k2	5.778
k3	3.273
ka1	12.758
ka2	6.311
ka3	3.656

REFERENCES

Adams, P. M., Fabricant, J. D., and Legator, M. S. (1981). Cyclophosphamide-induced spermatogenic effects detected in the F_1 generation by behavioral testing. *Science* 211: 80–82.

Auroux, M. et al. (1990). Cyclophosphamide in the F_0 male rat — physical and behavioral changes in 3 successive adult generations. *Mutat. Res.* 229: 189–200.

Brock, N. and Kreybig, T. (1964). [Experimental data on testing of drugs for teratogenicity in laboratory rats]. *Naunyn Schmiedebergs Arch. Pharmacol.* 249: 117–145.

Dolnick, E. H., Lindahl, I. L., and Terrill, C. E. (1970). Treatment of pregnant ewes with cyclophosphamide. *J. Anim. Sci.* 31: 944–946.

Dulioust, E. J. et al. (1989). Cyclophosphamide in the male rat — new pattern of anomalies in the Third generation. *J. Androl.* 10: 296–303.

Enns, G. M. et al. (1999). Apparent cyclophosphamide (Cytoxan) embryopathy: A distinct phenotype? *Am. J. Med. Genet.* 86: 237–241.

Evenson, D. P. et al. (1984). Male reproductive capacity may recover following drug treatment with the L-10 protocol for acute lymphocytic leukemia. *Cancer* 53: 30–36.

Gerlinger, P. and Clavert, J. (1964). Action du cyclophosphamide injecte a des lapines gestantes sur les gonads embryonnaires. *C. R. Acad. Sci. [D](Paris)* 258: 2899–2901.

Gilchrist, D. M. and Friedman, J. M. (1989). Teratogenesis and iv cyclophosphamide. *J. Rheumatol.* 16: 1008.

Greenberg, L. H. and Tanaka, K. R. (1964). Congenital anomalies probably induced by cyclophosphamide. *JAMA* 188: 423–426.

Hackenberger, I. and Kreybig, T. (1965). Vergleichende teratologische untersuchungen bei der maus und der ratte. *Arzneimittelforschung* 15: 1456–1460.

Hales, B. F., Smith, S., and Robaire, B. (1986). Cyclophosphamide in the seminal fluid of treated males: Transmission to females by mating and effect on pregnancy outcome. *Toxicol. Appl. Pharmacol.* 84: 423–430.

Kirshon, B. et al. (1988). Teratogenic effects of 1st trimester cyclophosphamide therapy. *Obstet. Gynecol.* 72: 462–464.

Lacy, C. F. et al. (2004). *Drug Information Handbook (Pocket). 2004–2005,* Lexi-Comp. Inc., Hudson, OH.

Lergier, J. E. et al. (1974). Normal pregnancy in multiple myeloma treated with cyclophosphamide. *Cancer* 34: 1018–1022.

Matalon, S. T., Ornoy, A., and Lishner, M. (2004). Review of the potential effects of three commonly used antineoplastic and immunosuppressive drugs (cyclophosphamide, azathioprine, doxorubicin) on the embryo and placenta. *Reprod. Toxicol.* 18: 219–230.

Meirow, D. et al. (2001). Administration of cyclophosphamide at different stages of follicular maturation in mice: Effects on reproductive performance and fetal malformations. *Human Reprod.* 16: 632–637.

Mirkes, P. E. (1985). Cyclophosphamide teratogenesis: A review. *Teratog. Carcinog. Mutag.* 5: 75–88.

Murphy, M. L. (1962). Teratogenic effects in rats of growth inhibiting chemicals, including studies on thalidomide. *Clin. Proc. Child. Hosp.* 18: 307–322.

Murray, C. L. et al. (1984). Multimodal cancer therapy for breast cancer in the first trimester of pregnancy. A case report. *JAMA* 252: 2607–2608.

Mutchinick, O., Aizpuru, E., and Grether, P. (1992). The human teratogenic effect of cyclophosphamide. *Teratology* 45: 329.

Nicholson, H. O. (1968). Cytotoxic drugs in pregnancy. Review of reported cases. *J. Obstet. Gynaecol. Br. Commonw.* 75: 307–312.

Paladini, D. et al. (2004). Prenatal detection of multiple fetal anomalies following inadvertent exposure to cyclophosphamide in the first trimester of pregnancy. *BDR (A)* 70: 99–100.

Russell, J. A., Powles, R. L., and Oliver, R. T. D. (1976). Conception and congenital abnormalities after chemotherapy of acute myelogenous leukemia in two men. *Br. Med. J.* 1: 1508.

Schardein, J. L. (2000). *Chemically Induced Birth Defects*, Third ed., Marcel Dekker, New York, pp. 581, 585.

Selevan, S. G. et al. (1985). A study of occupational exposure to antineoplastic drugs and fetal loss in nurses. *N. Engl. J. Med.* 313: 1173–1178.

Sosa Munoz, J. L. et al. (1983). [Acute leukemia and pregnancy]. *Rev. Invest. Clin.* 35: 55–58.

Toledo, T. M., Harper, R. C., and Moser, R. H. (1971). Fetal effects during cyclophosphamide and irradiation therapy. *Ann. Intern. Med.* 74: 87–91.

Vaux, K. K., Kahole, N., and Jones, K. L. (2002). Pattern of malformation secondary to prenatal exposure to cyclophosphamide. *Teratology* 65: 297.

Wilk, A. L., McClure, H. M., and Horigan, E. A. (1978). Induction of craniofacial malformations in the rhesus monkey with cyclophosphamide. *Teratology* 17: 24A.

Zemlickis, D. et al. (1993). Teratogenicity and carcinogenicity in a twin exposed *in utero* to cyclophosphamide. *Teratog. Carcinog. Mutag.* 13: 139–143.

4 Methotrexate

Chemical name: *N*-[4-[[(2,4-Diamino-6-pteridinyl)methyl]methylamino]benzoyl]-L-glutamic acid

Alternate name: Amethopterin

CAS #: 59-05-2

SMILES: c12c(ncc(n1)CN(c3ccc(cc3)C(NC(CCC(O)=O)C(O)=O)=O)C)nc(nc2N)N

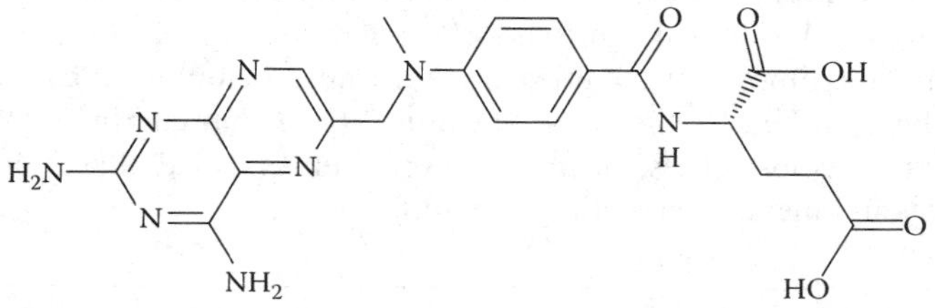

INTRODUCTION

Methotrexate is an antimetabolite (folate antagonist) antineoplastic and immunosuppressant drug. It is a methyl derivative of aminopterin, and it is used therapeutically in treating trophoblastic neoplasms, leukemias, psoriasis, and rheumatoid arthritis. It has largely replaced aminopterin in this therapeutic category; low dose therapy is indicated, however, as the drug, like aminopterin, has abortifacient properties. The drug is cell cycle specific for the S-phase of the cycle and acts by inhibiting DNA synthesis by irreversibly binding to dihydrofolate reductase, inhibiting the formation of reduced folates and thymidylate synthetase, resulting in inhibition of purine and thymidylic acid synthesis (Lacy et al., 2004). Methotrexate is available commercially as Rhematrex® and Trexall®, among others, and carries a pregnancy risk factor of X (contraindicated in pregnancy due to teratogenicity potential).

DEVELOPMENTAL TOXICOLOGY

ANIMALS

Methotrexate is teratogenic in animals when given by parenteral or oral routes of administration. Mice (Skalko and Gold, 1974), rats (Wilson and Fradkin, 1967), and rabbits (Jordan et al., 1970) have shown limb or digit defects and cleft palate. Effective doses ranged from 0.1 to 19.2 mg/kg by the intravenous route and 25 to 50 mg/kg by the intraperitoneal route. Cats had umbilical hernias and skull ossification defects in offspring of females treated with 0.5 mg/kg/day orally when given in 4-day cycles during organogenesis (Khera, 1976). Fetal mortality accompanied the malformations in all species. When given to primates at dosages of 3 to 4 mg/kg intravenously during various days of gestation, the drug did not cause malformations under the experimental conditions employed, but increased abortion rates were seen among the mothers, and there was fetal mortality (Wilson, 1971). Dogs were said to be resistant to the induction of malformations by methotrexate, but details of the study were incompletely reported, and further details were unobtainable (Esaki, 1978). Animal effect levels were generally higher than human therapeutic levels. Using a metabolic derivative of folinic acid, it was shown in the rabbit that the drug causes developmental toxicity by inhibition of dihydrofolate reductase (DeSesso and Goeringer, 1992).

HUMANS

In humans, methotrexate has been shown, from at least 18 case reports, to be an active teratogen and developmental toxicant (Table 1). Schardein (2000) detailed the embryopathy observed in the earlier cases, and Buckley et al. (1997) defined the range of features of the syndrome to include central nervous system abnormalities, including spina bifida, mental retardation, hydrocephaly, and anencephaly; skeletal abnormalities, including synostosis of lambdoid sutures, partial or absent ossification of bones, micrognathia, high or cleft palate, short extremities, wide-set eyes, syndactyly of fingers, absent digits, club foot, large fontanelles, and wide nasal bridge; and, in some cases, dextrocardia. Skull and limb abnormalities are the most common congenital malformations observed from analysis of the case histories shown in Table 1. Intrauterine growth retardation was an associated feature in most cases, death in fewer numbers, and functional deficits, such as developmental delay and mental retardation, in still fewer cases.

TABLE 1
Developmental Toxicity Profile of Methotrexate in Humans

Case Number	Malformations	Growth Retardation	Death	Functional Deficit	Ref.
1	Multiple: skull, digits, ears, face, ribs	✔			Milunsky et al., 1968; Holmes et al., 1972
2[a]	Multiple: skull, face, palate, ears, testes, digits			✔	Hermann and Opitz, 1969
3	Multiple: skull, ears, digits, skin				Powell and Ekert, 1971
4	Multiple: brain, face, ears, genital (female), skull	✔			Diniz et al., 1978
5[a]	Multiple: skull, face, ears, palate, testes, limbs, digits				Reich et al., 1978
6	Multiple: brain, axial skeleton, digits, heart	✔	✔		Buckley et al., 1997
7	Multiple: face, ears, digits, skin, skull (case #1)	✔			Bawle et al., 1998
8	Multiple: face, brain, ears, skin (case #2)	✔		✔	Bawle et al., 1998
9	Multiple: skull, face, ears, palate (case #3)	✔			Bawle et al., 1998
10	Typical embryopathy: skull, face, nipples, abdominal closure, genital, limbs, digits	✔		✔	delCampo et al., 1999
11	Multiple: brain, face, limbs		✔		Lloyd et al., 1999
12, 13	None		✔		Giacalone et al., 1999
14, 15, 16	None	✔			Giacalone et al., 1999
17	Multiple: craniofacial, axial skeletal, cardiopulmonary, gastrointestinal	✔	✔		Nguyen et al., 2002
18	Multiple: brain, face, digits, muscles	✔			Wheeler et al., 2002

[a] Drug intake uncertain — may be aminopterin treated. Face abnormality components include the jaw, eyes, nose, philtrum, and hair.

The dosage required to elicit the syndrome would appear to be on the order of 5 to 7.5 mg/day po (minimum 12.5 mg/week), and all reported treatment intervals were from prior to conception through the first 12 weeks of gestation. Therapeutic doses are in the range of 15 to 20 mg/m^2/2× per week po. These estimates are slightly different from those stated critically as >10 mg/week at 6 to 8 weeks of gestation (Feldkamp and Carey, 1993). Based on the large number of normal infants born following first trimester exposure to methotrexate (Frenkel and Meyers, 1960; Okun et al., 1979; Ayhan et al., 1990; Nantel et al., 1990; Aviles et al., 1991; Green et al., 1991; Giacalone et al., 1999; see Schardein, 2000), the risk for developmental toxicity would appear to be approximately 2.5%. The teratogenic risk is said to be moderate to high by one group of experts (Friedman and Polifka, 2000).

For further information on methotrexate developmental toxicity, see the literature (Christophidis, 1984; Lloyd et al., 1999; McElhatton, 2000).

CHEMISTRY

As the methylated analog of aminopterin, methotrexate has similar properties (relatively large size, hydrophilic, capable of engaging in hydrogen bonding). The calculated physicochemical and topological properties of methotrexate are as follows.

PHYSICOCHEMICAL PROPERTIES

Parameter	Value
Molecular weight	454.446 g/mol
Molecular volume	379.19 A^3
Density	1.452 g/cm^3
Surface area	465.17 A^2
LogP	−3.326
HLB	16.704
Solubility parameter	31.567 J$^{(0.5)}$/cm$^{(1.5)}$
Dispersion	26.141 J$^{(0.5)}$/cm$^{(1.5)}$
Polarity	8.714 J$^{(0.5)}$/cm$^{(1.5)}$
Hydrogen bonding	15.400 J$^{(0.5)}$/cm$^{(1.5)}$
H bond acceptor	3.40
H bond donor	1.88
Percent hydrophilic surface	78.97
MR	122.652
Water solubility	−1.940 log (mol/M^3)
Hydrophilic surface area	367.36 A^2
Polar surface area	220.02 A^2
HOMO	−8.965 eV
LUMO	−1.408 eV
Dipole	6.005 debye

TOPOLOGICAL PROPERTIES (UNITLESS)

Parameter	Value
x0	24.121
x1	15.634
x2	14.733
xp3	11.230

Continued.

Parameter	Value
xp4	8.927
xp5	7.114
xp6	4.975
xp7	3.445
xp8	2.074
xp9	1.646
xp10	1.008
xv0	17.595
xv1	9.750
xv2	7.200
xvp3	4.721
xvp4	3.027
xvp5	1.917
xvp6	1.072
xvp7	0.614
xvp8	0.321
xvp9	0.204
xvp10	0.105
k0	48.907
k1	27.585
k2	12.808
k3	8.000
ka1	24.060
ka2	10.351
ka3	6.227

REFERENCES

Aviles, A., Diaz-Moques, J. C., and Talavera, A. (1991). Growth and development of children of mothers treated with chemotherapy during pregnancy. Current status of 43 children. *Am. J. Hematol.* 36: 243–248.

Ayhan, A. et al. (1990). Pregnancy after chemotherapy for gestational trophoblastic disease. *J. Reprod. Med.* 35: 522–524.

Bawle, E. V., Conard, J. V., and Weiss, L. (1998). Adult and two children with fetal methotrexate syndrome. *Teratology* 57: 51–55.

Buckley, L. M. et al. (1997). Multiple congenital anomalies associated with weekly low dose methotrexate treatment of the mother. *Arthritis Rheu.* 40: 971–973.

Christophidis, N. (1984). Methotrexate. *Clin. Rheum. Dis.* 10: 401–415.

delCampo, M. et al. (1999). Developmental delay in fetal aminopterin/methotrexate syndrome. *Teratology* 60: 10–12.

DeSesso, J. M. and Goeringer, G. C. (1992). Methotrexate-induced developmental toxicity in rabbits is ameliorated by 1-(p-tosyl)-3,4,4-trimethylimidazolidine, a functional analog for tetrahydrofolate-mediated one-carbon transfer. *Teratology* 45: 271–283.

Diniz, E. M. et al. (1978). [Effect, on the fetus, of methotrexate (amethopterin) administered to the mother. Presentation of a case]. *Rev. Hosp. Clin. Fac. Sao Paolo* 33: 286–290.

Esaki, K. (1978). The beagle dog in embryotoxicity tests. *Teratology* 18: 129–130.

Feldkamp, M. and Carey, J. C. (1993). Clinical teratology counseling and consultation case report: Low dose methotrexate exposure in the early weeks of pregnancy. *Teratology* 7: 533–539.

Frenkel, E. P. and Meyers, M. C. (1960). Acute leukemia and pregnancy. *Ann. Intern. Med.* 53: 656–671.

Friedman, J. M. and Polifka, J. E. (2000). *Teratogenic Effects of Drugs. A Resource for Clinicians (TERIS)*, Second ed., Johns Hopkins University Press, Baltimore, MD.

Giacalone, P. L., Laffargue, F., and Benos, P. (1999). Chemotherapy for breast carcinoma during pregnancy. *Cancer* 86: 2266–2272.

Green, D. M. et al. (1991). Congenital anomalies in children of patients who received chemotherapy for cancer in childhood and adolescence. *N. Engl. J. Med.* 325: 141–146.

Hermann, J. and Opitz, J. M. (1969). An unusual form of acrocephalosyndactyly. *Birth Defects* 5: 39–42.

Holmes, L. B. et al. (1972). *Mental Retardation: An Atlas of Diseases with Associated Physical Abnormalities.* Macmillan, New York, p. 134.

Jordan, R. L., Terapane, J. F., and Schumacher, H. J. (1970). Studies on the teratogenicity of methotrexate in rabbits. *Teratology* 3: 203.

Khera, K. S. (1976). Teratogenicity studies with methotrexate, aminopterin, and acetylsalicylic acid in domestic cats. *Teratology* 14: 21–28.

Lacy, C. F. et al. (2004). *Drug Information Handbook (Pocket), 2004–2005*, Lexi-Comp., Inc., Hudson, OH.

Lloyd, M. E. et al. (1999). The effects of methotrexate on pregnancy, fertility and lactation. *Q. J. Med.* 92: 551–563.

McElhatton, P. R. (2000). A review of the reproductive toxicity of methotrexate in human pregnancy. *Reprod. Toxicol.* 14: 549.

Milunsky, A., Graef, J. W., and Gaynor, M. F. (1968). Methotrexate-induced congenital malformations. *J. Pediatr.* 72: 790–795.

Nantel, S., Parboosingh, J., and Poon, M. C. (1990). Treatment of an aggressive non-Hodgkin's lymphoma during pregnancy with MACOP-B chemotherapy. *Med. Pediatr. Oncol.* 18: 143–145.

Nguyen, C. et al. (2002). Multiple anomalies in a fetus exposed to low-dose methotrexate in the first trimester. *Obstet. Gynecol.* 99: 599–602.

Okun, D. B. et al. (1979). Acute leukemia in pregnancy. Transient neonatal myelosuppression after combination chemotherapy in the mother. *Med. Pediatr. Oncol.* 7: 315–319.

Powell, H. R. and Ekert, H. (1971). Methotrexate-induced congenital malformations. *Med. J. Aust.* 2: 1076–1077.

Reich, E. W. et al. (1978). Recognition in adult patients of malformations induced by folic acid antagonists. *Birth Defects* 14: 139–160.

Schardein, J. L. (2000). *Chemically Induced Birth Defects*, Third ed., Marcel Dekker, New York, pp. 586–587.

Skalko, R. G. and Gold, M. P. (1974). Teratogenicity of methotrexate in mice. *Teratology* 9: 159–164.

Wheeler, M., O'Meara, P., and Stanford, M. (2002). Fetal methotrexate and misoprostol exposure: The past revisited. *Teratology* 66: 73–76.

Wilson, J. G. (1971). Use of rhesus monkeys in teratological studies. *Fed. Proc.* 30: 104–109.

Wilson, J. G. and Fradkin, R. (1967). Interrelations of mortality and malformations in rats. In *Absts. Seventh Annu. Mtg. Teratology Society* pp. 57–58.

5 Chlorambucil

Chemical name: 4-[*Bis*(2-chloroethyl)amino]benzene butanoic acid

Alternate name: Chloraminophene

CAS #: 305-03-3

SMILES: c1(ccc(cc1)CCCC(O)=O)N(CCCl)CCCl

INTRODUCTION

Chlorambucil is an alkylating antineoplastic agent used therapeutically in the management of chronic lymphocytic leukemia, Hodgkin's and non-Hodgkin's lymphoma and several other malignancies. It is a derivative of mechlorethamine, another human developmental toxicant. As with other alkylators, chlorambucil interferes with DNA replication and RNA transcription by alkylation and cross-linking DNA strands (Lacy et al., 2004). The drug is commercially available as Leukeran® and has a pregnancy risk factor of D (labeling states "can cause fetal harm when administered to a pregnant woman: it is probably teratogenic in humans").

DEVELOPMENTAL TOXICOLOGY

ANIMALS

The drug has been tested in the laboratory for developmental toxicity in rodents. As expected, the drug is teratogenic in mice (Didcock et al., 1956) and rats (Murphy et al., 1958) when given intraperitoneally either once or twice during organogenesis at doses ranging from 6 to 40 mg/kg/day. Digit, limb, and central nervous system defects and cleft palate were produced. The drug is also embryolethal and causes stunting in rodents at doses in the range of 5 to 10 mg/kg by the same route (Murphy et al., 1958; Tanimura et al., 1965).

HUMANS

Chlorambucil is also teratogenic in humans: Four cases of congenital malformation were identified from published accounts in the literature, as tabulated in Table 1. Doses administered resulting in

TABLE 1
Developmental Toxicity Profile of Chlorambucil in Humans

Case Number	Malformations	Growth Retardation	Death	Functional Deficit	Ref.
1	None		✔		Revol et al., 1962
2	Kidney and ureter				Shotton and Monie, 1963
3	Eye		✔		Rugh and Skaredoff, 1965
4	None		✔		Nicholson, 1968
5	Kidney and ureter		✔		Steeger and Caldwell, 1980
6	Heart		✔		Thompson and Conklin, 1983

this toxicity ranged from 4 to 24 mg/day orally, and treatments ranged from conception through the 20th week of gestation. These levels are greater than the usual therapeutic doses of 2 to 4 mg/day po. There does not appear to be a syndrome of malformations other than for near-identical malformations of the urogenital system in two of the four cases. Interestingly, similar if not identical defects were produced in rats treated with the drug (Monie, 1961). Case # 5 was a twin pregnancy, and the other infant was spared the defect. The malformations were accompanied in five of the six cases by infant death. No intrauterine growth retardation was recorded, contrary to a review of treatment with antineoplastic drug therapy in humans, including chlorambucil, which was said to result in 40% of infants having low birth weight (Nicholson, 1968). There were no postnatal functional alterations reported in the single surviving infant. Only four nonmalformed infants were reported after chlorambucil use (Baynes et al., 1968; Nicholson, 1968; Jacobs et al., 1981; Zuazu et al., 1991); therefore, based on this published information, the risk to developmental toxicity is high, on the order of 60%.

CHEMISTRY

Chlorambucil is an aniline mustard of near average size. The compound is hydrophobic and of low polarity. Chlorambucil can participate in hydrogen bonding. The calculated physicochemical and topological properties are as follows.

PHYSICOCHEMICAL PROPERTIES

Parameter	Value
Molecular weight	304.216 g/mol
Molecular volume	265.58 A^3
Density	1.210 g/cm^3
Surface area	340.92 A^2
LogP	2.911
HLB	3.293
Solubility parameter	22.992 $J^{(0.5)}/cm^{(1.5)}$
Dispersion	20.768 $J^{(0.5)}/cm^{(1.5)}$
Polarity	5.261 $J^{(0.5)}/cm^{(1.5)}$
Hydrogen bonding	8.347 $J^{(0.5)}/cm^{(1.5)}$
H bond acceptor	0.68
H bond donor	0.29
Percent hydrophilic surface	20.665
MR	80.194

Continued.

Parameter	Value
Water solubility	-2.360 log (mol/M^3)
Hydrophilic surface area	70.45 A^2
Polar surface area	43.70 A^2
HOMO	-9.280 eV
LUMO	-0.005 eV
Dipole	1.610 debye

TOPOLOGICAL PROPERTIES (UNITLESS)

Parameter	Value
x0	14.088
x1	9.168
x2	7.443
xp3	5.271
xp4	4.120
xp5	3.078
xp6	2.052
xp7	1.275
xp8	0.843
xp9	0.525
xp10	0.286
xv0	12.330
xv1	7.416
xv2	5.035
xvp3	3.204
xvp4	2.299
xvp5	1.510
xvp6	0.877
xvp7	0.436
xvp8	0.283
xvp9	0.151
xvp10	0.067
k0	21.286
k1	17.053
k2	9.834
k3	6.817
ka1	16.444
ka2	9.319
ka3	6.389

REFERENCES

Baynes, T. L. S., Crickmay, G. F., and Vaughan Jones, R. (1968). Pregnancy in a case of chronic lymphatic leukemia. *Br. J. Obstet. Gynaecol.* 75: 1165–1168.

Didcock, K. A., Jackson, D., and Robson, J. M. (1956). The action of some nucleotoxic substances in pregnancy. *Br. J. Pharmacol.* 11: 437–441.

Jacobs, C. et al. (1981). Management of the pregnant patient with Hodgkin's disease. *Ann. Intern. Med.* 95: 669–675.

Lacy, C. F. et al. (2004). *Drug Information Handbook (Pocket), 2004–2005*, Lexi-Comp. Inc., Hudson, OH.

Monie, I. W. (1961). Chlorambucil-induced abnormalities of urogenital system of rat fetuses. *Anat. Rec.* 139: 145.

Murphy, M. L., Moro, A. D., and Lacon, C. (1958). Comparative effects of five polyfunctional alkylating agents on the rat fetus, with additional notes on the chick embryo. *Ann. NY Acad. Sci.* 68: 762–782.

Nicholson, H. O. (1968). Cytotoxic drugs in pregnancy. Review of reported cases. *J. Obstet. Gynaecol. Br. Commonw.* 75: 307–312.

Revol, L. et al. (1962). [Hodgkin's disease, lymphosarcoma, reticulosarcoma and pregnancy]. *Nouv. Rev. Fr. Hematol.* 2: 311–325.

Rugh, R. and Skaredoff, L. (1965). Radiation and radiomimetic chlorambucil and the fetal retina. *Arch. Ophthalmol.* 74: 382–393.

Shotton, D. and Monie, I. W. (1963). Possible teratogenic effect of chlorambucil on a human fetus. *JAMA* 186: 74–75.

Steege, J. F. and Caldwell, D. S. (1980). Renal agenesis after first trimester exposure to chlorambucil. *South. Med. J.* 73: 1414–1415.

Tanimura, T. et al. (1965). [Comparison of teratogenic effects between single and repeated administration of chemical agents]. *Kaibogaku Zasshi* 40: 13.

Thompson, J. and Conklin, K. A. (1983). Anesthetic management of a pregnant patient with scleroderma. *Anesthesiology* 59: 69–71.

Zuazu, J. et al. (1991). Pregnancy outcome in hematologic malignancies. *Cancer* 67: 703–709.

6 Mechlorethamine

Chemical name: 2-Chloro-*N*-(2-chloroethyl)-*N*-methylethanamine

Alternate names: Nitrogen mustard, chlormethine

CAS #: 51-75-2

SMILES: N(CCCl)(CCCl)C

INTRODUCTION

Mechlorethamine is an alkylating antineoplastic drug that has therapeutic utility in combination therapy for Hodgkin's disease and non-Hodgkin's lymphoma and other malignant lymphomas. The drug inhibits DNA and RNA synthesis via formation of carbonium ions by cross-linking strands of DNA, causing miscoding, breakage, and failure of replication. While the drug is not cell-phase specific, its effect is most pronounced in the S-phase, and cell proliferation is arrested in the G_2 phase (Lacy et al., 2004). The drug used commercially has the trade name Mustargen®, among others, and it has a pregnancy risk factor of D (labeling states "can cause fetal harm when administered to a pregnant woman").

DEVELOPMENTAL TOXICOLOGY

ANIMALS

Mechlorethamine is teratogenic in all laboratory species tested. Parenteral routes of administration were used. In the mouse, subcutaneous or intraperitoneal injection caused digit anomalies and hydrocephalus, as well as growth retardation and embryolethality (Danforth and Center, 1954; Thalhammer and Heller-Szollosy, 1955). In rats, mechlorethamine elicited multiple malformations, death, and growth retardation following subcutaneous administration (Haskin, 1948). Malformations were also produced in rabbits after intravenous dosing early in gestation (Gottschewski, 1964). In a seldom-used animal species, the ferret, malformations were induced in high incidence upon injection of mechlorethamine (Beck et al., 1976). Developmental toxicity was produced in animals at parenteral doses ranging from 1 μg/g/day in mice, 0.1 mg/kg/day in rabbits, 0.5 mg/kg/day in ferrets, to 1 mg/kg/day in rats, in decreasing order of sensitivity.

TABLE 1
Developmental Toxicity Profile of Mechlorethamine in Humans

Case Number	Malformations	Growth Retardation	Death	Functional Deficit	Ref.
1	None		✔		Revol et al., 1962
2	None		✔		Nicholson, 1968
3	None		✔		Nicholson, 1968
4	None		✔		Nicholson, 1968
5	Multiple: bone, digits, brain, ear				Garrett, 1974
6	Renal		✔		Mennuti et al., 1975
7	Heart	✔	✔		Thomas and Peckham, 1976
8	Palate				McKeen et al., 1979
9	Brain				McKeen et al., 1979
10	Inner ear				McKeen et al., 1979
11	None			✔	McKeen et al., 1979
12	None		✔		McKeen et al., 1979
13	Digits				Thomas and Andes, 1982
14	Brain		✔		Zemlickis et al., 1993

HUMANS

The drug has been associated with congenital malformation in the human as well. Recorded in the literature were at least eight cases from first trimester exposure as well as other developmental toxicity as shown in Table 1. The malformations recorded are diverse, having similarities in two cases with digit abnormalities and three with brain defects. The digit defects were similar to some recorded animal malformations. Further, in most cases, mechlorethamine was accompanied by combined antineoplastic drug treatment (especially procarbazine and vinblastine/vincristine as MOPP); thus, the teratogenic effect of this drug cannot be established with certainty. Malformations were accompanied by dysmaturity in one case and learning disability in a single case, neither of which are considered significant biological effects in the developmental toxicity parameter of this agent. Death or abortion occurred in the majority of the cases and was considered an associated feature of the developmental toxicity profile of mechlorethamine. Doses recorded in the cases, when stated, were 4 to 6 mg/m^2 po, and all cases are believed to have been limited to treatment in the first trimester. Therapeutic doses of this drug are much lower, 0.4 mg/kg (single dose) or 0.1 mg/kg (repeated doses) by the iv route, which are doses similar to the effect levels in animal studies. Based on the number of unaffected infants born after being exposed to mechlorethamine during the first trimester (Nicholson, 1968; Jones and Weinerman, 1979; McKeen et al., 1979; Whitehead et al., 1983; Andrieu and Ochoa-Molina, 1983; Green et al., 1991; see Schardein, 2000), the risk of developmental toxicity from this agent appears to be on the order of 22%. One group of experts stated the magnitude of teratogenic risk for this drug to be small to moderate (Friedman and Polifka, 2000).

For more information, see the review article by Dein et al. (1984) on the developmental toxicity of mechlorethamine and other antineoplastic drugs useful in treating Hodgkin's disease.

CHEMISTRY

Mechlorethamine is a nitrogen mustard of relatively small size. The chemical is hydrophobic. Its potential to engage in hydrogen bonding (as an acceptor) is relative low compared to the other human developmental toxicants. The calculated physicochemical and topological parameters for mechlorethamine are listed below.

PHYSICOCHEMICAL PROPERTIES

Parameter	Value
Molecular weight	156.054 g/mol
Molecular volume	134.15 A^3
Density	1.166 g/cm^3
Surface area	187.47 A^2
LogP	0.372
HLB	0.000
Solubility parameter	20.372 J$^{(0.5)}$/cm$^{(1.5)}$
Dispersion	17.539 J$^{(0.5)}$/cm$^{(1.5)}$
Polarity	8.087 J$^{(0.5)}$/cm$^{(1.5)}$
Hydrogen bonding	6.483 J$^{(0.5)}$/cm$^{(1.5)}$
H bond acceptor	0.13
H bond donor	0.00
Percent hydrophilic surface	0.70
MR	38.964
Water solubility	1.044 log (mol/M^3)
Hydrophilic surface area	1.30 A^2
Polar surface area	3.24 A^2
HOMO	−9.695 eV
LUMO	0.867 eV
Dipole	1.816 debye

TOPOLOGICAL PROPERTIES (UNITLESS)

Parameter	Value
x0	6.406
x1	3.808
x2	2.682
xp3	1.563
xp4	1.130
xp5	0.289
xp6	0.144
xp7	0.000
xp8	0.000
xp9	0.000
xp10	0.000
xv0	6.543
xv1	3.683
xv2	2.437
xvp3	1.270
xvp4	0.978
xvp5	0.254
xvp6	0.144
xvp7	0.000
xvp8	0.000
xvp9	0.000

Continued.

Parameter	Value
xvp10	0.000
k0	5.418
k1	8.000
k2	5.143
k3	5.000
ka1	8.540
ka2	5.673
ka3	5.540

REFERENCES

Andrieu, J. M. and Ochoa-Molina, M. E. (1983). Menstrual cycle, pregnancies and offspring before and after MOPP therapy for Hodgkin's disease. *Cancer* 52: 435–438.

Beck, F. et al. (1976). Comparison of the teratogenic effects of mustine hydrochloride in rats and ferrets. The value of the ferret as an experimental animal in teratology. *Teratology* 13: 151–160.

Danforth, C. H. and Center, E. (1954). Nitrogen mustard as a teratogenic agent in the mouse. *Proc. Soc. Exp. Biol. Med.* 86: 705–707.

Dein, R. A. et al. (1984). The reproductive potential of young men and women with Hodgkin's disease. *Obstet. Gynecol. Surv.* 39: 474–482.

Friedman, J. M. and Polifka, J. E. (2000). *Teratogenic Effects of Drugs. A Resource for Clinicians (TERIS)*, Second ed., Johns Hopkins University Press, Baltimore, MD.

Garrett, M. J. (1974). Teratogenic effects of combination chemotherapy. *Ann. Intern. Med.* 80: 667.

Gottschewski, G. H. M. (1964). Mammalian blastopathies due to drugs. *Nature* 201: 1232–1233.

Green, D. M. et al. (1991). Congenital anomalies in children of patients who received chemotherapy for cancer in childhood and adolescence. *N. Engl. J. Med.* 325: 141–146.

Haskin, D. (1948). Some effects of nitrogen mustard on the development of external body form in the fetal rat. *Anat. Rec.* 102: 493–511.

Jones, R. T. and Weinerman, B. H. (1979). MOPP (nitrogen mustard, vincristine, procarbazine and prednisone) given during pregnancy. *Obstet. Gynecol.* 54: 477.

Lacy, C. F. et al. (2004). *Drug Information Handbook (Pocket) 2004–2005*, Lexi-Comp, Inc., Hudson, OH.

McKeen, E. A. et al. (1979). Pregnancy outcome in Hodgkin's disease. *Lancet* 2: 590.

Mennuti, M. T., Shepard, T. H., and Mellman, W. J. (1975). Fetal renal malformation following treatment of Hodgkin's disease during pregnancy. *Obstet. Gynecol.* 46: 194–196.

Nicholson, H. O. (1968). Cytotoxic drugs in pregnancy. Review of reported cases. *J. Obstet. Gynaecol. Br. Commonw.* 75: 307–312.

Revol, L. et al. (1962). [Hodgkin's disease, lymphosarcoma, reticulosarcoma and pregnancy]. *Nouv. Rev. Fr. Hematol.* 2: 311–325.

Schardein, J. L. (2000). *Chemically Induced Birth Defects*, Third ed., Marcel Dekker, New York, p. 585.

Thalhammer, O. and Heller-Szollosy, E. (1955). Exogene Bildungfehler ("Missbildungen") durch Lostinjekion bei der graviden Maus (Ein Beitrag zur Pathogenese von Bildungsfehlern). *Z. Kinderheilk.* 76: 351.

Thomas, L. and Andes, W. A. (1982). Fetal anomaly associated with successful chemotherapy for Hodgkin's disease during the first trimester of pregnancy. *Clin. Res.* 30: 424A.

Thomas, P. R. M. and Peckham, M. J. (1976). The investigation and management of Hodgkin's disease in the pregnant patient. *Cancer* 38: 1443–1451.

Whitehead, E. et al. (1983). The effect of combination chemotherapy on ovarian function in women treated for Hodgkin's disease. *Cancer* 52: 988–993.

Zemlickis, D. et al. (1993). Teratogenicity and carcinogenicity in a twin exposed *in utero* to cyclophosphamide. *Teratog. Carcinog. Mutag.* 13: 139–143.

7 Cytarabine

Chemical name: 1-β-D-Arabinofuranosylcytosine

Alternate names: Ara-C, cytosine arabinoside, aracytidine

CAS #: 147-94-4

SMILES: C1(N2C(N=C(C=C2)N)=O)OC(C(C1O)O)CO

INTRODUCTION

Cytarabine is a purine antimetabolite used therapeutically as an antineoplastic agent, as it is active in treating leukemia and lymphoma. Its mechanism of action is by inhibition of DNA synthesis, through conversion to its active compound, aracytidine triphosphate, which is incorporated into DNA, inhibiting DNA polymerase and resulting in decreased DNA synthesis and repair; it is rapidly metabolized (Lacy et al., 2004). The drug is specific for the S phase of the cell cycle. Commercially available as Cytosar®, it has a pregnancy risk factor of D. (This category would indicate that the drug can cause fetal harm when administered to a pregnant woman.)

DEVELOPMENTAL TOXICOLOGY

ANIMALS

Among animal studies, cytarabine is teratogenic, and it increased fetal mortality and inhibited fetal body weight when given to mice during the organogenesis period of gestation (Puig et al., 1991). Cleft palate, renoureteral agenesis or hypoplasia, and poly- or oligodactyly in association with maternal toxicity were observed at intraperitoneal (IP) dose levels of 2 and 8 mg/kg/day, and resorption and decreased fetal body weight were observed at the higher dose. In an earlier study in mice, researchers recorded microcephaly with microscopic central nervous system malformations at a higher dose of 30 mg/kg/day ip (Kasubuchi et al., 1973). In another study, researchers observed the full pattern of developmental toxicity at an intravenous dose of 1.5 mg/kg/day during organogenesis in the same species (Nomura et al., 1969). In the rat, IP doses over a wide range (20 to 800 mg/kg/day) during 4 days of organogenesis produced cleft palate, limb, tail, and digit malformations, and fetal death in the offspring (Chaube et al., 1968). Toxicity was also recorded in the

TABLE 1
Developmental Toxicity Profile of Cytarabine in Humans

Cas Number	Malformations	Growth Retardation	Death	Functional Deficit	Ref.
1–11	None		✔		Newcomb et al., 1978; O'Donnell et al., 1979; Homer et al., 1979; Pizzuto et al., 1980; Taylor and Blom, 1980; DeSouza et al., 1982; Plows, 1982; Cantini and Yanes, 1984; Fassas et al., 1984; Volkenandt et al., 1987; Juarez et al., 1988
12	None	✔			Pizzuto et al., 1980
13	Multiple: ear, bone, digits				Wagner et al., 1980
14	Digits				Schafer, 1981
15	None	✔			Juarez et al., 1988
16	Multiple: face, digits, bone, brain			✔	Artlich et al., 1994

2-day-old neonatal rat at doses of 4 mg/kg for 5 days by the IP route (Gough et al., 1982). The toxicity was manifested by weight gain suppression, delayed hair growth, toxic clinical signs, cerebellar hypoplasia, retinal dysplasia, and delayed nephrogenesis.

HUMANS

There are a few recorded cases of malformation in humans (Table 1). Of some 16 cases illustrating developmental toxicity in humans with cytarabine (and usually combined antineoplastic therapy), 3 had malformations, all with digit defects, accompanied by large bone (leg) malformations in 2 of the cases. Dosage was not specified except in one case, at 160 mg/day intravenously in the first 2 months of pregnancy. Therapeutic doses are in the range of 100 mg to 3 g/m²/day. Intrauterine growth retardation was recorded in two cases in the published literature; intrauterine death in many cases; and functional deficit, defined as a slight retardation in postnatal motor milestones, in a single case (in this case the patient also had multiple anatomic malformations). Neither growth retardation nor functional deficits are considered representative characteristics of the developmental toxicity profile of cytarabine based on the few cases reported. In addition to the developmental effects, chromosomal abnormalities were also reported in several case reports (Maurer et al., 1971; Schleuning and Clemm, 1987). Paternal use of cytarabine combined with other antineoplastic drugs prior to conception was said to result in congenital anomalies (Russell et al., 1976). Based on the number of published cases of unaffected infants born following first trimester exposure to cytarabine (Lilleyman et al., 1977; Catanzarite and Ferguson, 1984; Reynoso et al., 1987; Juarez et al., 1988; see Schardein, 2000), the risk for developmental toxicity in the human associated with cytarabine is rather high, especially due to intrauterine death, at approximately 64%. The teratogenic risk of cytarabine is considered by one group of experts to be small to moderate in extent (Friedman and Polifka, 2000). Several thorough reviews of cytarabine combined therapy and pregnancy outcome were published (Catanzarite and Ferguson, 1984; Caliguri and Mayer, 1989).

CHEMISTRY

Cytarabine is a hydrophobic chemical of near average size as compared with the other human developmental toxicants. It is polar and capable of engaging in donor/acceptor hydrogen bonding interactions. The calculated physicochemical and topological properties are as follows.

PHYSICOCHEMICAL PROPERTIES

Parameter	Value
Molecular weight	243.219 g/mol
Molecular volume	196.16 A^3
Density	1.368 g/cm^3
Surface area	249.02 A^2
LogP	−0.959
HLB	21.540
Solubility parameter	36.455 J$^{(0.5)}$/cm$^{(1.5)}$
Dispersion	23.285 J$^{(0.5)}$/cm$^{(1.5)}$
Polarity	12.664 J$^{(0.5)}$/cm$^{(1.5)}$
Hydrogen bonding	25.028 J$^{(0.5)}$/cm$^{(1.5)}$
H bond acceptor	2.10
H bond donor	1.21
Percent hydrophilic surface	100.00
MR	57.692
Water solubility	4.405 log (mol/M^3)
Hydrophilic surface area	249.02 A^2
Polar surface area	133.99 A^2
HOMO	−9.505 eV
LUMO	−0.568 eV
Dipole	5.984 debye

TOPOLOGICAL PROPERTIES (UNITLESS)

Parameter	Value
x0	12.577
x1	8.041
x2	7.346
xp3	6.395
xp4	5.026
xp5	3.584
xp6	2.412
xp7	1.351
xp8	0.765
xp9	0.460
xp10	0.196
xv0	8.801
xv1	5.014
xv2	3.771
xvp3	2.643
xvp4	1.729
xvp5	1.027
xvp6	0.530
xvp7	0.251
xvp8	0.119
xvp9	0.056
xvp10	0.018
k0	20.918
k1	13.432
k2	5.325
k3	2.560
ka1	12.301
ka2	4.607
ka3	2.136

REFERENCES

Artlich, A. et al. (1994). Teratogenic effects in a case of maternal treatment for acute myelocytic leukaemia-neonatal and infantile course. *Eur. J. Pediatr.* 153: 488–491.

Caliguri, M. A. and Mayer, R. J. (1989). Pregnancy and leukemia. *Semin. Oncol.* 16: 388–396.

Cantini, E. and Yanes, B. (1984). Acute myelogenous leukemia in pregnancy. *South. Med. J.* 77: 1050–1051.

Catanzarite, V. A. and Ferguson, J. E. (1984). Acute leukemia and pregnancy: A review of management and outcome. *Obstet. Gynecol. Surv.* 39: 663–678.

Chaube, S. et al. (1968). The teratogenic effect of 1-β-D-arabinofuranosylcytosine in the rat. Protection by deoxycytidine. *Biochem. Pharmacol.* 17: 1213–1216.

DeSouza, J. J. L. et al. (1982). Acute leukaemia in pregnancy. *S. Afr. Med. J.* 62: 295–296.

Fassas, A. et al. (1984). Chemotherapy for acute leukemia during pregnancy: Five case reports. *Nouv. Rev. Fr. Hematol.* 26: 19–24.

Friedman, J. M. and Polifka, J. E. (2000). *Teratogenic Effects of Drugs. A Resource for Clinicians (TERIS),* Second ed., Johns Hopkins University Press, Baltimore, MD.

Gough, A. W. et al. (1982). Comparison of the neonatal toxicity of two antiviral agents — vidarabine phosphate and cytarabine. *Toxicol. Appl. Pharmacol.* 66: 143–152.

Homer, J. W., Beard, E. J., and Duff, G. B. (1979). Pregnancy complicated by acute myeloid leukaemia. *Aust. NZ J. Med.* 89: 212–213.

Juarez, S. (1988). Association of leukemia and pregnancy: Clinical and obstetric aspects. *Am. J. Clin. Oncol.* 11: 159–165.

Kasubuchi, Y. et al. (1973). Cytosine arabinoside induced microcephaly in mice. *Teratology* 8: 96.

Lacy, C. F. et al. (2004). *Drug Information Handbook (Pocket), 2004–2005,* Lexi-Comp., Inc., Hudson, OH.

Lilleyman, J. S., Hill, A. S., and Anderton, K. J. (1977). Consequences of acute myelogenous leukemia in early pregnancy. *Cancer* 40: 1300–1303.

Maurer, L. H. (1971). Fetal group C trisomy after cytosine arabinoside and thioguanine. *Ann. Intern. Med.* 75: 809–810.

Newcomb, M. et al. (1978). Acute leukemia in pregnancy: Successful delivery after cytarabine and doxorubicin. *JAMA* 239: 2691–2692.

Nomura, A. et al. (1969). [Teratogenic effects of 1-β-D-arabinofuranosyl-cytosine (AC-1075) in mice and rats]. *Gendai No Rinsho* 3: 758.

O'Donnell, R., Costigon, C., and O'Donnell, L. G. (1979). Two cases of acute leukemia in pregnancy. *Acta Haematol.* 61: 298–300.

Pizzuto, J. et al. (1980). Treatment of acute leukemia in pregnancy: Presentation of nine cases. *Cancer Treat. Rep.* 64: 679–683.

Plows, C. W. (1982). Acute myelomonocytic leukemia in pregnancy: Report of a case. *Am. J. Obstet. Gynecol.* 143: 41–43.

Puig, M. et al. (1991). Embryotoxic and teratogenic effects of cytosine arabinoside in mice. *Toxicologist* 11: 341.

Reynoso, E. E. et al. (1987). Acute leukemia during pregnancy: The Toronto Leukemia Study Group experience with long term follow-up of children exposed *in utero* to chemotherapeutic agents. *J. Clin. Oncol.* 5: 1098–1106.

Russell, J. A., Powles, R. L., and Oliver, R. T. D. (1976). Conception and congenital abnormalities after chemotherapy of acute myelogenous leukaemia in two men. *Br. Med. J.* 1: 1508.

Schafer, A. I. (1981). Teratogenic effects of antileukemic chemotherapy. *Arch. Intern. Med.* 141: 514–515.

Schardein, J. L. (2000). *Chemically Induced Birth Defects,* Third ed., Marcel Dekker, New York, pp. 591, 593.

Schleuning, M. and Clemm, C. (1987). Chromosomal aberrations in a newborn whose mother received cytotoxic treatment during pregnancy. *N. Engl. J. Med.* 317: 1666–1667.

Taylor, G. and Blom, J. (1980). Acute leukemia in pregnancy. *South. Med. J.* 73: 1314–1315.

Volkenandt, M. et al. (1987). Acute leukemia during pregnancy. *Lancet* 2: 1521–1522.

Wagner, V. M. et al. (1980). Congenital abnormalities in baby born to cytarabine treated mother. *Lancet* 2: 98–99.

8 Tretinoin

Chemical name: *All-trans* retinoic acid

Alternate name: Vitamin A acid

CAS #: 302-79-4

SMILES: C1(C(CCCC=1C)(C)C)C=CC(=CC=CC(=CC(O)=O)C)C

INTRODUCTION

Tretinoin is a vitamin A derivative with antipsoriatic properties when applied topically as a come-dolytic agent with utility in acne vulgaris and in photodamaged skin and in treating premalignant skin conditions. The drug is also used therapeutically by administration through the oral route in treating acute promyelocytic leukemia. It, and other retinoids, function to normalize the maturation of follicular epithelium, reduce inflammation, and enhance the penetration of other topical medications (Hardman et al., 2001). Retinoic acid is apparently identical to the body's own growth factor present in all cells and bound to specific retinoid receptors (Schaefer, 2001). The drug is available commercially as Renova®, Retin-A®, or by several other trade names and has a pregnancy risk factor of C (topical) or D (systemic).

DEVELOPMENTAL TOXICOLOGY

ANIMALS

In the laboratory, tretinoin was tested for developmental toxicity by the topical route, and the only consistent results have been obtained in the rabbit (Zbinden, 1975) and hamster (Sharma et al., 1990). In neither species were there malformations induced at doses of 0.1% or 30 mg/kg/day, respectively. However, fetal growth retardation and fetal death were observed in topically treated rabbits (Christian et al., 1997). The package label for the drug indicates the same for the rat. Studies by the oral route have shown teratogenicity in the rat (Collins et al., 1994), pig (Jorgensen, 1994), mouse (Kochhar, 1967), pigtail monkey (Newell-Morris et al., 1980), hamster (Shenefelt, 1972), and ferret (Hoar et al., 1988). Embryolethality was a common accompanying feature. Also according to the package label, tretinoin given orally to rabbits was said to be teratogenic. The malformations produced in all species by the oral route of administration were multiple and were described as being typical of those defects induced by other retinoids, including face, ear, eye, palate, limb, neural, and heart defects. Effect levels of oral dosing ranged from 3 mg/kg/day over 25 days of gestation for the pig, up to 50 mg/kg/day for 2 days in the organogenesis period. These dose ranges exceed the usual applied doses in humans. Postnatal behavioral effects were reported in rats following low doses (Nolen, 1986). The mouse has been a good "model" for retinoid-induced developmental effects (Padmanabhan et al., 1990).

HUMANS

In the human, four case reports identified tretinoin as a cause of congenital malformations following topical administration during pregnancy. In the first case, a growth-retarded infant with a unilateral external ear defect was born from a mother reported to have received treatment with 0.05% drug from before conception through 11 weeks of pregnancy (Camera and Pregliasco, 1992). In the second case, multiple malformations consisting of exomphalos, diaphragmatic hernia, heart, and unilateral limb defects were reported in an infant of a mother also receiving 0.05% tretinoin during the first 5 weeks of pregnancy (Lipson et al., 1993). In the third case, aortic, digit, and ear defects were described in a child whose mother received 0.05% tretinoin during the first 2 months of pregnancy (Navarre-Belhassen et al., 1998). The fourth case involved a woman treated topically in the first trimester with 0.025% of tretinoin. The infant had cerebral dysmorphology and an absence of an ear and external auditory canal (Selcen et al., 2000). The therapeutic dose for the drug is 0.01 to 0.05% (topical) and 45 to 200 mg/m^2/day (oral). Other developmental toxicity was reported: One group of investigators reported four spontaneous abortions (Johnson et al., 1994). A case of intrauterine growth retardation was reported in an infant born to a mother treated with tretinoin in the third trimester (Terada et al., 1997).

The case reports of malformations associated with drug administration were countered by several studies: First, in a prospective study comprised of 64 pregnant women who were exposed to the drug during pregnancy, no major malformations were observed (Johnson et al., 1994). A second prospective study of 94 tretinoin-exposed cases and 133 controls also found no excess malformations in the treatment group, and researchers concluded that tretinoin was not teratogenic in humans (Shapiro et al., 1997). A retrospective study comprised of 215 women exposed to tretinoin in early pregnancy compared to 430 controls found that the number of malformations in the exposed group was significantly less than in the control group (Jick et al., 1993). Still another more recent study of 107 first trimester exposures versus 389 controls also found no relation to major structural defects, abortions, or lowered birth weight compared to the controls (Lourerio et al., 2005). Further, the prevalence of retinoic acid-specific minor malformations did not differ significantly between the two groups. Rosa et al. (1994) reported that a specific brain malformation (holoprosencephaly) was found in a number of tretinoin-exposed cases reported to the Food and Drug Administration, but these could not be confirmed by others (DeWals et al., 1991). It has been said that the drug is not teratogenic topically in 0.1 to 0.5% concentration (Kligman, 1988).

Martinez-Frias and Rodriguez-Pinilla (1999) were critical of the above conclusions that tretinoin is not teratogenic because of the limitations of the cited studies. They concluded from the four positive studies that first trimester exposure to topical tretinoin may not be safe, and that we cannot exclude that it may imply a risk, and they recommend that the drug be contraindicated for use in pregnancy. It should also be stated that the recorded malformations in the positive studies are not dissimilar from the malformations induced by isotretinoin and etretinate, related retinoids considered human teratogens. One group of experts considers that it is unlikely there is a teratogenic risk from topical exposure (topical exposures are poorly absorbed), but that there is probably a substantial risk of developmental toxicity with systemic administration (Friedman and Polifka, 2000), conclusions not supported by the published reports. It seems to this writer that topical exposure to tretinoin is likely to be teratogenic based on the retinoid-like defects reported for it. And, despite an absence of positive case reports of systemic exposure causing congenital malformation, it seems that tretinoin is also likely to carry teratogenic risk, although risk-to-benefit ratio considerations to its use in cancer therapy apply. At any rate, tretinoin is clearly a developmental toxicant. Unfortunately, animal studies appear to have little relevance to risk issues in humans, either with respect to overall response or to dosage.

The mechanism of teratogenicity by retinoids has been studied perhaps more thoroughly than any other teratogen; the reader is referred to the published review by the NRC (2000) on the retinoic acids and their teratogenicity mechanisms. Several useful reviews on this subject were published (Rosa et al., 1986; Nau, 1993; Cohen, 1993; Kochhar and Christian, 1997; Collins and Mao, 1999).

CHEMISTRY

Tretinoin is a relatively large conjugated chemical that is highly hydrophobic as compared to the other compounds. Hydrogen bonding interactions can occur through the carboxylic acid portion of the molecule. The calculated physicochemical and topological properties for tretinoin are as follows.

PHYSICOCHEMICAL PROPERTIES

Parameter	Value
Molecular weight	300.441 g/mol
Molecular volume	310.17 A^3
Density	0.847 g/cm^3
Surface area	407.56 A^2
LogP	6.164
HLB	2.205
Solubility parameter	18.436 $J^{(0.5)}/cm^{(1.5)}$
Dispersion	17.428 $J^{(0.5)}/cm^{(1.5)}$
Polarity	1.432 $J^{(0.5)}/cm^{(1.5)}$
Hydrogen bonding	5.840 $J^{(0.5)}/cm^{(1.5)}$
H bond acceptor	0.52
H bond donor	0.31
Percent hydrophilic surface	15.94
MR	93.294
Water solubility	−4.127 log (mol/M^3)
Hydrophilic surface area	64.94 A^2
Polar surface area	40.46 A^2
HOMO	−7.849 eV
LUMO	−1.531 eV
Dipole	5.762 debye

TOPOLOGICAL PROPERTIES (UNITLESS)

Parameter	Value
x0	16.751
x1	10.220
x2	9.813
xp3	6.409
xp4	5.058
xp5	2.824
xp6	2.272
xp7	0.972
xp8	0.697
xp9	0.377

Continued.

Parameter	Value
xp10	0.324
xv0	14.441
xv1	7.867
xv2	6.761
xvp3	4.125
xvp4	2.875
xvp5	1.499
xvp6	1.101
xvp7	0.346
xvp8	0.199
xvp9	0.100
xvp10	0.082
k0	28.931
k1	20.046
k2	9.333
k3	7.422
ka1	18.379
ka2	8.092
ka3	6.330

REFERENCES

Camera, G. and Pregliasco, P. (1992). Ear malformation in baby born to mother using tretinoin cream. *Lancet* 339: 687.

Christian, M. S. et al. (1997). A developmental toxicity study of tretinoin emollient cream (Renova) applied topically to New Zealand white rabbits. *J. Am. Acad. Dermatol.* 36: S67–S76.

Cohen, M. (1993). Tretinoin: A review of preclinical toxicological studies. *Drug Dev. Res.* 30: 244–251.

Collins, M. D. and Mao, E. (1999). Teratology of retinoids. *Ann. Rev. Pharmacol. Toxicol.* 39: 399–430.

Collins, M. D. et al. (1994). Comparative teratology and transplacental pharmacokinetics of all-*trans* retinoic acid, 13-*cis* retinoic acid, and retinyl palmitate following daily administration to rats. *Toxicol. Appl. Pharmacol.* 127: 132–144.

DeWals, P. et al. (1991). Association between holoprosencephaly and exposure to topical retinoids. Results of the EUROCAT survey. *Paediatr. Perinat. Epidemiol.* 5: 445–447.

Friedman, J. M. and Polifka, J. E. (2000). *Teratogenic Effects of Drugs. Resource for Clinicians (TERIS)*, Second ed., Johns Hopkins University Press, Baltimore, MD.

Hardman, J. G., Limbird, L. E., and Gilman, A. G. (Eds.). (2001). *Goodman and Gilman's The Pharmacological Basis of Therapeutics*, 10th ed., McGraw-Hill, New York, p. 1809.

Hoar, R. M. et al. (1988). Similar dose sensitivity (mg/kg) of dogs and ferrets in a study of developmental toxicity. *Absts. 9th Ann. Mtg. ACT*, p. 21.

Jick, S. S., Terris, B. Z., and Jick, H. (1993). First trimester topical treatinoin and congenital disorders. *Lancet* 341: 1181–1182.

Johnson, K. A. et al. (1994). Pregnancy outcome in women prospectively ascertained with Retin-A exposures: An ongoing study. *Teratology* 49: 375.

Jorgensen, K. D. (1994). Teratogenic activity of tretinoin in the Gottingen mini-pig. *Teratology* 50: 26A–27A.

Kligman, A. M. (1988). Is topical tretinoin teratogenic? *JAMA* 259: 2918.

Kochhar, D. M. (1967). Teratogenic activity of retinoic acid. *Acta Pathol. Microbiol. Scand.* 70: 398–404.

Kochhar, D. M. and Christian, M. S. (1997). Tretinoin: A review of the nonclinical developmental toxicology experience. *J. Am. Acad. Derm.* 36: S47–S59.

Lipson, A. H., Collins, F., and Webster, W. S. (1993). Multiple congenital defects associated with maternal use of topical tretinoin. *Lancet* 341: 8856.

Lourerio, K. D. et al. (2005). Minor malformations characteristic of the retinoic acid embryopathy and other birth outcomes in children of women exposed to topical tretinoin during early pregnancy. *Am. J. Med. Genet.* 136A:117–121.

Martinez-Frias, M. L. and Rodriguez-Pinilla, E. (1999). First-trimester exposure to topical tretinoin: Its safety is not warranted. *Teratology* 60: 5.

Nau, H. (1993). Embryotoxicity and teratogenicity of topical retinoic acid. *Skin Pharmacol.* 6 (Suppl. 1): 35–44.

Navarre-Belhassen, C. et al. (1998). Multiple congenital malformations associated with topical tretinoin. *Ann. Pharmacother.* 32: 505–506.

Newell-Morris, L. et al. (1980). Teratogenic effects of retinoic acid in pigtail monkeys *(Macaca nemestrina).* II. Craniofacial features. *Teratology* 22: 87–101.

Nolen, G. A. (1986). The effects of prenatal retinoic acid on the viability and behavior of the offspring. *Neurobehav. Toxicol. Teratol.* 8: 643–654.

NRC (National Research Council). (2000). *Scientific Frontiers in Developmental Toxicology and Risk Assessment.* National Academy Press, Washington, D.C., pp. 75–80.

Padmanabhan, R., Vaidya, H. R., and Abu-Alatta, A. A. F. (1990). Malformations of the ear induced by maternal exposure to retinoic acid in the mouse fetuses. *Teratology* 42: 25A.

Rosa, F. W., Wilk, A. L., and Kelsey, F. O. (1986). Teratogen update: Vitamin A congeners. *Teratology* 33: 355–364.

Rosa, F., Piazza-Hepp, T., and Goetsch, R. (1994). Holoprosencephaly with 1st trimester topical tretinoin. *Teratology* 49: 418–419.

Schaefer, C. (Ed.). (2001). *Drugs during Pregnancy and Lactation. Handbook of Prescription Drugs and Comparative Risk Assessment*, Elsevier, Amsterdam, p. 109.

Selcen, D., Seidman, S., and Nigro, M. A. (2000). Otocerebral anomalies associated with topical tretinoin use. *Brain Dev.* 22: 218–220.

Shapiro, L. et al. (1997). Safety of first trimester exposure to topical tretinoin: Prospective cohort study. *Lancet* 350: 1143–1144.

Sharma, R. P. et al. (1990). Dose-dependent pharmacokinetics and teratogenic activity of topical retinoids. *Toxicologist* 10: 237.

Shenefelt, R. E. (1972). Morphogenesis of malformations in hamsters caused by retinoic acid: Relation to dose and stage of treatment. *Teratology* 5: 103–118.

Terada, Y. et al. (1997). Fetal arrhythmia during treatment of pregnancy-associated acute promyelocytic leukemia with all-*trans* retinoic acid and favorable outcome. *Leukemia* 11: 454–455.

Zbinden, G. (1975). Investigations on the toxicity of tretinoin administered systemically to animals. *Acta Dermatol. Venereol. Suppl. (Stockh.)* 74: 36–40.

9 Propranolol

Chemical name: 1-[(1-Methylethyl)amino]-3-(1-naphthalenyloxy-2-propanol)

CAS #: 525-66-6

SMILES: c12c(OCC(CNC(C)C)O)cccc1cccc2

INTRODUCTION

Propranolol is a β-blocking drug with therapeutic utility as an antianginal and antiarrhythmic agent (Class II) and an antihypertensive and antimigraine agent. Nonselective β-adrenergic blocking drugs, numbering about 15 including propranolol, competitively block response to β- and β-adrenergic stimulation, which results in decreases in heart rate, myocardial contractility, blood pressure, and myocardial oxygen demand (Lacy et al., 2004). The hydrochloride salt form of propranolol is available commercially as the prescription drug Inderal®, among other trade names, and it has a pregnancy risk factor of C (this label infers that potential benefits may outweigh the potential risk, because well-controlled human studies are lacking, and animal studies have shown a risk to the fetus or are lacking as well).

DEVELOPMENTAL TOXICOLOGY

ANIMALS

In animal studies, propranolol had little adverse developmental effect. In mice, intravenous doses on 1 day of organogenesis at 10 mg/kg produced no developmental toxicity (Fujii and Nishimura, 1974). In rats, the drug administered in drinking water during gestation and lactation at doses of 25 to 150 mg/kg/day elicited reduced litter size at birth and reduced neonatal growth but no malformations (Schoenfeld et al., 1985).

HUMANS

Authors of published studies on the developmental effects in humans caused by propranolol reported few adverse findings. A solitary study reported a case of tracheoesophageal fistula and intrauterine growth retardation in an infant whose mother was treated with 80 mg/day of the drug throughout the first trimester (Campbell, 1985). The therapeutic dose of propranolol is 80 to 320 mg/day orally. In a number of publications, researchers have alluded to intrauterine growth retardation (IUGR) and low birth weights in case reports of studies of women treated with the drug during pregnancy,

---

TABLE 1
**Representative Reports of Intrauterine Growth
Retardation (IUGR) in Infants of Women Treated
with Propranolol during Pregnancy**

Reed et al., 1974
Fiddler, 1974
Gladstone et al., 1975
Cottrill et al., 1977
Habib and McCarthy, 1977
Sabom et al., 1978
Lieberman et al., 1978
Eliahou et al., 1978
Pruyn et al., 1979
Oakley et al., 1979
Redmond, 1982
Paran et al., 1995

---

and they suggest that this effect may be related to treatment. A representative number of these reports published over time and including at least 185 births are shown in Table 1. In one publication, 23 reports were reviewed that involved 167 live-born infants exposed to chronic propranolol *in utero* and reported 14% with IUGR (Briggs et al., 2005). Several reports have cautioned against use of the drug in pregnancy on this account (Couston, 1982; Boice, 1982). The package label for the drug, in fact, states that growth retardation has been reported in neonates whose mothers received propranolol during pregnancy (*PDR*, 2002). The mechanisms possibly causing this effect were reviewed (Redmond, 1982). No consistent adverse effects including malformations, viability, or function were established with the drug.

As pointed out by Friedman and Polifka (2000), it is difficult in most studies cited to separate the action of the drug from an effect of the disease being treated. It appears that no postnatal studies were conducted to distinguish whether the growth deficiency recorded has any relevance to head circumference, as emphasized in one study (Pruyn et al., 1979), or to brain deficiency. It appears to this writer that propranolol treatment during pregnancy is associated with a reduction in fetal weight as an indication of developmental toxicity. In a recent review of the use of β-blockers in pregnancy, the drug was considered relatively safe, but the authors conceded that some drugs of this class, including propranolol, may cause IUGR and reduced placental weight, with treatment early in the second trimester resulting in the greatest effect (Frishman and Chesner, 1988).

CHEMISTRY

Propranolol is a hydrophobic molecule with average size in comparison to the other human developmental toxicants. It can participate in hydrogen bonding, both as acceptor and donor. The calculated physicochemical and topological properties are listed below.

PHYSICOCHEMICAL PROPERTIES

Parameter	Value
Molecular weight	259.347 g/mol
Molecular volume	253.21 A^3

Continued.

Parameter	Value
Density	1.025 g/cm^3
Surface area	311.27 A^2
LogP	2.434
HLB	6.328
Solubility parameter	22.636 J$^{(0.5)}$/cm$^{(1.5)}$
Dispersion	19.644 J$^{(0.5)}$/cm$^{(1.5)}$
Polarity	2.607 J$^{(0.5)}$/cm$^{(1.5)}$
Hydrogen bonding	10.941 J$^{(0.5)}$/cm$^{(1.5)}$
H bond acceptor	0.73
H bond donor	0.44
Percent hydrophilic surface	33.86
MR	76.123
Water solubility	–0.140 log (mol/M^3)
Hydrophilic surface area	105.39 A^2
Polar surface area	41.49 A^2
HOMO	–8.072 eV
LUMO	–0.108 eV
Dipole	3.681 debye

TOPOLOGICAL PROPERTIES (UNITLESS)

Parameter	Value
x0	13.665
x1	9.165
x2	8.053
xp3	5.990
xp4	4.691
xp5	4.018
xp6	2.338
xp7	1.638
xp8	1.132
xp9	0.790
xp10	0.403
xv0	11.466
xv1	6.686
xv2	5.005
xvp3	2.943
xvp4	1.938
xvp5	1.380
xvp6	0.635
xvp7	0.369
xvp8	0.221
xvp9	0.131
xvp10	0.058
k0	23.694
k1	15.390
k2	7.695
k3	4.795
ka1	13.999
ka2	6.661
ka3	4.028

REFERENCES

Boice, J. L. (1982). Propranolol during pregnancy. *JAMA* 248: 1834.

Briggs, G. G., Freeman, R. K., and Yaffe, S. J. (2005). *Drugs in Pregnancy and Lactation. A Reference Guide to Fetal and Neonatal Risk*, Seventh ed., Lippincott Williams & Wilkins, Philadelphia.

Campbell, J. W. (1985). A possible teratogenic effect of propranolol. *N. Engl. J. Med.* 313: 518.

Cottrill, C. M. et al. (1977). Propranolol therapy during pregnancy, labor, and delivery: Evidence for transplacental drug transfer and impaired neonatal drug disposition. *J. Pediatr.* 91: 812–814.

Couston, D. (1982). Antiarrhythmic agents during pregnancy. *JAMA* 247: 303.

Eliahou, H. E. et al. (1978). Propranolol for the treatment of hypertension in pregnancy. *Br. J. Obstet. Gynaecol.* 85: 431–436.

Fiddler, G. I. (1974). Propranolol and pregnancy. *Lancet* 2: 722–723.

Friedman, J. M. and Polifka, J. E. (2000). *Teratogenic Effects of Drugs. A Resource for Clinicians (TERIS)*, Second ed., Johns Hopkins University Press, Baltimore, MD.

Frishman, W. H. and Chesner, M. (1988). Beta-adrenergic blockers in pregnancy. *Am. Heart J.* 115: 147–152.

Fujii, T. and Nishimura, H. (1974). Reduction in frequency of fetopathic effects of caffeine in mice by pretreatment with propranolol. *Teratology* 10: 149–152.

Gladstone, G. R., Hordof, A., and Gersony, W. M. (1975). Propranolol administration during pregnancy: Effects on the fetus. *J. Pediatr.* 86: 962–964.

Habib, A. and McCarthy, J. S. (1977). Effects on the neonate of propranolol administered during pregnancy. *J. Pediatr.* 91: 808–811.

Lacy, C. F. et al. (2004). *Drug Information Handbook (Pocket), 2004–2005*, Lexi-Comp. Inc., Hudson, OH.

Lieberman, B. A. et al. (1978). The possible adverse effect of propranolol on the fetus in pregnancies complicated by severe hypertension. *Br. J. Obstet. Gynaecol.* 85: 678–683.

Oakley, G. D. G. et al. (1979). Management of pregnancy in patients with hypertrophic cardiomyopathy. *Br. Med. J.* 1: 1749–1750.

Paran, E. et al. (1995). β-adrenergic blocking agents in the treatment of pregnancy-induced hypertension. *Int. J. Clin. Pharmacol. Ther.* 33: 119–123.

PDR® (*Physicians' Desk Reference*®). (2002). Medical Economics Co., Inc., Montvale, NJ.

Pruyn, S. C., Phelan, J. P., and Buchanan, G. C. (1979). Long-term propranolol therapy in pregnancy: Maternal and fetal outcome. *Am. J. Obstet. Gynecol.* 135: 485–489.

Redmond, G. P. (1982). Propranolol and fetal growth retardation. *Semin. Perinatol.* 6: 142–147.

Reed, R. L. et al. (1974). Propranolol therapy throughout pregnancy: A case report. *Anesth. Analg.* 53: 214–218.

Sabom, M. M., Curry, R. C., and Wise, D. E. (1978). Propranolol therapy during pregnancy in a patient with idiopathic hypertrophic subaortic stenosis: Is it safe? *South. Med. J.* 71: 328–329.

Schoenfeld, N. et al. (1985). Effects of propranolol during pregnancy and development of rats. 2. Adverse effects on development. *Europ. J. Pediatr.* 143: 194–195.

10 Penicillamine

Chemical name: 3-Mercapto-D-valine

CAS #: 52-67-5

SMILES: C(C(C)(C)S)(C(O)=O)N

INTRODUCTION

Penicillamine has therapeutic utility as an antidote for copper and lead toxicity and is used in the treatment of Wilson's disease and cystinuria and as an adjunct in the treatment of rheumatoid arthritis. Mechanistically, penicillamine chelates with a number of heavy metals to form stable, soluble complexes that are excreted in urine. It also depresses circulating IgM rheumatoid factor and T cell but not B cell activity, and it combines with cystine to form a more soluble compound, thus preventing cystine calculi (Lacy et al., 2004). The drug is available by prescription as Cuprimine®, among other trade names, and it carries a pregnancy risk factor of D. The package label states that although normal outcomes have been reported (in pregnant women), characteristic congenital cutis laxa and associated birth defects have been reported in infants born of mothers who received therapy with penicillamine during pregnancy (see below; also see *PDR*, 2002).

DEVELOPMENTAL TOXICOLOGY

ANIMALS

Laboratory animal studies were conducted with the drug in mice, hamsters, and rats, and it is developmentally toxic in all three species. Given by the oral route, mice demonstrated cleft palate, increased abortion and resorptions, and decreased fetal body weight at high doses of 3.2 g/kg when administered 1 or 3 days during organogenesis (Myint, 1984). Similar doses in hamsters given on 1 day during organogenesis elicited fetal death, decreased fetal body weight, malformations of the central nervous system, and skeletal defects of the ribs and limbs (Wiley and Joneja, 1978). In rats, penicillamine given either by oral gavage or fed in the diet during organogenesis or throughout gestation produced malformations (palate and skeletal defects), reduced fetal body weight, and increased resorptions in the range of 360 to 1000 mg/kg (gavage) or 0.8% and higher (diet) in several studies (Steffek et al., 1972; Yamada et al., 1979; Mark-Savage et al., 1981). The doses used in these experiments were multiple those used in human therapy (see below).

TABLE 1
Congenital Malformation of the Skin in Penicillamine-Exposed Women

Case Number	Malformations	Growth Retardation	Death	Functional Deficit	Ref.
1	Skin, gastrointestinal, vessels, bones		✔		Mjolnerod et al., 1971
2	Skin, abdomen	✔	✔		Solomon et al., 1977
3	Skin				Linares et al., 1979
4	Skin, abdomen		✔		Beck et al., 1981
5	Skin, abdomen, jaw, ears				Harpey et al., 1983, 1984
6	Skin, brain, limbs, jaw			✔	Pinter et al., 2004

HUMANS

Developmental toxicity in the human is largely manifested as congenital malformation of the connective tissue of the skin, as tabulated in Table 1. Six cases of this disorder, termed *cutis laxa*, were described. Schardein (2000) described the defect in detail. In the cases reported, the general condition of the infants appeared normal, except for the generalized senescence of the skin, with extensive wrinkling and folding, having the appearance of too much skin for the body. However, three of the patients died in infancy. Intrauterine growth retardation was recorded in a single case, and a single case of developmental delay was reported. Neither effect is considered a significant parameter in the developmental toxicity profile of the drug. Clinically, the defect is apparently reversible: In the three surviving infants, the skin returned to normal externally within 4 months, with normal physical and neurological development in two of the cases. In each of the six cases, doses of 750 to 2000 mg/day orally had been administered, all in at least the first trimester. These doses are close to the recommended therapeutic drug dosage of 900 mg to 2 g/day orally. Interestingly, cutis laxa has been produced in an animal model — the rat (Hurley et al., 1982).

Six other cases of malformations were published in the literature but are not considered pertinent to this discussion. Rosa (1986) reported brain, eye and digits, brain and limb, and limb and digits defects among four cases known to the U.S. Food and Drug Administration. A single case of cleft lip/palate was recorded in another case report (Martinez-Frias et al., 1998). Another case, a patient with multiple malformations consisting of congenital contractures, hydrocephalus, and muscle dysfunction, was also reported (Gal and Ravenel, 1984). These malformations are dissimilar from the skin disorder recognized as a teratogenic finding and are largely dissimilar from each other; thus, they are not considered to be causally related to penicillamine administration.

Approximately 90 normal infants born of women treated during pregnancy with the drug were reported (Gregory and Mansell, 1983; Gal and Ravenel, 1984; Dupont et al., 1990; Hartard and Kunze, 1994; Berghella et al., 1997; see Schardein, 2000). The apparent risk for malformation appears to be about 5%. The skin defects are considered by one group of experts to have a small to moderate teratogenic risk (Friedman and Polifka, 2000). Several reviews of penicillamine developmental toxicity were published (Endres, 1981; Roubenoff et al., 1988; Domingo, 1998; Sternlieb, 2000).

CHEMISTRY

Penicillamine is a hydrophilic chemical of relatively small size. It is of average polarity as compared to the other chemicals, and it can participate in donor/acceptor hydrogen bonding interactions. Its calculated properties are as follows.

PHYSICOCHEMICAL PROPERTIES

Parameter	Value
Molecular weight	149.213 g/mol
Molecular volume	135.15 A^3
Density	1.092 g/cm^3
Surface area	191.40 A^2
LogP	−1.108
HLB	12.196
Solubility parameter	25.421 J$^{(0.5)}$/cm$^{(1.5)}$
Dispersion	19.739 J$^{(0.5)}$/cm$^{(1.5)}$
Polarity	7.843 J$^{(0.5)}$/cm$^{(1.5)}$
Hydrogen bonding	13.969 J$^{(0.5)}$/cm$^{(1.5)}$
H bond acceptor	1.16
H bond donor	0.82
Percent hydrophilic surface	59.38
MR	39.671
Water solubility	2.720 log (mol/M^3)
Hydrophilic surface area	113.65 A^2
Polar surface area	66.48 A^2
HOMO	−9.215 eV
LUMO	0.320 eV
Dipole	3.572 debye

TOPOLOGICAL PROPERTIES (UNITLESS)

Parameter	Value
x0	7.655
x1	3.854
x2	4.399
xp3	2.366
xp4	1.000
xp5	0.000
xp6	0.000
xp7	0.000
xp8	0.000
xp9	0.000
xp10	0.000
xv0	6.071
xv1	2.869
xv2	3.260
xvp3	1.218
xvp4	0.378
xvp5	0.000
xvp6	0.000
xvp7	0.000
xvp8	0.000
xvp9	0.000
xvp10	0.000
k0	7.986
k1	9.000
k2	2.722
k3	2.880
ka1	8.810
ka2	2.597
ka3	2.740

REFERENCES

Beck, R. B. et al. (1981). Ultrastructural findings in fetal penicillamine syndrome. In *Abstracts from the 14th Annual March of Dimes Birth Defects Conference*, San Diego, CA.

Berghella, V. et al. (1997). Successful pregnancy in a neurologically impaired woman with Wilson's disease. *Am. J. Obstet. Gynecol.* 176: 712–714.

Domingo, J. L. (1998). Developmental toxicity of metal chelating agents. *Reprod. Toxicol.* 12: 499–510.

Dupont, P., Irion, O., and Beguin, F. (1990). Pregnancy in a patient with treated Wilson's disease: A case report. *Am. J. Obstet. Gynecol.* 163: 1527–1528.

Endres, W. (1981). D-Penicillamine in pregnancy — to ban or not to ban. *Klin. Wochenschr.* 59: 535–538.

Friedman, J. M. and Polifka, J. E. (2000). *Teratogenic Effects of Drugs. A Resource for Clinicians (TERIS)*, Second ed., Johns Hopkins University Press, Baltimore, MD.

Gal, P. and Ravenel, S. D. (1984). Contractures and hydrocephalus with penicillamine and maternal hypotension. *J. Clin. Dysmorphol.* 2: 9–12.

Gregory, M. C. and Mansell, M. A. (1983). Pregnancy and cystinuria. *Lancet* 2: 1158–1160.

Harpey, J. P. et al. (1983). Cutis laxa and low serum zinc after neonatal exposure to penicillamine. *Lancet* 2: 858.

Harpey, J. P. et al. (1984). Neonatal cutis laxa due to D-penicillamine treatment during pregnancy. Hypozincaemia in the infant. *Teratology* 29: 29A.

Hartard, C. and Kunze, K. (1994). Pregnancy in a patient with Wilson's disease treated with D-penicillamine and zinc sulfate. *Eur. Neurol.* 34: 337–340.

Hurley, L. S. et al. (1982). Reduction by copper supplementation of teratogenic effects of D-penicillamine and triethylenetetramine. *Teratology* 25: 51A.

Lacy, C. F. et al. (2004). *Drug Information Handbook (Pocket), 2004–2005*, Lexi-Comp., Inc., Hudson, OH.

Linares, A. et al. (1979). Reversible cutis laxa due to maternal D-penicillamine treatment. *Lancet* 2: 43.

Mark-Savage, P. et al. (1981). Teratogenicity of D-penicillamine in rats. *Teratology* 23: 50A.

Martinez-Frias, M. L. et al. (1998). Prenatal exposure to penicillamine and oral clefts: Case report. *Am. J. Med. Genet.* 76: 274–275.

Mjolnerod, O. K. et al. (1971). Congenital connective-tissue defect probably due to D-penicillamine treatment in pregnancy. *Lancet* 1: 673–675.

Myint, B. (1984). D-Penicillamine-induced cleft palate in mice. *Teratology* 30: 333–340.

PDR® (Physicians' Desk Reference®). (2002). Medical Economics Co., Montvale, NJ.

Pinter, R., Hogge, W. A., and McPherson, E. (2004). Infant with severe penicillamine embryopathy born to a woman with Wilson disease. *Am. J. Med. Genet.* 128A: 294–298.

Rosa, F. W. (1986). Teratogen update: Penicillamine. *Teratology* 33: 127–131.

Roubenoff, R. et al. (1988). Effects of anti-inflammatory and immunosuppressive drugs on pregnancy and fertility. *Sem. Arthritis Rheum.* 18: 88–110.

Schardein, J. L. (2000). *Chemically Induced Birth Defects*, Third ed., Marcel Dekker, New York, pp. 640–641.

Solomon, L. et al. (1977). Neonatal abnormalities associated with D-penicillamine treatment during pregnancy. *N. Engl. J. Med.* 296: 54–55.

Steffek, A. J., Verrusio, A. C., and Watkins, C. A. (1972). Cleft palate in rodents after maternal treatment with various lathrogenic agents. *Teratology* 5: 33–40.

Sternlieb, I. (2000). Wilson's disease and pregnancy. *Hepatology* 31: 531–532.

Wiley, M. J. and Joneja, M. G. (1978). Neural tube lesions in the offspring of hamsters given single oral doses of lathrogens early in gestation. *Acta Anat.* 100: 347–353.

Yamada, T. et al. (1979). Reproduction studies of D-penicillamine in rats. 2. Teratogenicity study. *Oyo Yakuri* 18: 561–569.

11 Vitamin A

Chemical name: 3,7-Dimethyl-9-(2,6,6-trimethyl-1-cyclohexen-1-yl)-2,4,6,8-nonatetraen-1-ol

Alternate names: Oleovitamin A, retinol

CAS #: 68-26-8

SMILES: C1(C(CCCC=1C)(C)C)C=CC(=CC=CC(=CCO)C)C

INTRODUCTION

Vitamin A is a fat-soluble essential vitamin available from natural as well as synthetic sources. The vitamin promotes bone growth, tooth development, and reproduction; helps form and maintain healthy skin, hair, and mucous membranes; and builds the body's resistance to respiratory infections. It aids in the treatment of many eye disorders, and helps treat acne, impetigo, boils, carbuncles, and open ulcers when applied externally. It is also used therapeutically in the treatment and prevention of vitamin A deficiency. It has a long half-life and bioaccumulates (Hathcock et al., 1990). It is available commercially as an over-the-counter (OTC) preparation with the trade names Aquasol A® and Palmitate-A® among many other names. Vitamin A has a package label with contrasting pregnancy risk factors varying from A to X, the latter if used in excess of the recommended dietary allowance (RDA) doses (~1000 to 5000 IU/day) (Griffith, 1988). The RDA for pregnant women, depending on the source of information, is ~2700 (NRC, 1989) to 8000 IU/day (U.S. Teratology Society, 1987).

DEVELOPMENTAL TOXICOLOGY

ANIMALS

The studies described below are those related to excess vitamin A, as deficiency states of the vitamin also have developmental toxicity properties. Many studies conducted with different objectives were published for laboratory animals: The emphasis here is on representative responses by species, by the oral route (the same as that mainly used therapeutically in humans). The topical route has not been explored in this respect. The response in animals is best shown as tabulated in Table 1. A multitude of different malformations were recorded in these studies, but craniofacial, central nervous system, and skeletal defects appeared most commonly, according to one observer (Friedman and Polifka, 2000). In addition to structural malformations, learning skills and fine motor changes and

TABLE 1
Developmental Toxicity in Animals Administered Oral Vitamin A

Species	Developmental Toxic Dose (IU[a])	Toxicity Reported	Treatment Interval in Gestation (days)	Ref.
Mouse	3,000–10,000	Multiple M[b]	8–13 various	Kalter and Warkany, 1959; Giroud and Martinet, 1959
Rat	35,000–160,000	Craniofacial and brain M, postnatal behavioral changes	4–18 various	Cohlan, 1953; Hutchings et al., 1973; Kutz et al., 1985
Guinea pig	50,000	Jaw and tongue defects, D[c]	10–13	Giroud and Martinet, 1959
Hamster	20,000	Multiple M	7–10	Marin-Padilla and Ferm, 1965
Rabbit	41,000	Multiple M, D	5–14	Giroud and Martinet, 1958
Cat	1,000,000–2,000,000	Multiple M, D	(5 breedings)	Freytag and Morris, 1997
Dog	125,000	Multiple M	17–22	Wiersig and Swenson, 1967
Pig	3,000,000–10,000,000	Eye M	12–42 various	Palludan, 1966
Cyno monkey	7,500–80,000	Multiple M, D (maternal toxicity)	16–27	Hendrickx et al., 1997, 2000

[a] International units — a unit of measurement based on measured biological activities. For vitamin A, 1 IU = 0.3 mcg.
[b] Malformations.
[c] Death.

other behavioral abnormalities were also observed following large doses of vitamin A in rats (Hutchings et al., 1973).

Humans

A number of malformations in humans have been reported in case reports, as tabulated in Table 2. Approximately 23 cases were recorded. As with most other toxicologic dose relationships, all malformations have occurred at megadoses, on the order of 30,000 IU/day or greater, according to several sources; doses of 10,000 IU/day or less are apparently considered safe during pregnancy (Miller et al., 1998; Weigand et al., 1998). Transport to the fetus is by passive diffusion (Wild et al., 1974), and there is little or no difference between maternal and fetal blood levels, irrespective of when administered (Briggs et al., 2002). Most all developmentally effective doses in laboratory animals are many times greater than dietary and supplemental human doses. An important result in primates was a no observed effect level (NOEL) (7500 IU) that would correspond to a dose of 300,000 IU/day in humans. It appears that the rabbit is a good animal model for displaying similar defects as those shown in humans (Tzimas et al., 1997).

No discrete pattern of malformations is obvious from the recorded data given in Table 2. Variation in intake and patterns of ingestion may account for some of the differences in malformations. However, ear, limb, craniofacial, urinary, heart and blood vessels, cleft lip/palate, and brain abnormalities occurred most commonly in decreasing order (Rosa, 1993). These share a number of similarities to those reported in animals. The pattern of malformations is said by several investigators (Lungarotti et al., 1987; Rosa, 1991) to be a phenocopy of those defects induced by the vitamin A congener, isotretinoin, a recognized potent human teratogen and developmental toxicant.

These case reports are supported by at least one major epidemiological study — a prospective analysis of 22,748 pregnancies of women who consumed dietary or supplemental vitamin A during

TABLE 2
Developmental Toxicity Profile of Oral Vitamin A in Humans

Case Number	Malformations	Growth Retardation	Death	Functional Deficit	Ref.
1	Urinary tract				Pilotti and Scorta, 1965
2	Kidney				Bernhardt and Dorsey, 1974
3	[Goldenhar's syndrome]				Mounoud et al., 1975
4	Multiple: brain, kidney, adrenals, jaw		✔		Stange et al., 1978
5	Multiple: limbs, ears, face				Von Lennep et al., 1985
6	Brain				Vallet et al., 1985
7, 8	Ear				Vallet et al., 1985
9	[Vater's syndrome], ear				Vallet et al., 1985
10	Multiple: ears, jaw, eye				Vallet et al., 1985
11	Vessels				Vallet et al., 1985
12	Multiple: face, ears, palate				Rosa et al., 1986 (FDA case)
13	Ears, lip/palate				Rosa et al., 1986 (FDA case)
14	Lip				Rosa et al., 1986 (FDA case)
15	Heart, brain				Rosa et al., 1986 (FDA case)
16	Multiple: ears, vertebrae, limbs, digits				Rosa et al., 1986 (FDA case)
17	Multiple: lip/palate, jaw, face, eye				Rosa et al., 1986 (cited)
18	Multiple: ears, skull, nose, lip, jaw, tongue, skin, digits, gastrointestinal, heart, kidney, liver	✔	✔		Lungarotti et al., 1987
19–21	None		✔		Zuber et al., 1987
22	Eye				Evans and Hickey-Dwyer, 1991
23, 24	Brain				Miller et al., 1998 (manufacturer's cases)
25	Club feet				Miller et al., 1998 (manufacturer's case)
26	[Turner's syndrome]				Miller et al., 1998 (manufacturer's case)

their pregnancies in quantities of 5000 to >15,000 IU/day (Rothman et al., 1995). Of this cohort, there were 339 (1.5%) infants born with malformations, 121 of whom had defects occurring in sites that originated in the cranial neural crest, primarily craniofacial and cardiac defects, abnormalities commonly induced by retinoids in general. For women taking >10,000 IU/day, the relative risk was 4.8 (95% confidence interval [CI], 2.2 to 10.5) and 2.2 (95% CI, 1.3 to 3.8) for all malformations, regardless of origin. The apparent threshold was near 10,000 IU/day of supplemental vitamin A. These data supported the conclusion that high dietary intake of vitamin A appeared to be teratogenic, especially among women who had consumed these levels before the seventh gestational week. The authors concluded that about 1 infant in 57 exposed to vitamin A supplemented at these levels had a malformation attributable to it.

In contrast, a number of other fairly recent epidemiological studies comprising over 43,000 pregnancies do not support the premise that vitamin A has teratogenic properties, but the limiting factor may be that dosages in the studies reported were in the range of 8000 to ~10,000 IU/day (Martinez-Frias and Salvador, 1990; Werler et al., 1990; Shaw et al., 1997; Mills et al., 1997; Czeizel and Rockenbaur, 1998; Khoury et al., 1998; Mastroiacovo et al., 1999). Doses of this magnitude are generally considered safe and not teratogenic (Miller et al., 1998; Wiegand et al.,

1998). For one study of this group (Dudas and Czeizel, 1992), researchers reported dose administration of only 6000 IU/day, which would not be expected to be active. Two other studies of the group contained subsets of women who received higher doses (40,000 to 50,000 IU/day) and who did not illustrate an enhanced number of malformations (Martinez-Frias and Salvador, 1990; Mastroiacovo et al., 1999). However, too few subjects were evaluated to make significant statements related to safety. The U.S. Teratology Society (1987) has officially sanctioned doses of 8000 IU/day as being safe during pregnancy and considers doses of 25,000 IU/day and higher as potentially teratogenic.

It appears from analysis of these data that vitamin A supplementation or dietary intake during pregnancy of approximately 10,000 IU/day or less is a safe procedure with respect to teratogenic potential, and that quantities in excess of that dosage offer some risk of toxicity. One group of experts indicates a similar risk, and suggests further that doses of >25,000 IU/day have an undetermined (but perhaps real teratogenic risk) (Friedman and Polifka, 2000). It does not appear that other classes of developmental toxicity are affected by excessive quantities of the vitamin, only structural malformation.

A number of pertinent reviews addressing the toxicity of vitamin A excess in animals as well as humans were published (Gal et al., 1972; Geelen, 1979; Bendich and Lanseth, 1989; Hathcock et al., 1990; Pinnock and Alderman, 1992; Rosa, 1993; Monga, 1997; Miller et al., 1998).

CHEMISTRY

Vitamin A, structurally similar to tretinoin, is a highly hydrophobic compound that is larger in size in comparison to the other toxicants within this compilation. The compound contains a network of conjugated double bonds within its structure. It is of relatively low polarity. The calculated physicochemical and topological properties are as follows.

PHYSICOCHEMICAL PROPERTIES

Parameter	Value
Molecular weight	286.458 g/mol
Molecular volume	308.54 A^3
Density	0.813 g/cm^3
Surface area	406.37 A^2
LogP	5.753
HLB	0.269
Solubility parameter	18.673 $J^{(0.5)}/cm^{(1.5)}$
Dispersion	16.701 $J^{(0.5)}/cm^{(1.5)}$
Polarity	1.673 $J^{(0.5)}/cm^{(1.5)}$
Hydrogen bonding	8.182 $J^{(0.5)}/cm^{(1.5)}$
H bond acceptor	0.40
H bond donor	0.29
Percent hydrophilic surface	7.52
MR	91.550
Water solubility	−3.849 log (mol/M^3)
Hydrophilic surface area	30.54 A^2
Polar surface area	20.23 A^2
HOMO	−7.453 eV
LUMO	−1.004 eV
Dipole	1.511 debye

Topological Properties (Unitless)

Parameter	Value
x0	15.880
x1	9.864
x2	8.972
xp3	6.317
xp4	4.772
xp5	2.751
xp6	2.218
xp7	0.953
xp8	0.638
xp9	0.361
xp10	0.316
xv0	14.240
xv1	7.875
xv2	6.665
xvp3	4.187
xvp4	2.844
xvp5	1.500
xvp6	1.100
xvp7	0.352
xvp8	0.196
xvp9	0.100
xvp10	0.082
k0	27.164
k1	19.048
k2	9.209
k3	6.743
ka1	17.711
ka2	8.188
ka3	5.887

REFERENCES

Bendich, A. and Lanseth, L. (1989). Safety of vitamin A. *Am. J. Clin. Nutr.* 49: 358–371.

Bernhardt, I. B. and Dorsey, D. J. (1974). Hypervitaminosis A and congenital renal anomalies in a human infant. *Obstet. Gynecol.* 43: 750–755.

Briggs, G. G., Freeman, R. K., and Yaffe, S. J. (2002). *Drugs in Pregnancy and Lactation. A Reference Guide to Fetal and Neonatal Risk*, Sixth ed., Lippincott Williams & Wilkins, Philadelphia.

Cohlan, S. Q. (1953). Excessive intake of vitamin A during pregnancy as a cause of congenital anomalies in the rat. *Am. J. Dis. Child.* 86: 348–349.

Czeizel, A. E. and Rockenbaur, M. (1998). Prevention of congenital abnormalities of vitamin A. *Int. J. Vitam. Nutr. Res.* 68: 219–231.

Dudas, I. and Czeizel, A. E. (1992). Use of 6000 IU vitamin A during early pregnancy without teratogenic effect. *Teratology* 45: 335–336.

Evans, K. and Hickey-Dwyer, M. U. (1991). Cleft anterior segment with maternal hypervitaminosis A. *Br. J. Ophthalmol.* 75: 691–692.

Freytag, T. L. and Morris, J. G. (1997). Chronic administration of excess vitamin A in the domestic cat results in low teratogenicity. *FASEB* 11: A412.

Friedman, J. M. and Polifka, J. E. (2000). *Teratogenic Effects of Drugs. A Resource for Clinicians (TERIS)*, Second ed., Johns Hopkins University Press, Baltimore, MD.

Gal, I., Sharman, I. M., and Pryse-Davis, J. (1972). Vitamin A in relation to human congenital malformations. *Adv. Teratol.* 5: 143–159.

Geelen, J. A. G. (1979). Hypervitaminosis A induced teratogenesis. *CRC Crit. Rev. Toxicol.* 7: 351–375.

Giroud, A. and Martinet, M. (1958). Repercussions de l'hypervitaminose a chez l'embryon de lapin. *C. R. Soc. Biol. (Paris)* 152: 931–932.

Giroud, A. and Martinet, M. (1959). Teratogenese par hypervitaminose a chez le rat, la souris, le cobaye, et le lapin. *Arch. Fr. Pediatr.* 16: 971–975.

Griffith, H. W. (1988). *Complete Guide to Vitamins, Minerals and Supplements*, Fisher Books, Tucson, AZ, p. 23.

Hathcock, J. N. et al. (1990). Evaluation of vitamin-A toxicity. *Am. J. Clin. Nutr.* 52: 183–202.

Hendrickx, A. G., Hummler, H., and Oneda, S. (1997). Vitamin A teratogenicity and risk assessment in the cynomolgus monkey. *Teratology* 55: 68.

Hendrickx, A. G. et al. (2000). Vitamin A teratogenicity and risk assessment in the macaque retinoid model. *Reprod. Toxicol.* 14: 311–323.

Hutchings, D. E., Gibbon, J., and Kaufman, M. A. (1973). Maternal vitamin A excess during the early fetal period: Effects on learning and development in the offspring. *Dev. Psychobiol.* 6: 445–457.

Kalter, H. and Warkany, J. (1959). Teratogenic action of hypervitaminosis A in strains of inbred mice. *Anat. Rec.* 133: 396–397.

Khoury, M. J., Moore, C. A., and Mulinare, J. (1998). Do vitamin supplements in early pregnancy increase the risk of birth defects in the offspring? A population-based case-control study. *Teratology* 53: 91.

Kutz, S. A. et al. (1985). Vitamin A acetate: A behavioral teratology study in rats. II. *Toxicologist* 5: 106.

Lungarotti, M. S. et al. (1987). Multiple congenital anomalies associated with apparently normal maternal intake of vitamin A: A phenocopy of the isotretinoin syndrome. *Am. J. Med. Genet.* 27: 245–248.

Marin-Padilla, M. and Ferm, V. H. (1965). Somite necrosis and developmental malformations induced by vitamin A in the golden hamster. *J. Embryol. Exp. Morphol.* 13: 1–8.

Martinez-Frias, M. L. and Salvador, J. (1990). Epidemiological aspects of prenatal exposure to high doses of vitamin A in Spain. *Eur. J. Epidemiol.* 6: 118–123.

Mastroiacovo, P. et al. (1999). High vitamin A intake in early pregnancy and major malformations: A multicenter prospective controlled study. *Teratology* 59: 7–11.

Miller, R. K. et al. (1998). Periconceptual vitamin A use: How much is teratogenic? *Reprod. Toxicol.* 12: 75–88.

Mills, J. L. et al. (1997). Vitamin A and birth defects. *Am. J. Obstet. Gynecol.* 177: 31–36.

Monga, M. (1997). Vitamin A and its congeners. *Semin. Perinatol.* 21: 135–142.

Mounoud, R. L., Klein, D., and Weber, F. (1975). [A case of Goldenhar syndrome: Acute vitamin A intoxication in the mother during pregnancy]. *J. Genet. Hum.* 23: 135–154.

NRC (National Research Council). (1989). *Recommended Dietary Allowances*, 10th ed., Washington, D.C., National Academy Press.

Palludan, B. (1966). Swine in teratological research. In *Swine in Biomedical Research*, L. K. Bustad and R. O. McClellan, Eds., Battelle Memorial Institute, Columbus, OH, pp. 51–78.

Pilotti, G. and Scorta, A. (1965). Hypervitaminosis A during pregnancy and neonatal malformations of the urinary system. *Minerva Gynecol.* 17: 1103–1108.

Pinnock, C. B. and Alderman, C. P. (1992). The potential for teratogenicity of vitamin-A and its congeners. *Med. J. Aust.* 157: 804–809.

Rosa, F. (1991). Detecting human retinoid embryopathy. *Teratology* 43: 419.

Rosa, F. W. (1993). Retinoid embryopathy in humans. In *Retinoids in Clinical Practice*, G. Koren, Ed., Marcel Dekker, New York, pp. 77–109.

Rosa, F. W., Wilk, A. L., and Kelsey, F. O. (1986). Teratogen update: Vitamin A congeners. *Teratology* 33: 355–364.

Rothman, K. J. et al. (1995). Teratogenicity of high vitamin A intake. *N. Engl. J. Med.* 333: 1369–1373.

Shaw, G. M. et al. (1997). Periconceptual intake of vitamin A among women and risk of neural tube defect-affected pregnancies. *Teratology* 55: 132–133.

Stange, L., Carlstrom, K., and Erikkson, M. (1978). Hypervitaminosis A in early human pregnancy and malformations of the central nervous system. *Acta Obstet. Gynecol. Scand.* 57: 289–291.

Tzimas, G., Elmazar, M. M. A., and Nau, H. (1997). Why is the rat not an appropriate species to be used for teratogenic risk assessment of high vitamin A intake by humans. *Teratology* 56: 390.

U.S. Teratology Society. (1987). Position Paper. Guest Editorial: Vitamin A during pregnancy. *Teratology* 35: 267–268.

Vallet, H. L. et al. (1985). Isotretinoin (Accutane®), vitamin A, and human teratogenicity. In *Abstracts of the 113th American Public Health Association Meeting*, Washington, D.C.

Von Lennep, E. et al. (1985). A case of partial sirenomelia and possible vitamin A teratogenesis. *Prenat. Diagn.* 5: 35–40.

Werler, M. M. et al. (1990). Maternal vitamin A supplementation in relation to selected birth defects. *Teratology* 42: 497–503.

Wiegand, U. W., Hartmann, S., and Hummler, H. (1998). Safety of vitamin A: Recent results. *Int. J. Vitam. Nutr. Res.* 68: 411–416.

Wiersig, D. and Swenson, M. J. (1967). Teratogenicity of vitamin A in the canine. *Fed. Proc.* 26: 486.

Wild, J., Schorah, C. J., and Smithells, R. W. (1974). Vitamin A, pregnancy, and oral contraceptives. *Br. Med. J.* 1: 57–59.

Zuber, C., Librizzi, R. J., and Vogt, B. L. (1987). Outcomes of pregnancies exposed to high dose vitamin A. *Teratology* 35: 42A.

12 Carbamazepine

Chemical name: 5H-Dibenz[b,f]azepine-5-carboxamide

CAS #: 298-46-4

SMILES: N1(c2c(cccc2)C=Cc3c1cccc3)C(N)=O

INTRODUCTION

Carbamazepine is a tricyclic anticonvulsant drug that has activity against partial seizures of complex symptomology, generalized tonic-clonic seizures, and mixed seizure patterns, and provides pain relief of trigeminal or glosspharyngeal neuralgia (Lacy et al., 2004). Therapeutic efficacy has been found for carbamazepine in the treatment of bipolar and other affective disorders, resistant schizophrenia, ethanol withdrawal, restless leg syndrome, and posttraumatic disorders. Its mechanism of action is not clearly understood, but it is related chemically to the tricyclic antidepressants, and its chemical moiety of a carbonyl group at the 5-position is essential for its potent antiseizure activity (Hardman et al., 2001). Carbamazepine is available commercially by prescription under the trade names Carbatrol®, Epitol®, and Tegretol®, among others, and it has a pregnancy risk category of D. Stated on the package label is that the drug "can cause fetal harm when administered to a pregnant woman" (*PDR*, 2002).

DEVELOPMENTAL TOXICOLOGY

ANIMALS

In laboratory animal studies, carbamazepine was developmentally toxic in both mice and rats when given orally during the organogenesis period of gestation. In mice, doses in the range of 40 to 240 mg/kg/day were teratogenic, inducing central nervous system defects (McElhatton and Sullivan, 1977), and in rats given 600 mg/kg/day, a maternally toxic dose, the drug elicited skeletal and visceral abnormalities, reduced fetal weight, and resorption (Vorhees et al., 1990). Dose levels used in rodents were many times greater than therapeutic doses in humans (see below).

HUMANS

It should be mentioned at the onset that studies of induction of malformations in the human by anticonvulsants is problematic in that treatment is usually in the form of combined therapy with

TABLE 1
Representative Developmental Toxicity Studies with Monotherapy of Carbamazepine in Humans

Author	Developmental Toxicity Reported
Hicks, 1979	Multiple malformations in a stillborn
Hiilesmaa et al., 1981	Microcephaly in 20 cases
Bertollini et al., 1987	Microcephaly, growth inhibition described
Van Allen et al., 1988	10/21cases had syndrome of effects
Jones et al., 1988	Intrauterine growth retardation (IUGR), microcephaly, developmental delay and heart defects in two cases; poor newborn performance and defective skin and nails in three newborns
Dow and Riopelle, 1989	Case report of malformations
Jones et al., 1989	Among ~35 exposed women, spontaneous abortion in 3, prenatal growth deficiency in 2, postnatal growth deficiency in 2, developmental delay in 5, microcephaly in 4, multiple malformations (face, heart, digits) in multiple cases
Vestermark and Vestermark, 1991	Facial malformations and developmental and mental retardation recorded in case report
Rosa, 1991	Spina bifida 1% risk among 36 cases known to U.S. Food and Drug Administration
Oakeshott and Hunt, 1991	Case report of spina bifida
Gladstone et al., 1992	2/23 cases of malformation: myelomeningocele and multiple malformations
Little et al., 1993	Case report of neural tube defect
Omtzigt et al., 1993	9/159 (5.7%) cases of malformations reported in large study
Kaneko et al., 1993	3/43 (6.5%) with congenital malformations in large study
Kallen, 1994	Reported six cases of spina bifida
Ornoy and Cohen, 1996	Facial malformations, mild mental retardation (low cognitive scores) reported among 6/30 cases
Nulman et al., 1997	Increased minor anomalies among 35 cases
Jick and Terris, 1997	Seven cases (6.2%) with multiple malformations in large study
Samren et al., 1997	22/280 (7.9%) with major malformations (including spina bifida) from analysis of five large prospective European studies
Sutcliffe et al., 1998	Eye malformations in four cases
Canger et al., 1999	Twelve severe malformations in large prospective study
Wide et al., 2000	IUGR and microcephaly with cognitive dysfunction in large prospective study
Holmes et al., 2000	Developmental delay among >200 exposed children
Moore et al., 2000	Behavior phenotype described for drug
Arpino et al., 2000	Significant spina bifida in large surveillance study
Diav-Citrin et al., 2001	Considered teratogenic in prospective study of 210 subjects treated first trimester (cardiac and craniofacial defects; relative risk [RR] = 2.24)
Matalon et al., 2002	Meta-analysis of 22 studies comprised of 1255 subjects from first trimester exposures compared to 3756 controls: Increased risk (6.7% versus 2.3%) for malformations (mainly neural tube defects, cardiovascular and urinary tract anomalies, and cleft palate)
Wide et al., 2004	Increased major malformations in large registry study

more than one drug; most of the drugs used in this combination therapy are active teratogens when considered singly; and the treated patient has epilepsy, a factor that has often been associated with malformations in offspring. In light of these factors, evaluation of the developmental toxicity of carbamazepine is perhaps best considered from data in which the drug was used in monotherapy. A representative sampling of these data is presented in Table 1. It appears convincingly from the above data that carbamazepine is a human teratogen; several hundred cases exist in the literature. It is active at usual therapeutic doses (400 to 1200 mg/day orally) in the first trimester. In addition, it demonstrates developmental toxicity of other classes, including growth and functional deficiencies. The principal

---

TABLE 2
Clinical Findings among 35 Patients Whose
Mothers Received Carbamazepine
Monotherapy during Pregnancy

Clinical Findings	Frequency (%)
Hypoplastic fingernails	26
Epicanthal folds	26
Developmental delay	20
Short nose, long philtrum	11
Upslanting palpebral fissures	11
Microcephaly	11
Prenatal or postnatal growth deficiency	6
Multiple cardiac defects	3

Source: Modified after Jones, K. L. et al., *N. Engl. J. Med.*, 320,
1661–1666, 1989, by Schardein, J. L., *Chemically Induced Birth
Defects*, Third ed., Marcel Dekker, New York, 2000, p. 205.

---

features of the syndrome appear to be minor craniofacial defects, nail hypoplasia, and developmental delay, as initially proposed by Jones et al. (1989), features similar to those reported with fetal hydantoin syndrome (Schardein, 2000). Clinical findings in 35 cases of monotherapy with carbamazepine are tabulated in Table 2. Spina bifida was also reported in a number of studies in an incidence as high as 1%. The frequency of malformations is said to be two to three times as great as that generally seen in normal populations, and similar to that determined in children born to epileptic women who were treated with other anticonvulsants (Friedman and Polifka, 2000). These investigators consider carbamazepine to have a small to moderate teratogenic risk. Singular case reports of malformations reported with carbamazepine that are not associated with the syndrome of defects include those with adrenogenital syndrome (Vestergard, 1969), abnormal genital organs (McMullin, 1971), cranial nerve agenesis (Robertson et al., 1983), and rib defects (Legido et al., 1991).

In contrast to the positive indications as discussed above, a number of publications have not confirmed the teratogenic effect of the drug, alone or in combination with other anticonvulsants (see below), although several studies provided data to support the contention that the drug has some enhancement of effects when combined with other anticonvulsants, expecially with valproic acid (Meijer, 1984; Lindhout et al., 1984; Shakir and Abdulwahab, 1991; Kaneko et al., 1993; Janz, 1994; Matalon et al., 2002). It was proposed in this regard that the epoxide form of the drug combines with the toxic epoxide metabolites also formed by other anticonvulsants and binds covalently to macromolecules, thereby producing teratogenicity (Lindhout et al., 1984). Pertinent large studies that did not clearly demonstrate the developmentally toxic effects of monotherapy with carbamazepine as alluded to above are as follows: Niebyl et al., 1979; Nakane et al., 1980; Bertollini et al., 1985; Gaily et al., 1988, 1990; Van der Pol et al., 1991; Czeizel et al., 1992; Scolnik et al., 1994.

More recent review articles on carbamazepine alone and its use in combination with other anticonvulsant drugs and developmental toxicity potential were published (Lindhout et al., 1984; Hernandez-Diaz et al., 2001; Iqbal et al., 2001; Holmes et al., 2001; Wide et al., 2004; Ornoy et al., 2004).

CHEMISTRY

Carbamazepine is near average in terms of size. It is a hydrophobic molecule with average polarity and hydrogen bonding capability. The calculated physicochemical and topological properties are listed below.

PHYSICOCHEMICAL PROPERTIES

Parameter	Value
Molecular weight	236.273 g/mol
Molecular volume	208.60 A^3
Density	1.242 g/cm^3
Surface area	235.96 A^2
LogP	2.200
HLB	8.199
Solubility parameter	26.091 J$^{(0.5)}$/cm$^{(1.5)}$
Dispersion	23.029 J$^{(0.5)}$/cm$^{(1.5)}$
Polarity	7.615 J$^{(0.5)}$/cm$^{(1.5)}$
Hydrogen bonding	9.614 J$^{(0.5)}$/cm$^{(1.5)}$
H bond acceptor	0.79
H bond donor	0.58
Percent hydrophilic surface	41.99
MR	70.707
Water solubility	−1.803 log (mol/M^3)
Hydrophilic surface area	99.09 A^2
Polar surface area	51.18 A^2
HOMO	−8.781 eV
LUMO	−0.363 eV
Dipole	3.553 debye

TOPOLOGICAL PROPERTIES (UNITLESS)

Parameter	Value
x0	12.535
x1	8.771
x2	7.816
xp3	6.603
xp4	5.937
xp5	5.114
xp6	3.424
xp7	2.424
xp8	1.748
xp9	1.150
xp10	0.762
xv0	9.706
xv1	5.729
xv2	4.125
xvp3	3.018
xvp4	2.211
xvp5	1.558
xvp6	0.860
xvp7	0.507
xvp8	0.300
xvp9	0.167
xvp10	0.090
k0	18.380
k1	13.005
k2	5.551
k3	2.525
ka1	10.895
ka2	4.217
ka3	1.791

REFERENCES

Arpino, C. et al. (2000). Teratogenic effects of antiepileptic drugs: Use of an international database on malformations and drug exposure (MADRE). *Epilepsia* 41: 1436–1443.

Bertollini, R., Mastroiacovo, P., and Segni, G. (1985). Maternal epilepsy and birth defects: A case-control study in the Italian Multicentric Registry of Birth Defects (IPIMC). *Eur. J. Epidemiol.* 1: 67–72.

Bertollini, R. et al. (1987). Anticonvulsant drugs in monotherapy: Effect on the fetus. *Eur. J. Epidemiol.* 3: 164–171.

Canger, R. et al. (1999). Malformations in offspring of women with epilepsy. A prospective study. *Epilepsia* 40: 1231–1236.

Czeizel, A. E., Bod, M., and Halasz, P. (1992). Evaluation of anticonvulsant drugs during pregnancy in a population-based Hungarian study. *Eur. J. Epidemiol.* 8: 122–127.

Diav-Citrin, O. et al. (2001). Is carbamazepine teratogenic? A prospective controlled study of 210 pregnancies. *Neurology* 57: 321–324.

Dow, K. E. and Riopelle, R. J. (1989). Teratogenic effects of carbamazepine. *N. Engl. J. Med.* 321: 1480–1481.

Friedman, J. M. and Polifka, J. E. (2000). *Teratogenic Effects of Drugs. A Resource for Clinicians (TERIS)*, Second ed., Johns Hopkins University Press, Baltimore, MD.

Gaily, E. K., Kantola-Sorsa, E., and Granstrom, M.-L. (1988). Intelligence of children of epileptic mothers. *J. Pediatr.* 113: 677–684.

Gaily, E. K., Kantola-Sorsa, E., and Granstrom, M.-L. (1990). Specific congenital dysfunction in children with epileptic mothers. *Dev. Med. Child. Neurol.* 32: 403–414.

Gladstone, D. J. et al. (1992). Course of pregnancy and fetal outcome following maternal exposure to carbamazepine and phenytoin: A prospective study. *Reprod. Toxicol.* 6: 257–261.

Hardman, J. G., Limbird, L. E., and Gilman, A. G. (Eds.). (2001). *Goodman & Gilman's The Pharmacological Basis of Therapeutics*, 10th ed., McGraw-Hill, New York, p. 533.

Hernandez-Diaz, S. et al. (2001). Neural tube defects in relation to use of folic acid antagonists during pregnancy. *Am. J. Epidemiol.* 153: 961–968.

Hicks, E. P. (1979). Carbamazepine in two pregnancies. *Clin. Exp. Neurol.* 16: 269–275.

Hiilesmaa, V. K. et al. (1981). Fetal head growth retardation associated with maternal antiepileptic drugs. *Lancet* 2: 165–166.

Holmes, L. B. et al. (2000). Intelligence and physical features of children of women with epilepsy. *Teratology* 61: 196–202.

Holmes, L. B. et al. (2001). The teratogenicity of anticonvulsant drugs. *N. Engl. J. Med.* 344: 1132–1138.

Iqbal, M. M., Sohhan, T., and Mahmud, S. Z. (2001). The effects of lithium, valproic acid, and carbamazepine during pregnancy and lactation. *J. Toxicol. Clin. Toxicol.* 39: 381–392.

Janz, D. (1994). Are antiepileptic drugs harmful when taken during pregnancy? *J. Perinat. Med.* 22: 367–375.

Jick, S. S. and Terris, B. Z. (1997). Anticonvulsants and congenital malformations. *Pharmacotherapy* 17: 561–564.

Jones, K. L. et al. (1988). Pregnancy outcome in women treated with Tegretol. *Teratology* 37: 468–469.

Jones, K. L. et al. (1989). Pattern of malformations in the children of women treated with carbamazepine during pregnancy. *N. Engl. J. Med.* 320: 1661–1666.

Kallen, A. J. B. (1994). Maternal carbamazepine and infant spina bifida. *Reprod. Toxicol.* 8: 203–205.

Kaneko, S. et al. (1993). Teratogenicity of antiepileptic drugs and drug specific malformations. *Jpn. J. Psychiatr. Neurol.* 37: 306–308.

Lacy, C. F. et al. (2004). *Drug Information Handbook (Pocket), 2004–2005*, Lexi-Comp., Inc., Hudson, OH.

Legido, A., Toomey, K., and Goldsmith, L. (1991). Congenital rib anomalies in a fetus exposed to carbamazepine. *Clin. Pediatr.* 30: 63–64.

Lindhout, D., Hoopener, R. J. E. A., and Meinardi, H. (1984). Teratogenicity of antiepileptic drug combinations with special emphasis on epoxidation (of carbamazepine). *Epilepsia* 25: 77–83.

Little, B. B. et al. (1993). Megadose carbamazepine during the period of neural tube closure. *Obstet. Gynecol.* 82: 705–708.

Matalon, S. et al. (2002). The teratogenic effect of carbamazepine: A meta-analysis of 1255 exposures. *Reprod. Toxicol.* 16: 9–17.

McElhatton, P. R. and Sullivan, F. M. (1977). Comparative teratogenicity of six antiepileptic drugs in the mouse. *Br. J. Pharmacol.* 59: 494P–495P.

McMullin, G. P. (1971). Teratogenic effects of anticonvulsants. *Br. Med. J.* 4: 430.

Meijer, J. W. A. (1984). Possible hazard of valpromide–carbamazepine combination therapy in epilepsy. *Lancet* 1: 802.

Moore, S. J. et al. (2000). A clinical study of 57 children with fetal anticonvulsant syndromes. *J. Med. Genet.* 37: 489–497.

Nakane, Y. et al. (1980). Multiinstitutional study on the teratogenicity and fetal toxicity of antiepileptic drugs: A report of the collaborative study group in Japan. *Epilepsia* 21: 663–680.

Niebyl, J. R. et al. (1979). Carbamazepine levels in pregnancy and lactation. *Obstet. Gynecol.* 53: 139–140.

Nulman, I. et al. (1997). Findings in children exposed *in utero* to phenytoin and carbamazepine monotherapy: Independent effects of epilepsy and medications. *Am. J. Med. Genet.* 68: 18–24.

Oakeshott, P. and Hunt, G. M. (1991). Carbamazepine and spina bifida. *Br. Med. J.* 303: 651.

Omtzigt, J. G. C. et al. (1993). The 10,11-epoxide-10,11-diol pathway of carbamazepine in early pregnancy in maternal serum, urine, and amniotic fluid: Effect of dose, comedication, and relation to outcome of pregnancy. *Ther. Drug Monit.* 15: 1–10.

Ornoy, A. and Cohen, E. (1996). Outcome of children born to epileptic mothers treated with carbamazepine during pregnancy. *Arch. Dis. Child.* 75: 517–520.

Ornoy, A. et al. (2004). The developmental effects of maternal antiepileptic drugs with special reference to carbamazepine. *Reprod. Toxicol.* 19: 248–249.

PDR® (Physicians' Desk Reference®). (2002). Medical Economics Co., Inc., Montvale, NJ.

Robertson, I. G., Donnai, D., and D'Souza, S. (1983). Cranial nerve agenesis in a fetus exposed to carbamazepine. *Dev. Med. Child. Neurol.* 25: 540–541.

Rosa, F. W. (1991). Spina bifida in infants of women treated with carbamazepine during pregnancy. *N. Engl. J. Med.* 324: 674–677.

Samren, E. B. et al. (1997). Maternal use of antiepileptic drugs and the risk of major congenital malformations: A joint European prospective study of human teratogenesis associated with maternal epilepsy. *Epilepsia* 38: 981–990.

Schardein, J. L. (2000). *Chemically Induced Birth Defects*, Third ed., Marcel Dekker, New York, p. 205.

Scolnik, D. et al. (1994). Neurodevelopment of children exposed *in utero* to phenytoin and carbamazepine monotherapy. *JAMA* 271: 767–770.

Shakir, R. A. and Abdulwahab, B. (1991). Congenital malformations before and after the onset of epilepsy. *Acta Neurol. Scand.* 84: 153–156.

Sutcliffe, A. G., Jones, R. B., and Woodruff, G. (1998). Eye malformations associated with treatment with carbamazepine during pregnancy. *Ophthalmic Genet.* 19: 59–62.

Van Allen, M. L. et al. (1988). Increased major and minor malformations in infants of epileptic mothers: Preliminary results of the pregnancy and epilepsy study. *Am. J. Hum. Gen.* 43: A73.

Van der Pol, M. C. et al. (1991). Antiepileptic medication in pregnancy: Late effects on the children's central nervous system development. *Am. J. Obstet. Gynecol.* 164: 121–128.

Vestergard, S. (1969). Congenital adrenogenital syndrome. Report of a case observed after treatment with Tegretol during pregnancy. *Ugeskr. Laeger.* 131: 1129–1131.

Vestermark, V. and Vestermark, S. (1991). Teratogenic effects of carbamazepine. *Arch. Dis. Child.* 66: 641–642.

Vorhees, C. V. et al. (1990). Teratogenicity of carbamazepine in rats. *Teratology* 41: 311–317.

Wide, K. et al. (2000). Psychomotor development and minor anomalies in children exposed to antiepileptic drugs *in utero*: A prospective population-based study. *Dev. Med. Child. Neurol.* 42: 87–92.

Wide, K., Winbladh, B., and Kallen, B. (2004). Major malformations in infants exposed to antiepileptic drugs *in utero*, with emphasis on carbamazepine and valproic acid: A nation-wide, population-based register study. *Acta Paediatr.* 93: 174–176.

13 Danazol

Chemical name: 17α-Ethynyl-17β-hydroxy-4-androsteno[2,3-*d*]isoxazole

CAS #: 17230-88-5

SMILES: C12C(C3C(CC1)(C(CC3)(C#C)O)C)CCC4C2(Cc5c(C=4)onc5)C

INTRODUCTION

Danazol is a synthetically modified androgen derived from ethisterone, and it has androgenic, antigonadotropic, and antiestrogenic properties. It is used therapeutically in the treatment of endometriosis, fibrocystic breast disease, and hereditary angioedema. The drug acts by suppressing the pituitary–ovarian axis (Weiner and Buhimschi, 2004). Danazol is available as a prescription drug with the trade name Danocrine®, among others. It has a pregnancy risk category of X, due to its androgenic virilizing effects on female infants (see below).

DEVELOPMENTAL TOXICOLOGY

ANIMALS

There are no published animal studies concerning use of danazol. However, stated on the package label (*PDR*, 2002) is that oral doses of up to 250 mg/kg/day in rats and up to 60 mg/kg/day in rabbits, doses 15- and fourfold human therapeutic doses, respectively, are developmentally toxic only to the extent of inhibiting fetal development in the rabbit.

HUMANS

As stated above, danazol has moderate androgenic properties in the human. These properties include vaginal atresia, urogenital sinus defect, clitoral hypertrophy, labial fusion, and ambiguous genitalia only in females, lesions commonly termed pseudohermaphroditism or virilization. Internal reproductive organs are normal. The 28 cases recorded in the literature are tabulated in Table 1. All cases occurred at therapeutic dose levels (200 to 800 mg/day orally), but most occurred at the high end of the dose range. Effective treatment periods were only after eight gestational weeks, coinciding with the embryological derivation of the external genital structures. There was only a single report mentioning growth retardation, and in approximately one third of the total cases, spontaneous abortion or miscarriage were recorded. However, it is generally considered that these were more

TABLE 1
Developmental Toxicity Profile of Danazol in Humans

Case Number	Malformations	Growth Retardation	Death	Functional Deficit	Ref.
1, 2	None		✔		Dmowski and Cohen, 1978
3–8[a]	Virilization, body wall (1)				Castro-Magana et al., 1981
9	Virilization				Duck and Katayama, 1981
10	Virilization				Peress et al., 1982
11	Virilization	✔			Schwartz, 1982
12	Virilization				Shaw and Farquhar, 1984
13–17[b]	Virilization				Rosa, 1984 (FDA cases)
18–25	None		✔		Rosa, 1984 (FDA cases)
26	Virilization				Quagliarello and Greco, 1985
27	Virilization				Kingsbury, 1985
28–30	Virilization				ADRAC, 1989
31–35[c]	None		✔		Brunskill, 1992 (manufacturer's data)
36–43[c]	Virilization				Brunskill, 1992 (manufacturer's data)

[a] One case, cites five other known to them.
[b] Less four cases cited earlier.
[c] Less cases cited earlier (except numbers 4 through 8).

probably due to endometriosis, the indication for treatment, rather than to danazol. A publication by Brunskill (1992) reviewed most of the above cases from the various sources, totaling 129, 94 of which were pregnant, with 24% virilized. The teratogenic risk factor for virilization of female fetuses is considered by one group of experts to be moderate (Friedman and Polifka, 2000).

CHEMISTRY

Danazol is larger than the average size of human developmental toxicants. It is a hydrophobic compound with lower polarity. The calculated physicochemical and topological properties are as follows.

PHYSICOCHEMICAL PROPERTIES

Parameter	Value
Molecular weight	337.462 g/mol
Molecular volume	323.59 A^3
Density	1.066 g/cm^3
Surface area	387.54 A^2
LogP	4.737
HLB	0.000
Solubility parameter	23.422 J$^{(0.5)}$/cm$^{(1.5)}$
Dispersion	20.813 J$^{(0.5)}$/cm$^{(1.5)}$
Polarity	3.870 J$^{(0.5)}$/cm$^{(1.5)}$
Hydrogen bonding	10.021 J$^{(0.5)}$/cm$^{(1.5)}$
H bond acceptor	0.80
H bond donor	0.48
Percent hydrophilic surface	5.81
MR	96.501

Continued.

Parameter	Value
Water solubility	-4.218 log (mol/M^3)
Hydrophilic surface area	22.53 A^2
Polar surface area	46.26 A^2
HOMO	-8.989 eV
LUMO	-0.368 eV
Dipole	3.886 debye

TOPOLOGICAL PROPERTIES (UNITLESS)

Parameter	Value
x0	17.449
x1	11.912
x2	11.957
xp3	11.727
xp4	9.805
xp5	7.805
xp6	6.347
xp7	5.039
xp8	3.871
xp9	2.762
xp10	2.023
xv0	15.216
xv1	9.760
xv2	9.394
xvp3	8.666
xvp4	7.136
xvp5	5.518
xvp6	4.229
xvp7	3.106
xvp8	2.234
xvp9	1.463
xvp10	0.936
k0	34.948
k1	17.122
k2	5.510
k3	2.121
ka1	15.852
ka2	4.869
ka3	1.826

REFERENCES

ADRAC (Australian Drug Reactions Advisory Committee). (1989). Danazol masculinization — a reminder. *Aust. Adv. Drug React. Bull.* November, Abst. 1.

Brunskill, P. J. (1992). The effects of fetal exposure to danazol. *Br. J. Obstet. Gynaecol.* 99: 212–215.

Castro-Magana, M. et al. (1981). Transient adrenogenital syndrome due to exposure to danazol *in utero. Am. J. Dis. Child.* 135: 1032–1034.

Dmowski, W. P. and Cohen, M. R. (1978). Antigonadotropin (danazol) in the treatment of endometriosis. Evaluation of posttreatment fertility and three-year follow-up data. *Am. J. Obstet. Gynecol.* 130: 41–48.

Duck, S. C. and Katayama, K. P. (1981). Danazol may cause female pseudohermaphroditism. *Fertil. Steril.* 35: 230–231.

Friedman, J. M. and Polifka, J. E. (2000). *Teratogenic Effects of Drugs. A Resource for Clinicians (TERIS)*, Second ed., Johns Hopkins University Press, Baltimore, MD.

Kingsbury, A. C. (1985). Danazol and fetal masculinization: A warning. *Med. J. Aust.* 143: 410–411.

PDR® (Physicians' Desk Reference®). (2002). Medical Economics Co., Inc., Montvale, NJ.

Peress, M. R. et al. (1982). Female pseudohermaphroditism with somatic chromosomal anomaly in association with *in utero* exposure to danazol. *Am. J. Obstet. Gynecol.* 142: 708–709.

Quagliarello, J. and Greco, M. A. (1985). Danazol and urogenital sinus formation in pregnancy. *Fertil. Steril.* 43: 939.

Rosa, F. (1984). Virilization of the female fetus with maternal danazol exposure. *Am. J. Obstet. Gynecol.* 149: 99–100.

Schwartz, R. P. (1982). Ambiguous genitalia in a term female infant due to exposure to danazol *in utero. Am. J. Dis. Child.* 136: 474.

Shaw, R. W. and Farquhar, J. W. (1984). Female pseudohermaphroditism associated with danazol exposure *in utero.* Case report. *Br. J. Obstet. Gynaecol.* 91: 386–389.

Weiner, C. P. and Buhimschi, C. (2004). *Drugs for Pregnant and Lactating Women.* Church Livingstone, Philadelphia.

14 Parramethadione

Chemical name: 5-Ethyl-3,5-dimethyl-2,4-oxazolidinedione

CAS #: 115-67-3

SMILES: C1(C(N(C(O1)=O)C)=O)(CC)C

INTRODUCTION

Parramethadione is an oxazolidinedione anticonvulsant used in the treatment of petit mal epilepsy, a condition generally requiring treatment only in childhood. It is chemically related to trimethadione, a drug that has now largely been abandoned in the marketplace and is not used due to its severe effects on the human fetus (fetal trimethadione syndrome). The two drugs were introduced into the markeplace in the mid-1940s. As will be seen later, paramethadione has only slightly less and similar developmental toxicity potential, and its clinical use has also been increasingly limited due to this toxicity in favor of less toxic anticonvulsants. However, its inclusion in this series is justified by virtue that its history is an interesting lesson in relation to the dione effects in clinical practice. The drug was available as a prescription drug under the trade name Paradione®, and it had a pregnancy category designation of D (infers that the drug "may cause fetal harm when administered to a pregnant woman").

DEVELOPMENTAL TOXICOLOGY

ANIMALS

Studies with paramethadione in laboratory animals have been limited. In rats, oral doses over the range of 16.5 to 790 mg/kg/day during organogenesis had adverse maternal effects at 527 mg/kg and higher, and developmental effects were manifested by increased fetal death, inhibited fetal growth, and increased skeletal developmental variations at doses of 264 mg/kg and higher. No malformations were elicited in this species (Buttar et al., 1976). Oral doses in the range of 300 to 600 mg/day during a period of 16 to 21 days during the critical period of gestation produced no maternal or developmental toxicity in a primate species (Poswillo, 1972).

HUMANS

In human subjects, it was established that paramethadione, like its congener (trimethadione), produces a syndrome of developmental toxicity termed "fetal trimethadione syndrome." With paramethadione, six cases (three cases in one family) as tabulated in Table 1 were identified. Together

TABLE 1
Developmental Toxicity Profile with Paramethadione in Humans

Case Number	Malformations	Growth Retardation	Death	Functional Deficit	Ref.
1	None		✔		German et al., 1970a, 1970b (HEAL family II.1)
2	Multiple: lip/palate, spine, genital, urinary, brain, heart, vessels		✔		German et al., 1970a, 1970b (HEAL family II.2)
3	Multiple: ears, digits, genital, heart, vessels, renal	✔	✔		German et al., 1970a, 1970b (HEAL family II.3)
4	Multiple: ears, genitals, face	✔		✔	German et al., 1970a, 1970b (HEAL family II.4)
5	None		✔		German et al., 1970a, 1970b (HEAL family II.5)
6	Heart		✔		Rutman, 1973
7	Eye, brain	✔		✔	Rutman, 1973
8	Heart, brain				Rutman, 1973

with its more toxic congener, at least 37 cases have been described in the literature (Schardein, 2000). In all cases, other drugs, including other anticonvulsant drugs, were given to the mothers along with the diones, and normal infants were born following removal from the dione treatment, suggesting strongly that the drug had been responsible for the toxicity in the earlier pregnancies. The clinical findings of the fetal trimethadione syndrome are given in Table 2.

Paramethadione induced developmental toxicity in addition to the syndrome of malformations: intrauterine- or postnatal growth retardation and failure to thrive in about one half of the cases, postnatal death or spontaneous abortion in five of the eight cases, and mental retardation or delayed mental and motor development in two cases accompanying multiple malformations. These facts clearly indicate that paramethadione is a significant developmental toxicant, displaying the full spectrum of developmental toxicity. Fortunately, there is little chance for further adverse pregnancy effects, now that the drug has very limited clinical use. The magnitude of teratogenic risk is high, according to one group of experts (Friedman and Polifka, 2000).

TABLE 2
Clinical Findings in 53 Offspring of Women Treated with Diones during Pregnancy

Clinical Findings	Frequency (%)
Speech impairment	62
Congenital heart disease	50
Delayed mental development	50
Malformed ears	42
Urogenital malformations	30
Cleft lip/palate	28
Skeletal malformations	25
High arched palate	18
Inguinal or umbilical hernias	15

Source: Various sources, after Schardein, J. L., *Chemically Induced Birth Defects*, Third ed., Marcel Dekker, New York, 2000, p. 207.

CHEMISTRY

Paramethadione is a smaller hydrophilic compound capable of acting as a hydrogen bond acceptor. It is of average polarity in comparison to the other human developmental toxicants. Paramethadione's calculated physicochemical and topological properties are as follows.

PHYSICOCHEMICAL PROPERTIES

Parameter	Value
Molecular weight	157.169 g/mol
Molecular volume	140.81 A^3
Density	0.944 g/cm^3
Surface area	189.40 A^2
LogP	−1.668
HLB	7.673
Solubility parameter	22.723 J$^{(0.5)}$/cm$^{(1.5)}$
Dispersion	18.131 J$^{(0.5)}$/cm$^{(1.5)}$
Polarity	9.093 J$^{(0.5)}$/cm$^{(1.5)}$
Hydrogen bonding	10.242 J$^{(0.5)}$/cm$^{(1.5)}$
H bond acceptor	0.54
H bond donor	0.00
Percent hydrophilic surface	39.71
MR	41.397
Water solubility	1.961 log (mol/M^3)
Hydrophilic surface area	75.21 A^2
Polar surface area	52.93 A^2
HOMO	−10.967 eV
LUMO	0.159 eV
Dipole	2.767 debye

TOPOLOGICAL PROPERTIES (UNITLESS)

Parameter	Value
x0	8.646
x1	5.010
x2	4.837
xp3	4.382
xp4	2.701
xp5	1.537
xp6	0.500
xp7	0.048
xp8	0.000
xp9	0.000
xp10	0.000
xv0	6.879
xv1	3.522
xv2	2.816
xvp3	2.018
xvp4	0.966
xvp5	0.466
xvp6	0.115
xvp7	0.007

Continued.

Parameter	Value
xvp8	0.000
xvp9	0.000
xvp10	0.000
k0	11.455
k1	9.091
k2	2.803
k3	1.322
ka1	8.358
ka2	2.390
ka3	1.080

REFERENCES

Buttar, H. S., Dupuis, I., and Khera, K. S. (1976). Fetotoxicity of trimethadione and paramethadione in rats. *Toxicol. Appl. Pharmacol.* 37: 126.

Friedman, J. M. and Polifka, J. E. (2000). *Teratogenic Effects of Drugs. A Resource for Clinicians (TERIS)*, Second ed., Johns Hopkins University Press, Baltimore, MD.

German, J. et al. (1970a). Possible teratogenicity of trimethadione and paramethadione. *Lancet* 2: 261–262.

German, J., Kowal, A., and Ehlers, K. H. (1970b). Trimethadione and human teratogenesis. *Teratology* 3: 349–361.

Poswillo, D. E. (1972). Tridone and paradione as suspected teratogens. An investigation in subhuman primates. *Ann. R. Coll. Surg. Engl.* 50: 367–370.

Rutman, J. Y. (1973). Anticonvulsants and fetal damage. *N. Engl. J. Med.* 289: 696–697.

Schardein, J. L. (2000). *Chemically Induced Birth Defects*, Third ed., Marcel Dekker, New York, p. 207.

15 Carbon Monoxide

Chemical name: CO

Alternate names: Carbonic oxide, exhaust gas, illuminating gas, flue gas

CAS #: 630-08-0

SMILES: [O+]#[C-]

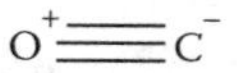

$$O^+ {\equiv\equiv} C^-$$

INTRODUCTION

Carbon monoxide (CO) is a highly poisonous, odorless, colorless, tasteless, flammable gas used as a reducing chemical in metallurgical operations, in organic synthesis of petroleum-type products, and in the manufacture of metal carbonyls. Its toxicity resides in its ability to combine with the hemoglobin in the blood to form carboxyhemoglobin, which disrupts oxygen transport and delivery throughout the body. Maternal smoking probably constitutes the most common source of (fetal) exposure to high concentrations of CO; measurements exceed 50,000 ppm in some cases (Robinson and Forbes, 1975). This source of the chemical will not be discussed in this section. Rather, exposures discussed here are in the context of human environmental atmospheric exposures. The threshold limit value (TLV) adopted for CO for the human is 25 ppm (time weighted average); its toxic activity is via anoxia to the cardiovascular, central nervous, and reproductive systems (ACGIH, 2005). We will discuss the latter two systems here, as developmental neurotoxicity is the primary manifestation of the effects of CO in the human (see below).

DEVELOPMENTAL TOXICOLOGY

ANIMALS

Carbon monoxide inhalation has not proven to be a consistent teratogen in laboratory animals. As a multitude of studies in a variety of species have been conducted, a tabulation of developmental effects by exposure level and response is provided in Table 1. The characteristic responses indicate that developmental toxicity in the form of embryolethality, growth retardation, and postnatal functional impairment is commonly induced in laboratory animals from CO exposures and, rarely, malformation is induced, only in the rat and guinea pig.

HUMANS

In the human, the pattern of toxicity recorded was mainly confined to the central nervous system, and representative historical references over the interval 1929 to contemporary times are provided in Table 2. Over 20 cases are recorded here. Exposures ranged from the first month of gestation

TABLE 1
Developmental Toxicity Profile of Carbon Monoxide (CO) in Laboratory Animals

Species	Characteristic Response	Gestational Exposure Level	Ref.
Mouse	Increased fetal mortality and decreased fetal weight, postnatal behavior effects	65–500 ppm	Singh and Scott, 1984; Singh, 1986
Rat	Postnatal behavioral effects, central nervous system abnormalities	150–1000 ppm	Daughtrey and Norton,1983; Mactutus and Fechter, 1984
Guinea pig	Limb malformations	0.42–0.48%	Giuntini and Corneli, 1955
Rabbit	Reduced fetal weight, increased fetal mortality	90–180 ppm	Astrup et al., 1972
Pig	Increased stillbirth	180–250 ppm	Wood, 1979; Dominick and Carson, 1983
Primate	Brain lesions (fetal hemorrhagic necrosis)	0.1–0.3%	Ginsberg and Myers, 1974

TABLE 2
Developmental Toxicity Profile of Carbon Monoxide (CO) in Humans

Case Number	Malformations	Growth Retardation	Death	Functional Deficit	Ref.
1	Brain		✔		Maresch, 1929
2	Brain		✔		Neuberger, 1935
3	Brain	✔			Brander, 1940
4	Jaw, tongue		✔	✔	Zourbas, 1947
5	Brain, pancreas		✔		Lombard, 1950
6	Eyes			✔	Lombard, 1950
7	None			✔	Lombard, 1950
8	None			✔	Desclaux et al., 1951
9	Limbs				Gere and Szekeres, 1955
10	Skeletal				Corneli, 1955
11	None			✔	Muller and Graham, 1955
12	Brain				Ingalls, 1956
13, 14	None			✔	Beau et al., 1956
15	Eyes	✔			Beau et al., 1956
16	None		✔	✔	Beau et al., 1956
17	Limbs, digits				Bette, 1957
18	Brain	✔	✔		Schwedenberg, 1959
19	None	✔			Nishimura, 1974
20	Multiple: brain, skull, ears, oral, genital, lungs, limbs	✔	✔		Caravati et al., 1988
21	None		✔		Caravati et al., 1988
22	None		✔		Caravati et al., 1988
23	Muscle				Buyse, 1990
24	Brain				Woody and Brewster, 1990
25	None			✔	Koren et al., 1991
26	Lip/palate, heart				Hennequin et al., 1993

until the ninth month or near-term. Carboxyhemoglobin levels ranging from chronic (5 to 20%) to acute (30 to 50%) to life-threatening (50 to 66%) to lethal (>66%) were cataloged (Aubard and Mogne, 2000). Growth retardation was an associated feature in 20%, but death and functional deficits of various descriptions (retarded psychomotor development, subnormal mentality, lack of reflexes, mental retardation, spasticity, cerebral palsy) were commonplace findings. As stated above, the developmental toxicity pattern has been primarily as a developmental neurotoxicant, characterized chiefly as anoxic encephalopathy and mortality. A number of useful reviews on carbon-monoxide-induced developmental toxicity are available. Included are home and vehicle exposures (Jaeger, 1981), workplace exposures (Norman and Halton, 1990), animal and human exposures (Annau and Fechter, 1994), and exposures in general (Longo, 1977; Barlow and Sullivan, 1982; Bailey, 2001).

CHEMISTRY

Carbon monoxide, a linear molecule, is one of the smallest human developmental toxicants. Its calculated physicochemical and topological properties are shown below.

PHYSICOCHEMICAL PROPERTIES

Parameter	Value
Molecular weight	28.010 g/mol
Molecular volume	28.12 A^3
Density	1.079 g/cm^3
Surface area	45.53 A^2
LogP	−1.270
HLB	21.540
Solubility parameter	26.923 J$^{(0.5)}$/cm$^{(1.5)}$
Dispersion	26.923 J$^{(0.5)}$/cm$^{(1.5)}$
Polarity	0.000 J$^{(0.5)}$/cm$^{(1.5)}$
Hydrogen bonding	0.000 J$^{(0.5)}$/cm$^{(1.5)}$
H bond acceptor	0.89
H bond donor	0.00
Percent hydrophilic surface	100.00
MR	7.027
Water solubility	4.209 log (mol/M^3)
Hydrophilic surface area	45.53 A^2
Polar surface area	19.90 A^2
HOMO	−12.362 eV
LUMO	2.175 eV
Dipole	0.806 debye

TOPOLOGICAL PROPERTIES (UNITLESS)

Parameter	Value
x0	2.000
x1	1.000
x2	0.000
xp3	0.000
xp4	0.000

Continued.

Parameter	Value
xp5	0.000
xp6	0.000
xp7	0.000
xp8	0.000
xp9	0.000
xp10	0.000
xv0	0.908
xv1	0.204
xv2	0.000
xvp3	0.000
xvp4	0.000
xvp5	0.000
xvp6	0.000
xvp7	0.000
xvp8	0.000
xvp9	0.000
xvp10	0.000
k0	0.602
k1	2.000
k2	0.000
k3	0.000
ka1	1.800
ka2	0.000
ka3	0.000

REFERENCES

ACGIH (American Conference of Government Industrial Hygienists). (2005). *TLVs® and BEIs®, Threshold Limit Values for Chemical Substances and Physical Agents & Biological Exposure Indices*, ACGIH, Cincinnati, OH, p. 18.

Annau, Z. and Fechter, L. D. (1994). The effects of prenatal exposure to carbon monoxide. In *Prenatal Exposure to Toxicants. Developmental Consequences*, H. L. Needleman and D. Bellinger, Eds., Johns Hopkins University Press, Baltimore, MD, pp. 249–267.

Astrup, P. et al. (1972). Effect of moderate carbon-monoxide exposure on fetal development. *Lancet* 2: 1220–1222.

Aubard, Y. and Mogne, I. (2000). Carbon monoxide poisoning in pregnancy. *Br. J. Obstet. Gynecol.* 107: 833–838.

Bailey, B. (2001). Carbon monoxide poisoning during pregnancy. In *Maternal–Fetal Toxicology. A Clinicians Guide*, Third ed., G. Koren, Ed., Marcel Dekker, New York, pp. 257–268.

Barlow, S. M. and Sullivan, F. M. (1982). *Reproductive Hazards of Industrial Chemicals. An Evaluation of Animal and Human Data*, Academic Press, New York, pp. 179–199.

Beau, A., Neimann, N., and Pierson, M. (1956). [The role of carbon monoxide poisoning during pregnancy on the genesis of neonatal encephalopathies. A propos of 5 observations]. *Arch. Fr. Pediatr.* 13: 130–143.

Bette, H. (1957). Extremitaten Missbildungen nach Leuchtgasvergiftung der Mutter, kasuistike Beitrag zur Missbildungsforschung. *Munch. Med. Wochenschr.* 99: 1246.

Brander, T. (1940). Microcephalus und Tetraplegie bei emem kinde nach Kohlenmonoxydvergiftung der Mutter wahrend der Schwangerschaft. *Acta Paediat.* 28 (Suppl. 1): 123–132.

Buyse, M. L. (Ed.). (1990). *Birth Defects Encyclopedia*, Center for Birth Defects Information Services, Dover, MA, Blackwell Scientific, St. Louis, pp. 697–699.

Caravati, E. M. et al. (1988). Fetal toxicity associated with maternal carbon monoxide poisoning. *Ann. Emerg. Med.* 17: 714–717.

Corneli, F. (1955). Contributo sperimentale all'azione teratogenica dell'ossido do carbonio nei mammiferi. *Ortop. Traumatol. Protez* 23: 261–271.

Daughtrey, W. C. and Norton, S. (1983). Caudate morphology and behavior of rats exposed to carbon monoxide in utero. *Exp. Neurol.* 80: 265–275.

Desclaux, P., Soulairac, A., and Morlon, C. (1951). Intoxication oxycarbonee au cours d'une gestation (5-mois). Arrieration mentale consecutive. *Arch. Fr. Pediatr.* 8: 316–317.

Dominick, M. A. and Carson, T. L. (1983). Effects of carbon monoxide exposure on pregnant sows and their fetuses. *Am. J. Vet. Res.* 44: 35–40.

Gere, K. and Szekeres, V. (1955). Ujabb adapt az embryopathiak pathogeneishez. *Kulonlenyomat a Gyermekgyogydszet* 8: 245–248.

Ginsberg, M. D. and Myers, R. E. (1974). Fetal brain damage following maternal carbon monoxide intoxication: An experimental study. *Acta Obstet. Gynecol. Scand.* 53: 309–317.

Giuntini, L. and Corneli, F. (1955). Nota preliminari sull'azione teratogenica dell' ossido di carbonio. *Bull. Soc. Ital. Biol. Sper.* 31: 258–260.

Hennequin, Y. et al. (1993). *In utero* carbon monoxide poisoning and multiple fetal abnormalities (letter). *Lancet* 341: 240.

Ingalls, T. H. (1956). Causes and prevention of developmental defects. *JAMA* 161: 1047–1051.

Jaeger, R. J. (1981). Carbon monoxide in houses and vehicles. *Bull. NY Acad. Sci.* 57: 860–872.

Koren, G. et al. (1991). A multicenter prospective study of fetal outcome following accidental carbon monoxide poisoning in pregnancy. *Reprod. Toxicol.* 5: 397–403.

Lombard, J. (1950). Du role de l'intoxication oxycarbonee au cours de la grossesse comme facteur de malformations. Thesis, Université Nancy, France.

Longo, L. D. (1977). The biological effects of carbon monoxide on the pregnant woman, fetus, and newborn infant. *Am. J. Obstet. Gynecol.* 129: 69–103.

Mactutus, C. F. and Fechter, L. D. (1984). Prenatal exposure to carbon monoxide: Learning and memory deficits. *Science* 223: 409–411.

Maresch, R. (1929). Uber emen Fall von Kohlenoxydgasschadigung der Kinder in der Gebarmutter. *Wien. Klin. Wochenschr.* 79: 454–456.

Muller, G. L. and Graham, S. (1955). Intrauterine death of the fetus due to accidental carbon monoxide poisoning. *N. Engl. J. Med.* 252: 1075–1078.

Neuburger, F. (1935). Uber emen intrauterinen Hirnschadigung nach emer Leuchtgasvergiftung der Mutter. *Beitr. Gerrichtl. Med.* 13: 85–95.

Nishimura, H. (1974). CO poisoning during pregnancy and microcephalic child. *Cong. Anom.* 14: 41–46.

Norman, C. A. and Halton, D. M. (1990). Is carbon monoxide a workplace teratogen? A review and evaluation of the literature. *Ann. Occupat. Hyg.* 4: 335–347.

Robinson, J. C. and Forbes, W. F. (1975). The role of carbon dioxide in cigarette smoking. I. Carbon monoxide yield from cigarettes. *Arch. Environ. Health* 30: 425–434.

Schwedenberg, T. H. (1959). Leukoencephalopathy following carbon monoxide asphyxia. *J. Neuropathol. Exp. Neurol.* 18: 597–608.

Singh, J. (1986). Early behavioral alterations in mice following prenatal carbon monoxide exposure. *Neurotoxicology* 7: 475–482.

Singh, J. and Scott, L. H. (1984). Threshold for carbon monoxide induced fetotoxicity. *Teratology* 30: 253–257.

Wood, E. N. (1979). Increased incidence of stillbirth in piglets associated with high levels of atmospheric carbon monoxide. *Vet. Rec.* 104: 283–284.

Woody, R. C. and Brewster, M. A. (1990). Telencephalic dysgenesis associated with presumptive maternal carbon monoxide intoxication in the first trimester of pregnancy. *Clin. Toxicol.* 28: 467–475.

Zourbas, M. (1947). Encephalopathie congenitale avec troubles du tonus neuromusculaire vraisemblablement consecutive a une intoxication par l'oxyde de carbone. *Arch. Fr. Pediatr.* 4: 513–515.

16 Formaldehyde

Alternate names: Formic aldehyde, methanal, methylene oxide, oxomethane, formalin (aqueous solution)

CAS #: 50-00-0

SMILES: C=O

$$O =\!=\!=$$

INTRODUCTION

Formaldehyde is a colorless gas used in the production of resins, wood products, plastics, fertilizers, and foam insulation. It also has utility as a textile finish, preservative, stabilizer, disinfectant, and antibacterial food additive. In solution as formalin (formol), it has uses as a disinfectant, and the total number of products containing formaldehyde exceeds 3000, any of which may give off formaldehyde vapors (Winter, 1992). Inhalational exposures are thus of major concern. In addition to its generic name, it is also available by several trade names, including BFV®, Formalith®, Ivalon®, and Lysoform®, among others. The threshold limit value (TLV) short-term exposure limit (STEL) for occupational exposure to formaldehyde vapor in the atmosphere is 0.3 ppm (ACGIH, 2005).

DEVELOPMENTAL TOXICOLOGY

ANIMALS

Laboratory animal studies by the inhalational route have been limited to the rat, and their relevance to human exposures is unknown. Microscopic changes in the liver and kidney were reported following exposure levels as high as 0.8 mg/m^3 (Gofmekler and Bonashevskaya, 1969), but levels of up to 5 mg/m^3 were said to produce only decreased postnatal activity of 30-day-old young following prenatal treatment of the dams (Sheveleva, 1976).

HUMANS

Studies in the human indicated developmental toxicity, manifested by malformation or spontaneous abortion, although there are contradictory results to report, as shown in Table 1. In addition, there is one poorly documented foreign report in which lower birth weights were said to be recorded among 446 females exposed to formaldehyde vapor at concentrations ranging from 1.2 to 3.6 ppm compared to 200 control women (Shumilina, 1975). In the absence of corroborating and better validated studies, this report is not included here as being valid.

In summary, it appears that there is evidence, in at least four published reports of variable quality, that formaldehyde or its aqueous counterpart, formalin, have the potential to induce spontaneous abortion or miscarriage in the human when exposures occur early in pregnancy. However, study quality and general absence of exposure concentrations leave much to be desired with respect

TABLE 1
Reported Associations to Developmental Toxicity with Formaldehyde or Formalin in Humans

Author	Malformations	Death
Axelsson et al., 1984	—	Increased spontaneous abortion (RR = 3.2, 95% CI 1.36 to 7.47) among 745 laboratory workers exposed to solvents including formalin
Ericson et al., 1984	No association among 76 laboratory workers	No association to stillbirths among 76 laboratory workers
Hemminki et al., 1985	No association among 34 nurses occupationally exposed in first trimester	No association to spontaneous abortion among 164 nurses occupationally exposed in first trimester
John, 1990	—	Weak association (twofold increase) with miscarriage among 61 cosmetologists exposed in first trimester
Taskinen et al., 1994	—	Weak association (RR = 3.5, 95% CI 1.1 to 11.2) with miscarriage among 206 laboratory workers exposed in first trimester
Saurel-Cubizolles et al., 1994	Significant increase in all congenital anomalies (but not major malformations) in cohort of 271 infants of operating room nurses exposed during first trimester	Significant association to spontaneous abortion among 316 operating room nurses exposed during first trimester

Note: RR is the relative risk; CI is the confidence interval.

to hazard estimation. It was shown in a recent review that there was some evidence of increased risk for spontaneous abortion (meta-relative risk = 1.4, 95% confidence interval [CI] 0.9–2.1), but study biases made it impossible for these investigators to assign significant risk for spontaneous abortion due to the chemical (Collins et al., 2001). With contradictory reports on the potential for this chemical to induce malformations, the data are tenuous at best, and it remains to be seen whether teratogenesis is, in fact, a real response. At this time, it appears that it is not. In addition, growth retardation and functional deficits have not been associated with pregnancy exposure outcomes. Several useful review articles on formaldehyde toxicity in pregnancy in both animals and humans were published (Ma and Harris, 1988; Collins et al., 2001).

CHEMISTRY

Formaldehyde is one of the smallest organic human developmental toxicants. It is hydrophilic and is capable of participating in hydrogen bonding interactions as an acceptor. The calculated physicochemical and topological properties of formaldehyde are shown below.

PHYSICOCHEMICAL PROPERTIES

Parameter	Value
Molecular weight	30.026 g/mol
Molecular volume	30.83 A^3
Density	0.821 g/cm^3

Continued.

Parameter	Value
Surface area	50.67 A^2
LogP	−0.980
HLB	21.540
Solubility parameter	24.178 J$^{(0.5)}$/cm$^{(1.5)}$
Dispersion	15.748 J$^{(0.5)}$/cm$^{(1.5)}$
Polarity	15.748 J$^{(0.5)}$/cm$^{(1.5)}$
Hydrogen bonding	9.412 J$^{(0.5)}$/cm$^{(1.5)}$
H bond acceptor	0.16
H bond donor	0.06
Percent hydrophilic surface	100.00
MR	8.562
Water solubility	3.901 log (mol/M^3)
Hydrophilic surface area	50.67 A^2
Polar surface area	20.23 A^2
HOMO	−10.489 eV
LUMO	1.511 eV
Dipole	1.739 debye

TOPOLOGICAL PROPERTIES (UNITLESS)

Parameter	Value
x0	2.000
x1	1.000
x2	0.000
xp3	0.000
xp4	0.000
xp5	0.000
xp6	0.000
xp7	0.000
xp8	0.000
xp9	0.000
xp10	0.000
xv0	1.115
xv1	0.289
xv2	0.000
xvp3	0.000
xvp4	0.000
xvp5	0.000
xvp6	0.000
xvp7	0.000
xvp8	0.000
xvp9	0.000
xvp10	0.000
k0	0.602
k1	2.000
k2	0.000
k3	0.000
ka1	1.670
ka2	0.000
ka3	0.000

REFERENCES

ACGIH (American Conference of Government Industrial Hygienists). (2005). *TLVs® and BEIs®. Threshold Limit Values for Chemical Substances and Physical Agents & Biological Exposure Indices*, ACGIH, Cincinnati, OH, p. 31.

Axelsson, G., Lutz, C., and Rylander, R. (1984). Exposure to solvents and outcomes of pregnancy in university laboratory employees. *Br. J. Ind. Med.* 41: 305–312.

Collins, J. J. et al. (2001). A review of adverse pregnancy outcomes and formaldehyde exposure in human and animal studies. *Regul. Toxicol. Pharmacol.* 34: 17–34.

Ericson, A. et al. (1984). Delivery outcome of women working in laboratories during pregnancy. *Arch. Environ. Health* 29: 5–10.

Gofmekler, V. A. and Bonashevskaya, T. I. (1969). Experimental study of the teratogenic action of formaldehyde from data obtained from morphological studies. *Gig. Sanit.* 34: 92–94.

Hemminki, K., Kyyronen, P., and Lindbohm, M.-L. (1985). Spontaneous abortions and malformations in the offspring of nurses exposed to anesthetic gases, cytostatic drugs, and other potential health hazards in hospitals based on registered information of outcome. *J. Epidemiol. Community Health* 39: 141–147.

John, E. M. (1990). Spontaneous abortion among cosmetologists. NTIS Report /PB 91-222703, National Technical Information Service, Springfield, VA.

Ma, T.-H. and Harris, M. M. (1988). Review of the genotoxicity of formaldehyde. *Mutat. Res.* 196: 37–57.

Saurel-Cubizolles, M. J., Hays, M., and Estryn-Behar, M. (1994). Work in operating rooms and pregnancy outcome among nurses. *Int. Arch. Occup. Environ. Health* 66: 235–241.

Sheveleva, G. A. (1976). Investigation of the specific effect of formaldehyde on the embryogenesis and progeny of white rats. *Toksikol. Nauykh. Orim. Khim. Veschestv.* 12: 78–86.

Shumilina, A. V. (1975). Menstrual and child-bearing functions of female workers occupationally exposed to the effects of formaldehyde. *Gigiena Truda I Prof. 'Nye Zabolevaniya* 12: 18–21.

Taskinen, H. et al. (1994). Laboratory work and pregnancy outcome. *J. Occup. Med.* 36: 311–319.

Winter, R. (1992). *A Consumer's Dictionary of Household, Yard and Office Chemicals*, Crown Publishers, New York, p. 142.

17 Isotretinoin

Chemical name: 13-*cis*-Retinoic acid

Alternate name: Neovitamin A acid

CAS #: 4759-48-2

SMILES: C1(C=CC(C)=CC=C/C(C)=C\C(=O)O)=C(C)CCCC1(C)C

INTRODUCTION

Isotretinoin, an analog of vitamin A, belongs to the group termed "retinoids" that includes the well-known developmental toxicants etretinate, tretinoin, and acitretin. It has therapeutic value in the treatment of severe, recalcitrant nodular acne unresponsive to conventional therapy. In this therapy, it reduces sebaceous gland size and sebum production and regulates cell proliferation and differentiation (Lacy et al., 2004). The mechanism for this action is via retinoic acid receptors (RARs) as discussed in recent publications, but it is not known whether the parent drug or its 4-oxo-metabolite is the active teratogen (Collins and Mao, 1999; see below). The drug is available commercially by prescription under the trade name Accutane® and several other names, and it has a pregnancy category of X. The package label contains a black box warning label stating that while not every fetus exposed to the drug has resulted in a deformed child, there is an extremely high risk that a deformed infant can result if pregnancy occurs while taking the drug in any amount, even for short periods of time; potentially any fetus exposed during pregnancy can be affected. Restrictive conditions apply for use in women of childbearing potential, and an "avoid pregnancy" icon exists on the label (*PDR*, 2002; Arnon et al., 2004).

DEVELOPMENTAL TOXICOLOGY

ANIMALS

Isotretinoin is a potent developmental toxicant, including teratogenicity, in every animal species tested. Positive effects by the oral route were observed in hamsters (Burk and Willhite, 1988), mice (Vannoy and Kwashigroch, 1987), rabbits (Kamm, 1982), rats (Henck et al., 1987; Collins et al., 1994), and cynomolgus monkeys (Hummler et al., 1990) when administered the drug during one or more days during organ formation in the respective species. Embryo death and decreased fetal weight at maternally toxic doses were observed in mice, rats, and primates. Effective dose levels were observed from 2.5 mg/kg/day in the primate, 10 mg/kg/day in the rabbit, 30 mg/kg/day in the rat, 50 mg/kg/day in the hamster, and 200 mg/kg/day in the mouse, in decreasing order of sensitivity (U.S. Teratology Society, 1991). In all species, these levels exceed the therapeutic (oral) dose in humans (0.5 to 2 mg/kg/day).

HUMANS

Isotretinoin is also a potent teratogen in humans, and it affects most classes of developmental toxicity as well. The history of its toxicity is of interest. It was the first (and perhaps only) drug introduced into the marketplace (in September, 1982) when it was already known to be teratogenic in laboratory species (rat and rabbit; see Kamm, 1982). Because its teratogenicity is universally accepted, a summary is provided below by class effects rather than by tabulation of all case and study reports.

Malformations

Within 6 months of the drug being placed on the market, in the literature in 1983, an abstract authored by an U.S. Food and Drug Administration (FDA) official attested to the knowledge of five cases of malformation known to the agency that were associated with the use of this drug in pregnancy (Rosa, 1983). By the end of the year, a total of 11 cases of malformation were reported to the agency (Rosa, 1984a, 1984b). The same year, the first case report of malformation associated with treatment with isotretinoin was published by scientific investigators (Braun et al., 1984). In the 1982–1985 interval, the manufacturer of the drug estimated that 160,000 women of childbearing age took the drug; the manufacturer allegedly had reports of 426 pregnancy exposures in the interval up to 1989. Additionally, the FDA estimated that 900 to 1300 babies were born with severe birth defects in the first 5 yr the drug was marketed (press accounts, April 1988). This is in contrast to an estimate made in 2000 that about 95 case reports had been published describing cases of malformation (Schardein, 2000). A number of additional cases have come to light since 2000, and the total number of cases of malformation reported in the medical literature up to the present is approximately 210 (Stern et al., 1984; Hersh et al., 1985; Bigby and Stern, 1988; Strauss et al., 1988; Coberly et al., 1996; Honein et al., 2001; Giannoulis et al., 2004; Arnon et al., 2004; Giannoulis et al., 2005). This is not surprising, based on the estimate that up to 60,000 female patients of childbearing age per year are treated with isotretinoin (Strauss et al., 1988), an estimate undoubtedly much greater today. The fact that it has been shown by the manufacturer that no contraception was used by 50% of the patients in a survey of pregnancy reports, in spite of the label warnings, substantiates this estimate. It was said editorially upon discovery of the developmental effects of the drug that there was a 100% risk of abortion or malformation if drug treatment occurred in the second month of gestation (Hall, 1984).

Isotretinoin must qualify as the most widely used teratogenic drug in this country at present. One group of experts considers the teratogenic risk of the drug to be high (Friedman and Polifka, 2000). Characteristic features of the syndrome include central nervous system malformations, microtia/anotia, micrognathia, cleft palate, cardiac and great vessel defects, thymic abnormalities, and eye malformations. A summary of malformation types observed in 61 cases of isotretinoin-exposed pregnancies is shown in Table 1. The cynomolgus monkey is considered a good animal model for human toxicity, demonstrating malformations in similar sites, and embryolethality at maternally toxic dose levels (Hummler et al., 1990, 1996).

Daily doses eliciting teratogenicity are in the range of 0.5 to 1.5 mg/kg, but doses as low as 0.2 mg/kg may be responsible for inducing malformation in some cases. The critical period is believed to be 3 to 5 weeks following conception. In a series of 88 prospectively ascertained pregnancies following 17 to 55 days after discontinuing drug treatment, there was a high rate of conception, but the outcomes included 8 spontaneous abortions, 1 abnormal birth, 75 normal liveborns, but only 4 (4.5%) with congenital malformations (Dai et al., 1989). Oddly enough, the

TABLE 1
Types of Malformations Observed among 61 Isotretinoin-Exposed Pregnancies

Defect	Percent (%) with Defects
Ear, absence or stricture of auditory canal, absence of auricle or microtia	71
Central nervous system (CNS): microcephalus, reduction deformities of brain or hydrocephalus	49
Cardiovascular system (CVS): common truncus, transposition of great vessels, tetralogy of Fallot, common ventricle, coarctation of aorta/aortic arch, or other aortic anomalies	33
Ear + CNS	39
Ear + CVS	25
CNS + CVS	23
Ear + CNS + CVS	18
Ear + (CNS or CVS)	46

Source: From Lynberg, M. C. et al., *Teratology*, 42, 513–519, 1990. With permission.

defects in the four cases included noncharacteristic isotretinoin malformations. Reproductive capacity is at least in part restored following discontinuation of treatment.

The mechanism of teratogenicity by the retinoids has been investigated thoroughly, and the reader is referred to the review article on retinoic acid metabolism by the National Research Council (NRC; 2000). It appears that the receptors for retinoids are of two types (RAR and retinoid X receptor [RXR]) of the nuclear hormone ligand-dependent, transcription-factor superfamily, and the receptor specificity of retinoids correlates with their teratogenic actions — RAR agonists are potent, while RXR agonists are ineffective. In both cases, the receptor, when activated by exogenously added retinoic acid, affects gene expression at abnormal times and sites, as compared with that done by endogenous retinoid. Additional details are available in the literature (NRC, 2000).

Growth Retardation

Of the four classes of developmental toxicity, growth retardation has not been a characteristic feature of the isotretinoin-induced syndrome of defects, with the exception of microcephaly recorded in some case reports in association with abnormalities of various types. Paradoxically however, the first case of a first-trimester-exposed child had intrauterine growth retardation (IUGR) and no malformations (Kassis et al., 1985). And in another of the initial descriptions of isotretinoin-induced malformations in 154 cases, only two infants were small for gestational age, and although there were 11 premature infants, only five were less than 35% gestational age (Lammer et al., 1985).

Death

Because of the high frequency of spontaneous abortions in women exposed to isotretinoin, one authoritative source (the Centers for Disease Control and Prevention [CDC]) stated that death may be a more common adverse outcome than malformations seen in liveborn infants (Anon., 1984). The FDA estimated that 700 to 1000 women had spontaneous abortions in the initial marketing period of 1982 to 1986, and that another 5000 to 7000 women had induced abortions in that same interval for fear of birth defects (press reports, April 1988).

In the 1982 to 1984 interval, of 154 pregnancies identified as those of women who received isotretinoin treatment, 12 had spontaneous abortions, 3 of 21 with major malformations were stillborn, and 9 died after birth (Lammer et al., 1985). In a later evaluation by these investigators, the outcomes of 57 more pregnancies included 9 more spontaneous abortions plus a stillborn with malformations (Lammer et al., 1987). Stillbirth was listed for one of two isotretinoin-exposed infants in a case report (Lancaster and Rogers, 1988). In a recent report, an incidence of 18% was observed for spontaneous abortion among 115 pregnancies (Dai et al., 1992).

Functional Deficit

Functional deficits of several forms have been associated with malformations and death in isotretinoin-induced developmental toxicity. In a follow-up study of 31 5-year old children born to women who were treated with isotretinoin during the first 60 days after conception, 15 (47%) performed in the subnormal range on standard intelligence tests. And of 12 children who had major malformations, all had low IQs in the range of <70 to >85 (Adams and Lammer, 1993).

Several central nervous system malformations that may have been the cause of, and are associated with, functional impairment were described in the syndrome in high incidence, including hydrocephalus, cortical and cerebellar defects, and spina bifida, to name a few. In one study of 61 cases of malformation, central nervous system involvement appeared in 49% of the cases (Lynberg et al., 1990). Review articles on isotretinoin and human pregnancy were published (Hall, 1984; Nygaard, 1988; Gollnick and Orfanos, 1989; Holmes and Wolfe, 1989; Thomson and Cordero, 1989; Lynberg et al., 1990; Chen et al., 1990; Collins and Mao, 1999).

CHEMISTRY

Isotretinoin is an isomer of tretinoin, with its terminal double bond in the Z configuration. It is large compared to the other compounds. The chemical is highly hydrophobic with low polarity. Hydrogen bonding can occur through the carboxylic acid. The calculated physicochemical and topological properties are listed below.

PHYSICOCHEMICAL PROPERTIES

Parameter	Value
Molecular weight	300.441 g/mol
Molecular volume	310.15 A^3
Density	0.847 g/cm^3
Surface area	407.53 A^2
LogP	6.164
HLB	2.204
Solubility parameter	18.436 J$^{(0.5)}$/cm$^{(1.5)}$
Dispersion	17.428 J$^{(0.5)}$/cm$^{(1.5)}$
Polarity	1.432 J$^{(0.5)}$/cm$^{(1.5)}$
Hydrogen bonding	5.840 J$^{(0.5)}$/cm$^{(1.5)}$
H bond acceptor	0.52
H bond donor	0.31
Percent hydrophilic surface	15.93
MR	93.294
Water solubility	−4.127 log (mol/M^3)
Hydrophilic surface area	64.92 A^2
Polar surface area	40.46 A^2
HOMO	−7.817 eV
LUMO	−1.514 eV
Dipole	5.615 debye

TOPOLOGICAL PROPERTIES (UNITLESS)

Parameter	Value
x0	16.751
x1	10.220
x2	9.813
xp3	6.409
xp4	5.058
xp5	2.824
xp6	2.272
xp7	0.972
xp8	0.697
xp9	0.377
xp10	0.324
xv0	14.441
xv1	7.867
xv2	6.761
xvp3	4.125
xvp4	2.875
xvp5	1.499
xvp6	1.101
xvp7	0.346
xvp8	0.199
xvp9	0.100
xvp10	0.082
k0	28.931
k1	20.046
k2	9.333
k3	7.422
ka1	18.379
ka2	8.092
ka3	6.330

REFERENCES

Adams, J. and Lammer, E. J. (1993). Neurobehavioral teratology of isotretinoin. *Reprod. Toxicol.* 7: 175–177.

Anon. (1984). Isotretinoin — a newly recognized human teratogen. *MMMR* 33: 171–173.

Arnon, J. et al. (2004). Pregnancy exposure to isotretinoin: A continuing problem. *Reprod. Toxicol.* 19: 258–259.

Bigby, M. and Stern, R. S. (1988). Adverse reactions to isotretinoin. A report from the adverse drug reaction reporting system. *J. Am. Acad. Dermatol.* 18: 543–552.

Braun, J. T. et al. (1984). Isotretinoin dysmorphic syndrome. *Lancet* 1: 506–507.

Burk, D. T. and Willhite, C. C. (1988). Inner ear malformations induced by isotretinoin. *Teratology* 37: 448.

Chen, D. T., Jacobson, M. M., and Kuntzman, R. G. (1990). Experience with the retinoids in human pregnancy. In *Basic Science in Toxicology*, G. N. Volans, Ed., Taylor & Francis, London, pp. 473–482.

Coberly, S., Lammer, E., and Alashari, M. (1996). Retinoic acid embryopathy: Case report and review of the literature. *Pediatr. Pathol. Lab. Med.* 16: 823–836.

Collins, M. D. and Mao, G. E. (1999). Teratogenicity of retinoids. *Ann. Rev. Pharmacol. Toxicol.* 39: 399–430.

Collins, M. D. et al. (1994). Comparative teratology and transplacental pharmacokinetics of all-*trans* retinoic acid, 13-*cis* retinoic acid, and retinyl palmitate following daily administration to rats. *Toxicol. Appl. Pharmacol.* 127: 132–144.

Dai, W. S., Hsu, M.-A., and Itri, L. M. (1989). Safety of pregnancy after discontinuation of isotretinoin. *Arch. Dermatol.* 125: 362–365.

Dai, W. S., LaBraico, J. M., and Stern, R. S. (1992). Epidemiology of isotretinoin exposure during pregnancy. *J. Am. Acad. Dermatol.* 26: 599–606.

Friedman, J. M. and Polifka, J. E. (2000). *Teratogenic Effects of Drugs. A Resource for Clinicians (TERIS)*, Second ed., Johns Hopkins University Press, Baltimore, MD.

Giannoulis, H. et al. (2004). Multiple teratogenesis after use of isotretinoin during the first trimester of pregnancy. *Reprod. Toxicol.* 18: 729.

Giannoulis, C. H. et al. (2005). Isotretinoin(Ro-Accutane)teratogenesis: A case report. *Clin. Exp. Obstet. Gynecol.* 32: 78.

Gollnick, H. and Orfanos, C. E. (1989). Treatment of acne by isotretinoin. Dosage and side-effects including teratogenicity. *Muench. Med. Wochenschr.* 131: 457–461.

Hall, J. G. (1984). Vitamin A — a newly recognized human teratogen — harbinger of things to come. *J. Pediatr.* 105: 583–584.

Henck, J. W. et al. (1987). Teratology of all-*trans* retinoic acid and 13-*cis* retinoic acid in two rat strains. *Teratology* 36: 26A.

Hersh, J. H. et al. (1985). Retinoic acid embryopathy: Timing of exposure amd effects on fetal development. *JAMA* 254: 909–910.

Holmes, A. and Wolfe, S. (1989). When a uniquely effective drug is teratogenic. The case for isotretinoin. *N. Engl. J. Med.* 321: 756–757.

Honein, M. A., Paulozzi, L. J., and Erickson, J. D. (2001). Continued occurrence of Accutane®-exposed pregnancies. *Teratology* 64: 142–147.

Hummler, H., Hendrickx, A. G., and Nau, H. (1996). Maternal toxicokinetics, metabolism, and embryo exposure following a teratogenic dosing regimen with 13-*cis*-retinoic acid (isotretinoin) in the cynomolgus monkey. *Teratology* 50: 184–193.

Hummler, H., Korte, R., and Hendrickx, A. G. (1990). Induction of malformations in the cynomolgus monkey with 13-*cis* retinoic acid. *Teratology* 42: 263–272.

Kamm, J. J. (1982). Toxicology, carcinogenicity, and teratogenicity of some orally administered retinoids. *J. Am. Acad. Dermatol.* 6: 652–659.

Kassis, I., Sunderji, S., and Abdul-Karim, R. (1985). Isotretinoin (Accutane) and pregnancy. *Teratology* 32: 145–146.

Lacy, C. F. et al. (2004). *Drug Information Handbook (Pocket), 2004–2005*, Lexi-Comp, Inc., Hudson, OH.

Lammer, E. J. et al. (1985). Retinoic acid embryopathy. *N. Engl. J. Med.* 313: 837–841.

Lammer, E. J. et al. (1987). Risk for major malformations among human fetuses exposed to isotretinoin (13-*cis* retinoic acid). *Teratology* 35: 68A.

Lancaster, A. L. and Rogers, J. G. (1988). Isotretinoin use in pregnancy. *Med. J. Aust.* 148: 654–655.

Lynberg, M. C. et al. (1990). Sensitivity, specificity, and positive predictive value of multiple malformations in isotretinoin embryopathy surveillance. *Teratology* 42: 513–519.

NRC (National Research Council). (2000). *Scientific Frontiers in Developmental Toxicology and Risk Assessment*, National Academy Press, Washington, D.C., pp. 75–80.

Nygaard, D. A. (1988). Accutane — is the drug a prescription for birth defects. *Trial* 24: 81–83.

PDR® (*Physicians' Desk Reference*®). (2002). Medical Economics Co., Inc., Montvale, NJ.

Rosa, F. W. (1983). Teratogenicity of isotretinoin. *Lancet* 2: 513.

Rosa, F. W. (1984a). A syndrome of birth defects with maternal exposure to a vitamin A congener: Isotretinoin. *J. Clin. Dysmorphol.* 2: 13–17.

Rosa, F. W. (1984b). Isotretinoin (13-*cis* retinoic acid) human teratogenicity. *Teratology* 29: 55A.

Schardein, J. L. (2000). *Chemically Induced Birth Defects*, Third ed., Marcel Dekker, New York, p. 779.

Stern, R. S., Rosa, F., and Baum, C. (1984). Isotretinoin and pregnancy. *J. Am. Acad. Dermatol.* 10: 851–854.

Strauss, J. S. et al. (1988). Isotretinoin and teratogenicity. *J. Am. Acad. Dermatol.* 19: 353–354.

Thomson, E. J. and Cordero, J. F. (1989). The new teratogens: Accutane and other vitamin-A analogs. *MCN* 14: 244–248.

U.S. Teratology Society. (1991). Recommendations for isotretinoin use in women of child-bearing potential. *Teratology* 44: 1–6.

Vannoy, J. and Kwashigroch, T. E. (1987). Accutane-induced congenital heart defects in the mouse. *Teratology* 35: 42A.

18 Captopril

Chemical name: 1-[2*S*-3-Mercapto-2-methyl-1-oxopropyl]-L-proline

CAS #: 62571-86-2

SMILES: N1(C(CCC1)C(O)=O)C(C(CS)C)=O

INTRODUCTION

Captopril is an antihypertensive agent used in the management of hypertension, the treatment of congestive heart failure and left ventricular dysfunction following myocardial infarction, and diabetic nephropathy. It is one of some 15 or so drugs that are classed as angiotensin-converting enzyme (ACE) inhibitors. These drugs are competitive inhibitors of ACE; they prevent conversion of angiotensin I to angiotensin II, a potent vasoconstrictor, resulting in lower levels of the latter, which causes an increase in plasma renin activity and a reduction in aldosterone secretion (Lacy et al., 2004). Captopril is available commercially under the trade name Captoten®, among other names, and it has variable pregnancy category risk factors. For use early in pregnancy, the category is C, but its unique toxicity in later (second and third trimesters) pregnancy warrants a pregnancy category of D. This is explained in a black box warning on the package label as follows: "When used in the second and third trimesters, ACE inhibitors can cause injury and even death in the developing fetus" (*PDR*, 2002; see below).

DEVELOPMENTAL TOXICOLOGY

ANIMALS

In animal studies, captopril given orally at relatively low doses of 3 mg/kg/day to rabbits late in gestation lengthened the gestation period and increased stillbirths (Pipkin et al., 1980). Sheep given the drug intraveneously late in gestation had no developmental toxicity other than a high incidence of stillborn lambs (Pipkin et al., 1980). In rats, doses in the range of 10 to 30 mg/kg/day given during organogenesis were maternally toxic and reduced implants and resulted in fetal growth retardation and decreased ossification in several sectors (Al-Shabanah et al., 1991). Teratogenesis was not induced in any of the three species.

TABLE 1
Developmental Toxicity Profile of Captopril in Humans

Case Number	Malformations	Growth Retardation	Death	Functional Deficit	Ref.
1	Renal			✔	Guignard et al., 1981
2	Skull, limbs		✔		Duminy and Burger, 1981
3	Multiple: lungs, skull, limbs	✔	✔		Rothberg and Lorenz, 1984
4	None	✔			Coen et al., 1985
5–9	None		✔		Kreft-Jais et al., 1988
10–24	None	✔			Kreft-Jais et al., 1988
25	Renal	✔	✔	✔	Knott et al., 1989
26	Multiple: skull, lungs, renal, vessels	✔		✔	Barr, 1990; Barr and Cohen, 1991
27	Brain	✔			Piper et al., 1992
28	Skull, renal, lungs	✔		✔	Pryde et al., 1992, 1993; Sedman et al., 1995

HUMANS

The ACE inhibitors (ACEIs) have unique properties in human development. They elicit a significant developmental toxicity termed "ACEI fetopathy" when administered in the second and third trimesters; the 26th week of gestation is said to be critical. The six cases of fetopathy (numbered 1–3, 25, 26, and 28 in Table 1) related to captopril administration are tabulated in Table 1.

Oligohydramnios, hypocalvaria (an unusual underdevelopment of skull bones), fetal growth retardation, neonatal renal failure, hypotension, pulmonary hypoplasia, joint contractures, and death were repeatedly observed after maternal treatment later in pregnancy. Therapeutic doses of up to 150 mg/day orally are typically administered. The mechanism for fetal calvarial hypoplasia is possibly related to the drug-induced oligohydramnios that allows the uterine musculature to exert direct pressure on the fetal skull. Combined with fetal hypotension, the result could be due to inhibition of peripheral perfusion and ossification of the calvaria (Brent and Beckman, 1991). Renal defects are probably also caused by decreased renal perfusion related to reduced renal blood flow (Martin et al., 1992). The most consistent renal findings are associated with a disruption of function, resulting in oligohydramnios and neonatal anuria accompanied by severe hypotension (Beckman et al., 1997), constituting functional deficit in the neonatal period.

Use of captopril during the first trimester of pregnancy does not appear to present a risk to the fetus; therefore, there is no reason not to use the drug in the first trimester (Brent and Beckman, 1991).

The usual laboratory species are inappropriate models for fetopathy in the human, because their renal development is postnatal according to one scientist (Barr, 1997). Identical late malformations are also observed with two other drugs in the class: published reports of cases with enalapril and with lisinopril exist (Mehta and Modi, 1989; Cunniff et al., 1990; Barr and Cohen, 1991; Bhatt-Mehta and Deluga, 1993; Lavorotti et al., 1997). The U.S. Food and Drug Administration (FDA) was aware of more than 50 cases resultant of fetopathy from ACEIs when considered over 10 years ago (FDA, 1992). A review some years ago of 56 cases of fetopathy from all ACEI sources from the literature indicated intrauterine growth retardation (IUGR) in 36%, oligohydramnios in 56%, hypotension anuria in 52%, and a mortality rate of 25% (Pryde et al., 1993). One group of investigators reviewed a large number of ACEI pregnancies from the literature, and while the data with captopril were too meager to provide information on malformations per se, the investigators suggested that because of the perinatal problems with the ACEIs, extreme caution should be applied in prescribing these drugs during pregnancy (Hanssens et al., 1991). Should a child be born with fetopathy, aggressive therapy with dialysis to remove the inhibitor may mitigate the profound hypotensive effects, according to one group of investigators (Sedman et al., 1995).

The magnitude of teratogenic risk is considered by one group of experts to be moderate (Friedman and Polifka, 2000). Several good reviews on the subject are available (Barr, 1994; Buttar, 1997).

CHEMISTRY

Captopril is a hydrophilic compound of average size. The compound is of average polarity and can act as both a hydrogen bond donor and acceptor. Captopril's calculated physicochemical and topological properties are as follows.

PHYSICOCHEMICAL PROPERTIES

Parameter	Value
Molecular weight	217.288 g/mol
Molecular volume	194.00 A^3
Density	1.307 g/cm^3
Surface area	255.57 A^2
LogP	−1.844
HLB	13.408
Solubility parameter	24.430 $J^{(0.5)}/cm^{(1.5)}$
Dispersion	20.149 $J^{(0.5)}/cm^{(1.5)}$
Polarity	7.903 $J^{(0.5)}/cm^{(1.5)}$
Hydrogen bonding	11.330 $J^{(0.5)}/cm^{(1.5)}$
H bond acceptor	0.83
H bond donor	0.34
Percent hydrophilic surface	64.64
MR	58.205
Water solubility	0.578 log (mol/M^3)
Hydrophilic surface area	165.22 A^2
Polar surface area	63.93 A^2
HOMO	−9.038 eV
LUMO	0.598 eV
Dipole	2.818 debye

TOPOLOGICAL PROPERTIES (UNITLESS)

Parameter	Value
x0	10.715
x1	6.575
x2	5.747
xp3	4.698
xp4	3.292
xp5	1.971
xp6	1.125
xp7	0.453
xp8	0.218
xp9	0.094
xp10	0.032
xv0	8.755

Continued.

Parameter	Value
xv1	5.151
xv2	3.862
xvp3	2.892
xvp4	1.756
xvp5	0.909
xvp6	0.503
xvp7	0.218
xvp8	0.081
xvp9	0.029
xvp10	0.008
k0	16.046
k1	12.071
k2	5.186
k3	2.750
ka1	11.554
ka2	4.816
ka3	2.505

REFERENCES

Al-Shabanah, O. A. et al. (1991). The effect of maternal administration of captopril on fetal development in the rat. *Res. Commun. Chem. Pathol. Pharmacol.* 73: 221–230.

Barr, M. (1990). Fetal effects of angiotensin converting enzyme inhibitor. *Teratology* 41: 536.

Barr, M. (1994). Teratogen update: Angiotensin-converting enzyme inhibitors. *Teratology* 50: 399–409.

Barr, M. (1997). Lessons from human teratogens: ACE inhibitors. *Teratology* 56: 373.

Barr, M. and Cohen, M. M. (1991). ACE inhibitor fetopathy and hypocalvaria: The kidney–skull connection. *Teratology* 44: 485–495.

Beckman, D. A., Fawcett, L. B., and Brent, R. L. (1997). Developmental toxicity. In *Handbook of Human Toxicity*, E. J. Massaro, Ed., J. L. Schardein, Sect. Ed., CRC Press, Boca Raton, FL, pp. 1007–1084.

Bhatt-Mehta, V. and Deluga, K. S. (1993). Fetal exposure to lisinopril: Neonatal manifestations and management. *Pharmacotherapy* 13: 515–518.

Brent, R. L. and Beckman, D. A. (1991). Angiotensin-converting enzyme inhibitors, an embryopathic class of drugs with unique properties: Information for clinical teratology counselors. *Teratology* 43: 543–546.

Buttar, H. S. (1997). An overview of the influence of ACE inhibitors on fetal-placental circulation and perinatal development. *Mol. Cell. Biochem.* 176: 61–71.

Coen, G. et al. (1985). Successful treatment of longlasting severe hypertension with captopril during a twin pregnancy. *Nephron* 40: 498–500.

Cunniff, C. et al. (1990). Oligohydramnios sequence and renal tubular malformation associated with maternal enalapril use. *Am. J. Obstet. Gynecol.* 162: 187–189.

Duminy, P. C. and Burger, P. du T. (1981). Fetal abnormality associated with the use of captopril during pregnancy. *S. Afr. Med. J.* 60: 805.

FDA (Food and Drug Administration). (1992). Dangers of ACE inhibitors during second and third trimesters of pregnancy. *Med. Bull.* 22: 2.

Friedman, J. M. and Polifka, J. E. (2000). *Teratogenic Effects of Drugs. A Resource for Clinicians (TERIS)*, Second ed., Johns Hopkins University Press, Baltimore, MD.

Guignard, J. P., Burgener, F., and Calame, A. (1981). Persistent anuria in neonate: A side effect of captopril. *Int. J. Pediatr. Nephrol.* 2: 133.

Hanssens, M. et al. (1991). Fetal and neonatal effects of treatment with angiotensin-converting enzyme inhibitors in pregnancy. *Obstet. Gynecol.* 78: 128–135.

Knott, P. D., Thorpe, S. S., and Lamont, C. A. R. (1989). Congenital renal dysgenesis possibly due to captopril. *Lancet* 1: 451.

Kreft-Jais, C. et al. (1988). Angiotensin-converting-enzyme inhibitors during pregnancy: A survey of 22 patients given captopril and none given enalapril. *Br. J. Obstet. Gynaecol.* 95: 420–422.

Lacy, C. F. et al. (2004). *Drug Information Handbook (Pocket), 2004–2005.* Lexi-Comp. Inc., Hudson, OH.

Lavoratti, G. et al. (1997). Neonatal anuria by ACE inhibitors during pregnancy. *Nephron* 76: 235–236.

Martin, R. A. et al. (1992). Effect of ACE inhibition on the fetal kidneys: Decreased renal blood flow. *Teratology* 46: 317–321.

Mehta, N. and Modi, N. (1989). ACE inhibitors in pregnancy. *Lancet* 2: 96.

PDR® (Physicians' Desk Reference®). (2002). Medical Economics Co., Inc., Montvale, NJ.

Piper, J. M., Ray, W. A., and Rosa, F. W. (1992). Pregnancy outcome following exposure to angiotensin-converting enzyme inhibitors. *Obstet. Gynecol.* 80: 429–432.

Pipkin, F. B., Turner, S. R., and Symonds, E. M. (1980). Possible risk with captopril in pregnancy. *Lancet* 1: 1256.

Pryde, P. G. et al. (1992). ACE inhibitor fetopathy. *Am. J. Obstet. Gynecol.* 166: 348.

Pryde, P. G. et al. (1993). Angiotensin-converting enzyme inhibitor fetopathy. *J. Am. Soc. Nephrol.* 3: 1575–1582.

Rothberg, A. D. and Lorenz, R. (1984). Can captopril cause fetal and neonatal renal failure? *Pediatr. Pharmacol* 4: 189–192.

Sedman, A. B., Kershaw, D. B., and Bunchman, T. E. (1995). Recognition and management of angiotensin converting enzyme inhibitor fetopathy. *Pediatr. Nephrol.* 9: 382–385.

19 Misoprostol

Chemical name: (11α,13E)-11,16-Dihydroxy-16-methyl-9-oxoprost-13-en-1-oic
acid methyl ester

CAS #: 59122-46-2

SMILES: C1(C(C(CC1O)=O)CCCCCCC(OC)=O)C=CCC(CCCC)(C)O

INTRODUCTION

Misoprostol is a synthetic prostaglandin E1 analog used therapeutically for the prevention of nonsteroidal anti-inflammatory drug (NSAID)-induced gastric ulcers (an antiulcerative agent). It has abortifacient properties as well, and is used in that manner orally for terminating pregnancies of less than 49 days in duration (usually in association with another abortifacient agent, mifepristone), as it has been shown to induce uterine contractions (Lacy et al., 2004). It is available commercially under the trade name Cytotec®, and it has a pregnancy category factor of X. The package label contains a black box warning stating that "misoprostol administration to women who are pregnant can cause abortion, premature birth, or birth defects" (*PDR*, 2002).

DEVELOPMENTAL TOXICOLOGY

ANIMALS

Studies in laboratory animals have been limited to rats and rabbits, and fetotoxic and teratogenic effects were not shown in these species. The package label of the drug states that doses (route unspecified) of up to 625 times (rats) or 63 times (rabbits) the human therapeutic dosage are innocuous (*PDR*, 2002). A single published study in the rat indicates that intravaginal doses of up

TABLE 1
**Malformations Attributed to Misoprostol in Published Reports
in Humans**

Number of Malformation Cases	Malformations Reported	Ref.
5	Skull	Fonseca et al., 1991, 1993; Schonhofer, 1991
7	Mobius, limbs	Gonzalez et al., 1993
1	Multiple: limbs, body wall, digits	Genest et al., 1994
4	Multiple: limbs, body wall, digits, face, lip/palate, skin	Castilla and Orioli, 1994
3	Limbs	Hall, 1996
4	Limbs, brain	Orioli and Castilla, 1997, 2000
17 (?)	Multiple: Mobius, digits, other	Vargas et al., 1997
42	Multiple: Mobius, brain, body wall	Gonzalez et al., 1998
1	Limb	Hofmeyr et al., 1998
39	Multiple: brain, body wall, face	Blanch et al., 1998
15	Limbs, other	Coelho et al., 2000
32	Multiple: Mobius, limbs, face, ears	Vargas et al., 2000
3	Mobius	Marques-Dias et al., 2003

to 1 mg/kg/day on gestation days 0 to 7 inhibited implantation, but when administered on gestation days 7 to 21 following implantation, no developmental toxicity was elicited (Ichikawa et al., 1982).

HUMANS

The drug has been marketed for use in humans since 1986. It was misused beginning in the early 1990s in Brazil as an abortifacient due to its availability over the counter; it was subsequently banned and fell into black market use (Schonhofer, 1991; Costa and Vessey, 1993; Luna-Coelho et al., 1993). As pointed out by Brent (1993), it is difficult to interpret the case reports emanating from its use by an exposed population that could number as high as 5 million. Nevertheless, over 170 case reports of malformations following failed abortion were described in publications in the 1990s to the present time. When given in early pregnancy, misoprostol must be considered developmentally toxic, with toxicity manifest as arthrogryposis and limb reduction defects, brain abnormalities, gastroschisis, and the rare Mobius sequence or syndrome (a functional maldevelopment of the sixth and seventh cranial nerves resulting in unbalanced movements [palsies] of the facial muscles). These cases are tabulated in Table 1. Mortality, of course, was a feature in many cases, and growth retardation was the only class of developmental toxicity not affected by misoprostol. Vascular disruption is thought to be the cause of such cases. In the case of the Mobius sequence, a possible mechanism proposed by one highly respected investigator was flexion of the embryo in the area of cranial nuclei 6 and 7 that results in vascular disruption of the region (Shepard, 1995). In reviews of induced malformations by misoprostol, 47 cases of Mobius syndrome were identified from case reports and a prospective cohort study (Pastuszak et al., 1997, 1998). Limb reduction defects of the hands were also recorded as associated abnormalities in these reports. A relative risk of >7 for congenital malformations, particularly for Mobius syndrome, was reported in another study (Vargas et al., 1997). Miscarriages and fetal death in high incidence were reported in several studies evaluated (Bugalho et al., 1994; Schuler et al., 1997). In addition to these developmental effects, a study of 60 placentas of misoprostol-exposed pregnancies found 16 to be abnormal microscopically (Vaux et al., 2004). However, data examined from several studies, including the first prospectively controlled study of 67 treated versus 81 nonexposed cohorts, did not support a potent teratogenic action of misoprostol during pregnancy (Schuler et al., 1992, 1999). In another study, skull defects were not reported in the initial published studies considered related to the drug by another investigator (Paumgartten et al., 1992). Where indicated, doses eliciting the recorded malformations ranged from 400 to 4000 mcg/day, doses exceeding the therapeutic doses of 400 to 800 mcg/day orally (gastic ulcers) or 25 mcg intravaginally for labor induction. Most commonly, the dose taken was 800 mcg, consisting of two 200-mcg tablets orally plus two 200-mcg tablets intravaginally, and some women took as much as 6000 or 9200 mcg/day, according to reports (Schonhofer, 1991; Bond and Zee, 1994). The critical period for teratogenesis appears to be in the first trimester, up to 12 weeks, or more specifically, 6 to 8 weeks postconception (Marques-Dias et al., 2003). No adverse effects were observed in the newborns of 966 women who were administered the drug for cervical ripening and labor induction near term who were evaluated in eight studies (Sanchez-Ramos et al., 1997). One group of experts determined the magnitude of teratogenic risk for vascular disruption to be small (Friedman and Polifka, 2000). Another investigator reported an increased risk of Mobius sequence and limb defects associated with misoprostol exposure in the first trimester in low frequency and the overall risk of increased major malformations to be low (Fawcett, 2005). Investigators in over 200 studies involving a total of over 16,000 women evaluated the drug's effectiveness in pregnancy, and results support its continued use (Goldberg et al., 2000).

CHEMISTRY

Misoprostol is a large hydrophobic compound. It can participate in hydrogen bonding both as an acceptor and a donor. It is of average polarity with respect to the other human developmental toxicants. The calculated physicochemical and topological properties of misoprostol are as shown in the following.

PHYSICOCHEMICAL PROPERTIES

Parameter	Value
Molecular weight	382.541 g/mol
Molecular volume	390.10 Å^3
Density	0.941 g/cm^3
Surface area	516.96 Å^2
LogP	1.892
HLB	3.378
Solubility parameter	21.454 J$^{(0.5)}$/cm$^{(1.5)}$
Dispersion	17.591 J$^{(0.5)}$/cm$^{(1.5)}$
Polarity	3.297 J$^{(0.5)}$/cm$^{(1.5)}$
Hydrogen bonding	11.829 J$^{(0.5)}$/cm$^{(1.5)}$
H bond acceptor	1.24
H bond donor	0.45
Percent hydrophilic surface	21.04
MR	109.310
Water solubility	−2.026 log (mol/M^3)
Hydrophilic surface area	108.74 Å^2
Polar surface area	90.15 Å^2
HOMO	−9.975 eV
LUMO	0.780 eV
Dipole	2.747 debye

Topological Properties (Unitless)

Parameter	Value
x0	20.286
x1	12.803
x2	11.183
xp3	8.035
xp4	5.804
xp5	4.029
xp6	2.375
xp7	1.784
xp8	1.164
xp9	0.791
xp10	0.557
xv0	17.284
xv1	10.470
xv2	8.183
xvp3	5.591
xvp4	3.820
xvp5	2.518
xvp6	1.446
xvp7	0.997
xvp8	0.636
xvp9	0.411
xvp10	0.259
k0	38.647
k1	25.037
k2	13.265
k3	9.846
ka1	23.998
ka2	12.425
ka3	9.131

REFERENCES

Blanch, G. et al. (1998). Embryonic abnormalities at medical termination of pregnancy with mifepristone and misoprostol during first trimester: Observational study. *Br. Med. J.* 316: 1712–1713.

Bond, G. R. and Zee, A. V. (1994). Overdosage of misoprostol in pregnancy. *Am. J. Obstet. Gynecol.* 171: 561–562.

Brent, R. L. (1993). Congenital malformation case reports: The editor's and reviewer's dilemma. *Am. J. Med. Genet.* 47: 872–874.

Bugalho, A. et al. (1994). Induction of labor with intravaginal misoprostol in intrauterine fetal death. *Am. J. Obstet. Gynecol.* 171: 538–541.

Castilla, E. E. and Orioli, I. M. (1994). Teratogenicity of misoprostol: Data from the Latin-American Collaborative Study of Congenital Malformations (ECLAMC). *Am. J. Med. Genet.* 51: 161–162.

Coelho, K. E. F. A. et al. (2000). Misoprostol embryotoxicity: Clinical evaluation of fifteen patients with arthrogryposis. *Am. J. Med. Genet.* 95: 297–301.

Costa, S. H. and Vessey, M. P. (1993). Misoprostol and illegal abortion in Rio de Janiero, Brazil. *Lancet* 341: 1258–1261.

Fawcett, L. B. (2005). Misoprostol. *Birth Defects Res. (A)* 73: 327.

Fonseca, W. et al. (1991). Misoprostol and congenital malformation. *Lancet* 338: 56.

Fonseca, W. et al. (1993). Congenital malformation of the scalp and cranium after failed first trimester abortion attempt with misoprostol. *Clin. Dysmorphol.* 2: 76–80.

Friedman, J. M. and Polifka, J. E. (2000). *Teratogenic Effects of Drugs. A Resource for Clinicians (TERIS)*, Second ed., Johns Hopkins University Press, Baltimore, MD.

Genest, D. R. et al. (1994). Limb defects and omphalocele in a 17 week fetus following first trimester misoprostol exposure. *Teratology* 49: 418.

Goldberg, A. B., Greenberg, M. B., and Dorney, P. D. (2000). Drug therapy: Misoprostol in pregnancy. *N. Engl. J. Med.* 344: 38–47.

Gonzalez, C. H. et al. (1993). Limb deficiency with or without Mobius sequence in seven Brazilian children associated with misoprostol use in the first trimester of pregnancy. *Am. J. Med. Genet.* 47: 59–64.

Gonzalez, C. H. et al. (1998). Congenital abnormalities in Brazilian children associated with misoprostol misuse in first trimester of pregnancy. *Lancet* 351: 1624–1627.

Hall, J. G. (1996). Arthrogryposis associated with unsuccessful attempts at termination of pregnancy. *Am. J. Med. Genet.* 63: 293–300.

Hofmeyr, G. J. et al. (1998). Limb reduction anomaly after failed misoprostol abortion. *S. Afr. Med. J.* 88: 566–567.

Ichikawa, Y. et al. (1982). Studies on the administration of 16,16-dimethyl-trans-δ-2-prostaglandin E1 in the pregnant rat. *Gendai Iryo* 14: 593–618.

Lacy, C. F. et al. (2004). *Drug Information Handbook (Pocket), 2004–2005*. Lexi-Comp., Inc., Hudson, OH.

Luna-Coelho, H. L. et al. (1993). Misoprostol and illegal abortion in Fortaleza, Brazil. *Lancet* 341: 1261–1263.

Marques-Dias, M. J., Gonzalez, C. H., and Rosemberg, S. (2003). Mobius sequence in children exposed *in utero* to misoprostol: Neuropathological study of three cases. *Birth Defects Res. (A)* 67: 1002–1007.

Orioli, I. and Castilla, E. (1997). Teratogenicity of misoprostol. *Teratology* 55: 161.

Orioli, I. M. and Castilla, E. E. (2000). Epidemiological assessment of misoprostol teratogenicity. *Br. J. Obstet. Gynaecol.* 107: 519–523.

Pastuszak, A. L. et al. (1997). Misoprostol use during pregnancy is associated with an increased risk for Mobius sequence. *Teratology* 55: 36.

Pastuszak, A. L. et al. (1998). Use of misoprostol during pregnancy and Mobius syndrome in infants. *N. Engl. J. Med.* 338: 1881–1885.

Paumgartten, F. J. R. et al. (1992). Risk assessment in reproductive toxicology as practiced in South America. In *Risk Assessment of Prenatally-Induced Adverse Health Effects*, D. Neubert, R. J. Kavlock, H.-J. Merker, and J. Klein, Eds., Springer-Verlag, Berlin, pp. 163–179.

PDR® (*Physicians' Desk Reference®*). (2002). Medical Economics Co., Inc., Montvale, NJ.

Sanchez-Ramos, L. et al. (1997). Misoprostol for cervical ripening and labor induction: A meta-analysis. *Obstet. Gynecol.* 89: 633–642.

Schonhofer, P. S. (1991). Brazil: Misuse of misoprostol as an abortifacient may induce malformations. *Lancet* 337: 1534–1535.

Schuler, L., Ashton, P. W., and Sanseverino, M. T. (1992). Teratogenicity of misoprostol. *Lancet* 339: 437.

Schuler, L. et al. (1997). Pregnancy outcome after abortion attempt with misoprostol. *Teratology* 55: 36.

Schuler, L. et al. (1999). Pregnancy outcome after exposure to misoprostol in Brazil: A prospective, controlled study. *Reprod. Toxicol.* 13: 147–151.

Shepard, T. H. (1995). Mobius syndrome after misoprostol: A possible teratogenic mechanism. *Lancet* 346: 780.

Vargas, F. R. et al. (1997). Investigation of the teratogenic potential of misoprostol. *Teratology* 55: 104.

Vargas, F. R. et al. (2000). Prenatal exposure to misoprostol and vascular disruption defects: A case-control study. *Am. J. Med. Genet.* 95: 302–306.

Vaux, K. K. et al. (2004). Placental abnormalities associated with misoprostol administration. *Birth Defects Res. (A)* 70: 259.

20 Streptomycin

Chemical name: *0*-2-Deoxy-2-(methylamino-α-L-glucopyranosyl-1(1→2)-*0*-5-deoxy-3-*C*-formyl-α-L-lyxofuranosyl-(1→4)-*N,N'*-*bis*(aminoiminomethyl)-D-streptamine

CAS #: 57-92-1

SMILES: C1(C(OC2C(C(C(C(C2O)O)NC(N)=N)O)NC(N)=N)OC(C1(O)C=O)C)OC3C(C(C(C(O3)CO)O)O)NC

INTRODUCTION

Streptomycin is an aminoglycoside antibiotic used therapeutically as an antitubercular agent. It is also used as part of combination therapy for treatment of streptococcal or enterococcal endocarditis, plague, tularemia, and brucellosis. It is produced by the soil actinomycete *Streptomyces griseus*, and several salt forms have been formulated for therapeutic use from synthesized material. The drug is used in both human and veterinary therapeutics. Its mechanism of action is by inhibition of bacterial protein synthesis by binding directly to the 30S ribosomal subunits, causing a faulty peptide sequence to form the protein chain (Lacy et al., 2004). The drug is known by its generic name as well as by a variety of trade names. It has a pregnancy category of D, due largely to its ototoxic properties (see below).

DEVELOPMENTAL TOXICOLOGY

ANIMALS

The drug has been studied by the pertinent human route (intramuscular) in the guinea pig, mouse, and rabbit. Guinea pigs injected with up to 100 mg/kg/day late in gestation evidenced no developmental toxicity (Riskaer et al., 1952). Mice given 500 mg/kg/day during 5 days of the organogenesis period had no overt developmental toxicity, but about 20% of the fetuses had subtle microscopic brain alterations (Ericson-Strandvik and Gyllensten, 1963). In rabbits, an unquantitated dose produced no developmental toxicity (Nurazyan, 1973). Inner ear damage pertinent to this discussion (see below) was recorded postnatally in mice given 250 mg/kg/day streptomycin on gestational days 12 to 18 by the intraperitoneal route (Nakamoto et al., 1985).

HUMANS

In the human, the aminoglycosides are well-established ototoxins in adults. Ototoxicity has also been recorded with streptomycin during pregnancy. Approximately 40 cases were published on this condition (a malformative and functional deficit), and the pertinent reports are tabulated in Table 1.

Hearing deficits resulted from lesions varying from vestibular dysfunction and cochlear damage to social hearing deficits related to structural damage of the eighth cranial nerve. Particularly affected was high-tone sensorineural hearing loss outside the speech frequencies. The deficit has no specific pregnancy-specific relationship, nor, apparently, an association with dose level (the therapeutic dose level ranges from 75 mg/kg/week up to 4 g/week maximum). No congenital malformations have been attributed to the drug from larger studies of antitubercular drugs (Marynowski and Sianoz-Ecka, 1972; Heinonen et al., 1977; Czeizel et al., 2000). Likewise, no other class of developmental toxicity (growth retardation or death) has been associated with the congenital eighth nerve lesion. Other aminoglycosides for which cases of fetal ototoxicity were recorded include dihydrostreptomycin and kanamycin, totaling about 28 cases (Schardein, 2000). One group of experts placed the magnitude of teratogenic risk (for deafness) due to streptomycin as being small (Friedman and Polifka, 2000). Other investigators placed the incidence of inner ear defects as 1:6 (Snider et al., 1980), as 1:10 (Ganguin and Rempt, 1970), and as 1:12 (Schardein, 2000) of those exposed. Reviews on the subject of aminoglycoside ototoxicity during development include those by Warkany (1979) and Snider et al. (1980).

TABLE 1
Hearing Deficits Recorded in Offspring Following
Maternal Treatment of Streptomycin during Pregnancy

Ref.

Leroux, 1950
Sakula, 1954
Kreibich, 1954
Bolletti and Croatto, 1958
Rebattu et al., 1960
Lenzi and Ancona, 1962
Kern, 1962
Robinson and Cambon, 1964
Conway and Birt, 1965
Matsushima, 1967
Rasmussen, 1969
Varpela et al., 1969
Khanna and Bhatia, 1969
Ganguin and Rempt, 1970
Nishimura and Tanimura, 1976
Heinonen et al., 1977
Donald and Sellers, 1981
Donald et al., 1991

CHEMISTRY

Streptomycin is a human developmental toxicant of very large size. It is highly hydrophilic with a high polar surface area. Streptomycin can act as both a hydrogen bond donor and acceptor. The calculated physicochemical and topological properties are shown in the following.

PHYSICOCHEMICAL PROPERTIES

Parameter	Value
Molecular weight	581.581 g/mol
Molecular volume	491.23 A^3
Density	1.316 g/cm^3
Surface area	640.42 A^2
LogP	−12.158
HLB	20.298
Solubility parameter	34.974 J$^{(0.5)}$/cm$^{(1.5)}$
Dispersion	22.944 J$^{(0.5)}$/cm$^{(1.5)}$
Polarity	6.332 J$^{(0.5)}$/cm$^{(1.5)}$
Hydrogen bonding	25.626 J$^{(0.5)}$/cm$^{(1.5)}$
H bond acceptor	6.64
H bond donor	4.37
Percent hydrophilic surface	94.60
MR	135.489
Water solubility	8.874 log (mol/M^3)
Hydrophilic surface area	605.85 A^2
Polar surface area	334.59 A^2
HOMO	−8.982 eV
LUMO	0.302 eV
Dipole	4.669 debye

TOPOLOGICAL PROPERTIES (UNITLESS)

Parameter	Value
x0	30.102
x1	18.708
x2	17.617
xp3	15.114
xp4	11.994
xp5	9.455
xp6	6.589
xp7	4.870
xp8	3.439
xp9	2.244
xp10	1.455
xv0	21.694
xv1	12.414
xv2	9.970
xvp3	7.456
xvp4	5.176
xvp5	3.440

Continued.

Parameter	Value
xvp6	2.090
xvp7	1.257
xvp8	0.744
xvp9	0.421
xvp10	0.238
k0	64.082
k1	34.490
k2	14.189
k3	7.059
ka1	32.879
ka2	13.125
ka3	6.420

REFERENCES

Bolletti, M. and Croatto, L. (1958). Deafness in a 5 year old girl resulting from streptomycin therapy during pregnancy. *Acta Paediatr. Lat.* 11: 1–15.

Conway, N. and Birt, B. D. (1965). Streptomycin in pregnancy: Effect on the foetal ear. *Br. Med. J.* 2: 260–263.

Czeizel, A. E. et al. (2000). A teratological study of aminoglycoside antibiotic treatment during pregnancy. *Scand. J. Infect. Dis.* 32: 309–313.

Donald, P. R. and Sellers, S. L. (1981). Streptomycin ototoxicity in the unborn child. *S. Afr. Med. J.* 60: 316.

Donald, P. R., Doherty, E., and Van Zyl, F. J. (1991). Hearing loss in the child following streptomycin administration during pregnancy. *Cent. Afr. J. Med.* 37: 268–271.

Ericson-Strandvik, B. and Gyllensten, L. (1963). The central nervous system of foetal mice after administration of streptomycin. *Acta Pathol. Microbiol. Scand.* 59: 292–300.

Friedman, J. M. and Polifka, J. E. (2000). *Teratogenic Effects of Drugs. A Resource for Clinicians (TERIS)*, Second ed., Johns Hopkins University Press, Baltimore, MD.

Ganguin, G. and Rempt, E. (1970). Streptomycin Behandlung in der Schwangerschaft und ihre Auswirkung auf das Gehor Kindes. *Z. Laryngol. Rhinol. Otol. Ihre. Grenzgeb.* 49: 496–503.

Heinonen, O. P., Slone, D., and Shapiro, S. (1977). *Birth Defects and Drugs in Pregnancy*, Publishing Sciences Group, Littleton, MA.

Kern, G. (1962). [On the problem of intrauterine streptomycin damage]. *Schweiz. Med. Wschr.* 92: 77–79.

Khanna, B. K. and Bhatia, M. L. (1969). Congenital deaf mutism following streptomycin therapy to mother during pregnancy. A case of streptomycin ototoxicity *in utero. Indian J. Chest Dis.* 11: 51–53.

Kreibich, H. (1954). Sind nach einer Streptomycin-behandlung Tuberculoser Schwangerer schadigung des Kindes zu erwarten? *Dtsch. Gesundheitswes.* 9: 177–181.

Lacy, C. F. et al. (2004). *Drug Information Handbook (Pocket), 2004–2005*, Lexi-Comp., Inc., Hudson, OH.

Lenzi, E. and Ancona, F. (1962). Sul problema delle lesioni dell'apparato uditivo da passaggio transplacentare di streptomicina. *Riv. Ital Ginecol.* 46: 115.

Leroux, L. (1950). Existe-t-ii une surdite congenitale acquise due a la streptomycina? *Ann. Otolaryngol. (Paris)* 67: 1194–1196.

Marynowski, A. and Sianoz-Ecka, E. (1972). [Comparison of the incidence of congenital malformations in neonates from healthy mothers and from patients treated for tuberculosis]. *Ginekol. Pol.* 43: 713–715.

Matsushima, M. (1967). A study of pulmonary tuberculosis of pregnant women. Report 7. Effects of chemotherapy during pregnancy on the fetus. *Kekkaku* 42: 463–464.

Nakamoto, Y., Otani, H., and Tanaka, O. (1985). Effects of aminoglycosides administered to pregnant mice on postnatal development of inner ear in their offspring. *Teratology* 32: 34B.

Nishimura, H. and Tanimura, T. (1976). *Clinical Aspects of the Teratogenicity of Drugs*, Excerpta Medica, New York, pp. 130, 131.

Nurazyan, A. G. (1973). [Distribution of antibiotics in the organism of a pregnant rabbit and its fetus]. *Antibiotiki* 18: 268–269.

Rasmussen, F. (1969). The ototoxic effect of streptomycin and dihydrostreptomycin on the foetus. *Scand. J. Resp. Dis.* 50: 61–67.

Rebattu, J. P., Lesne, G., and Megard, M. (1960). Streptomycin, barriere placentaire, troubles cochleovestib-ulaires. *J. Fr. Otolaryngol.* 9: 411.

Riskaer, N., Christensen, E., and Hertz, H. (1952). The toxic effects of streptomycin and dihydrostreptomycin in pregnancy, illustrated experimentally. *Acta Tuberc. Pneumol. Scand.* 27: 211–212.

Robinson, G. E. and Cambon, K. G. (1964). Hearing loss in infants of tuberculous mothers treated with streptomycin during pregnancy. *N. Engl. J. Med.* 271: 949–951.

Sakula, A. (1954). Streptomycin and the foetus. *Br. J. Tuberc.* 48: 69–72.

Schardein, J. L. (2000). *Chemically Induced Birth Defects*, Third ed., Marcel Dekker, New York, pp. 391–392.

Snider, D. E. et al. (1980). Treatment of tuberculosis during pregnancy. *Am. Rev. Respir. Dis.* 122: 65–79.

Varpela, E., Hietalahti, J., and Aro, M. J. T. (1969). Streptomycin and dihydrostreptomycin medication during pregnancy amd their effect on the child's inner ear. *Scand. J. Respir. Dis.* 50: 101–109.

Warkany, J. (1979). Antituberculous drugs. *Teratology* 20: 133–138.

21 Methimazole

Chemical name: 1,3-Dihydro-1-methyl-2H-imidazole-2-thione

Alternate names: Mercazolyl, thiamazole

CAS #: 60-56-0

SMILES: C1(N(C=CN1)C)=S

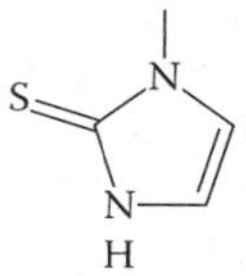

INTRODUCTION

Methimazole is a thioamide chemical used therapeutically as an antithyroid agent, given for the palliative treatment of hyperthyroidism and to control thyrotoxic crises that may accompany thyroidectomy. The drug inhibits the synthesis of thyroid hormones by blocking the oxidation of iodine in the thyroid gland, hindering its ability to combine with tyrosine to form thyroxine and triiodothyronine (Lacy et al., 2004). Methimazole is available as a prescription drug under the trade name Tapazole®, among other names. It has a pregnancy category risk factor of D. The package label carries a warning that the drug "can cause fetal harm when administered to a pregnant woman." The label goes on to state that it can induce goiter and even cretinism in the developing fetus, and, in addition, rare instances of congenital defects: aplasia cutis as manifested by scalp defects, esophageal atresia with tracheoesophageal fistula, and choanal atresia with absent/hypoplastic nipples (see below; see also *PDR*, 2002).

DEVELOPMENTAL TOXICOLOGY

ANIMALS

In animal studies, methimazole has not been shown to be teratogenic. However, in two species, the mouse and the rat, functional behavioral effects were produced following oral dosing of the drug late in gestation through postnatal day 10 (Comer and Norton, 1982; Rice et al., 1987). Administration of methimazole in low doses to the rabbit throughout the gestational period did not elicit any developmental or maternal toxicity (Zolcinski et al., 1964).

HUMANS

In humans, methimazole is associated with malformations as described above in the package label for the drug. Included is a peculiar, ulcer-like midline lesion of the scalp termed "aplasia cutis

TABLE 1
Developmental Toxicity Profile of Methimazole in Humans

Case Number	Malformations	Growth Retardation	Death	Functional Deficit	Ref.
1	Limbs	✔			Zolcinski and Heimrath, 1966
2–4	Scalp				Milham and Elledge, 1972
5, 6	Scalp, gastrointestinal, genital				Mujtaba and Burrow, 1975
7	Scalp				Bacharach and Burrow, 1984
8–14	Scalp				Milham, 1985
15, 16	Scalp, urinary				Milham, 1985
17	Scalp				Kalb and Grossman, 1986
18	Choana, face, nipples	✔		✔	Greenberg, 1987
19	Scalp				Van Dijke et al., 1987
20	Scalp				Farine et al., 1988
21	(DiGeorge syndrome)				Kawamura et al., 1989
22	(West syndrome), others				Shikii et al., 1989
23	Scalp				Tanaka et al., 1989
24	Gastrointestinal, thyroid	✔	✔		Ramirez et al., 1992
25	Gastrointestinal, heart, thyroid	✔	✔		Ramirez et al., 1992
26	Skin				Martinez-Frias et al., 1992
27	Scalp				Mandel et al., 1994
28	Scalp, ears, nipples				Sargent et al., 1994
29	Scalp				Vogt et al., 1995
30	Scalp, choana, heart, body wall, gastrointestinal	✔	✔		Johnsson et al., 1997
31	Choana, face, eye, renal				Hall, 1997
32	Scalp, choana, face, nipples	✔		✔	Wilson et al., 1998
33	Scalp, choana, face, palate, digits, gastrointestinal	✔		✔	Clementi et al., 1999
34	Scalp, digits, face				Martin-Denavit et al., 2000
35	Scalp, choana, body wall, face, limb	✔			Ferraris et al., 2003
36	Scalp, body wall	✔			Ferraris et al., 2003

congenita," and less commonly, esophageal atresia and tracheoesophageal fistulae (a gastrointestinal defect), choanal atresia, and athelia (absent nipple(s)). These findings are considered components of the "methimazole embryopathy," and there may be other associated anomalies as well. The reported cases are tabulated in Table 1. Of these, some 28 cases had single or multiple aplasia cutis, and several had choana, esophageal atresia and tracheoesophageal fistulae, and the absence of nipples. Other classes of developmental toxicity were occasionally associated; a number of cases of intrauterine growth retardation (IUGR) were recorded, as well as functional impairments (psychomotor retardation, developmental delay, and mental retardation) and death in three of the published cases. Even though the latter effect falls within normal frequency in pregnancy, the finding cannot be dismissed with certainty. Functional behavioral deficits occurred in animal studies as well, and this also cannot be dismissed as irrelevant. Thus, except for the rather rare malformation, aplasia cutis of the scalp, many of the affected cases appeared to be otherwise normal. The usual therapeutic dose of up to 40 mg/day orally was sufficient to induce the malformations, but the developmental timetable was less well defined: Most occurred in the first trimester, but at least one resulting case had been treated in the third trimester.

One large study evaluated 241 women who had prenatal exposure to methimazole compared to 1089 women who were exposed to nonteratogenic drugs (diGianantonio et al., 2001). They found no major malformations or abortions but a higher incidence of choana and esophageal atresia between the third and seventh gestational weeks in the methimazole-exposed group than in the controls.

It should be stated that in several large studies, researchers found no association of methimazole with scalp defects (Momotani et al., 1984; Van Dijke et al., 1987). Researchers who conducted another study found no effects on somatic growth, intellectual development, or thyroid function caused by use of methimazole (Messer et al., 1990). Lack of effect on intellectual development by the drug was also reported by other investigators (Eisenstein et al., 1992). One group of respected clinicians considered the scalp defects rare but definitely related to treatment (Shepard et al., 2002), and another group found the magnitude of teratogenic risk to be minimal to small (Friedman and Polifka, 2000). Goiters in the newborn have not been a major finding, although several cases were recorded (Warkany, 1971; Refetoff et al., 1974). The closely related drug and parent compound of methimazole, carbimazole, was also associated with similar malformations in several cases, and thyroid effects in a number of other reports (Schardein, 2000).

Several reviews exist of methimazole treatment and resulting developmental effects (Mandel et al., 1994; Wing et al., 1994; Clementi et al., 1999; Diav-Citrin and Ornoy, 2002).

CHEMISTRY

Methimazole is a small heterocyclic compound with a relatively low polar surface area. It is of average hydrophobicity compared to the other compounds within this compilation. It can participate in hydrogen bonding. The calculated physicochemical and topological properties are as follows.

PHYSICOCHEMICAL PROPERTIES

Parameter	Value
Molecular weight	114.171 g/mol
Molecular volume	95.83 A^3
Density	1.295 g/cm^3
Surface area	123.20 A^2
LogP	1.360
HLB	14.926
Solubility parameter	27.433 $J^{(0.5)}/cm^{(1.5)}$
Dispersion	21.281 $J^{(0.5)}/cm^{(1.5)}$
Polarity	13.097 $J^{(0.5)}/cm^{(1.5)}$
Hydrogen bonding	11.321 $J^{(0.5)}/cm^{(1.5)}$
H bond acceptor	0.29
H bond donor	0.27
Percent hydrophilic surface	71.24
MR	33.328
Water solubility	2.854 log (mol/M^3)
Hydrophilic surface area	87.77 A^2
Polar surface area	20.72 A^2
HOMO	−8.129 eV
LUMO	0.382 eV
Dipole	6.518 debye

TOPOLOGICAL PROPERTIES (UNITLESS)

Parameter	Value
x0	5.276
x1	3.304
x2	2.886
xp3	2.290
xp4	1.331
xp5	0.471
xp6	0.118
xp7	0.000
xp8	0.000
xp9	0.000
xp10	0.000
xv0	4.827
xv1	2.413
xv2	1.763
xvp3	1.231
xvp4	0.519
xvp5	0.166
xvp6	0.046
xvp7	0.000
xvp8	0.000
xvp9	0.000
xvp10	0.000
k0	5.916
k1	5.143
k2	1.852
k3	0.960
ka1	4.898
ka2	1.694
ka3	0.851

REFERENCES

Bacharach, L. K. and Burrow, G. N. B. (1984). Aplasia cutis congenita and methimazole. *Can. Med. Assoc. J.* 130: 1264.

Clementi, M. et al. (1999). Methimazole embryopathy: Delineation of the phenotype. *Am. J. Med. Genet.* 83: 43–46.

Comer, C. P. and Norton, S. (1982). Effects of perinatal methimazole exposure on a developmental test battery for neurobehavioral toxicity in rats. *Toxicol. Appl. Pharmacol.* 63: 133–141.

Diav-Citrin, O. and Ornoy, A. (2002). Teratogen update: Antithyroid drugs — methimazole, carbimazole, and propylthiouracil. *Teratology* 65: 38–44.

diGianantonio, E. et al. (2001). Adverse effects of prenatal methimazole exposure. *Teratology* 64: 262–266.

Eisenstein, Z. et al. (1992). Intellectual capacity of subjects exposed to methimazole or propylthiouracil *in utero*. *Eur. J. Pediatr.* 151: 558–559.

Farine, D. et al. (1988). Elevated α-fetoprotein in pregnancy complicated by aplasia cutis after exposure to methimazole. *Obstet. Gynecol.* 71: 996.

Ferraris, S. et al. (2003). Malformations following methimazole exposure *in utero*: An open issue. *Birth Defects Res. (A)* 67: 989–992.

Friedman, J. M. and Polifka, J. E. (2000). *Teratogenic Effects of Drugs. A Resource for Clinicians (TERIS)*, Second ed., Johns Hopkins University Press, Baltimore, MD.

Greenberg, F. (1987). Brief clinical report: Choanal atresia and athelia: Methimazole teratogenicity or a new syndrome? *Am. J. Med. Genet.* 28: 931–934.

Hall, B. D. (1997). Methimazole as a teratogenic etiology of choanal atresia/multiple congenital anomaly syndrome. *Am. J. Hum. Genet.* (Suppl. 61): A100.

Johnsson, E., Larsson, G., and Ljunggran, M. (1997). Severe malformations in infant born to hyperthyroid woman on methimazole. *Lancet* 350: 1520.

Kalb, R. E. and Grossman, M. E. (1986). The association of aplasia cutis congenita with therapy of maternal thyroid disease. *Perspect. Dermatol.* 3: 327–330.

Kawamura, M. et al. (1989). A case of partial DiGeorge syndrome born to a mother with familial Basedow disease and methimazole treatment during pregnancy. *Teratology* 40: 663.

Lacy, C. F. et al. (2004). *Drug Information Handbook (Pocket), 2004–2005*, Lexi-Comp. Inc., Hudson, OH.

Mandel, S. J., Brent, G. A., and Larsen, P. R. (1994). Review of antithyroid drug use during pregnancy and report of a case of aplasia cutis. *Thyroid* 4: 129–133.

Martin-Denavit, T. et al. (2000). Ectodermal abnormalities associated with methimazole intrauterine exposure. *Am. J. Med. Genet.* 94: 338–340.

Martinez-Frias, M. L. et al. (1992). Methimazole in animal feed and congenital aplasia cutis. *Lancet* 339: 742–743.

Messer, P. M., Houffa, B. P., and Olbricht, T. (1990). Antithyroid drug treatment of Grave's disease in pregnancy: Long-term effects on somatic growth, intellectual development and thyroid function of the offspring. *Acta Endocrinol. (Copenh.),* 123: 311–316.

Milham, S. (1985). Scalp defects in infants of mothers treated for hyperthyroidism with methimazole or carbimazole during pregnancy. *Teratology* 32: 321.

Milham, S. and Elledge, W. (1972). Maternal methimazole and congenital defects in children. *Teratology* 5: 125.

Momotani, N. et al. (1984). Maternal hyperthyroidism and congenital malformations in the offspring. *Clin. Endocrinol.* 20: 695–700.

Mujtaba, Q. and Burrow, G. N. (1975). Treatment of hyperthryoidism in pregnancy with propylthiouracil and methimazole. *Obstet. Gynecol.* 46: 282–286.

PDR® (Physicians' Desk Reference®). (2002). Medical Economics Co., Inc., Montvale, NJ.

Ramirez, A. et al. (1992). Esophageal atresia and tracheoesophageal fistula in two infants born to hyperthyroid women receiving methimazole (Tapazole) during pregnancy. *Am. J. Med. Genet.* 44: 200–202.

Refetoff, S. et al. (1974). Neonatal hypothyroidism and goiter of each of two sets of twins due to maternal therapy with antithyroid drugs. *J. Pediatr.* 85: 240–244.

Rice, S. A., Millan, D. P., and West, J. A. (1987). The behavioral effects of perinatal methimazole administration in Swiss Webster mice. *Fundam. Appl. Toxicol.* 8: 531–540.

Sargent, K. A. et al. (1994). Apparent scalp–ear–nipple (Findlay) syndrome in a neonate exposed to methimazole *in-utero. Am. J. Hum. Genet.* 55 (Suppl.): A312.

Schardein, J. L. (2000). *Chemically Induced Birth Defects*, Third ed., Marcel Dekker, New York, p. 468.

Shepard, T. H. et al. (2002). Update on new developments in the study of human teratogens. *Teratology* 65: 153–161.

Shikii, A. et al. (1989). A case of hydrops fetalis, minor anomalies and symptomatic West syndrome born to a mother with Basedow disease and thiamazole treatment. *Teratology* 40: 663.

Tanaka, S. et al. (1989). Three cases of neonatal congenital anomalies associated with maternal hyperthyroidism. *Teratology* 40: 673–674.

Van Dijke, C. P., Heydeendael, R. J., and de Kleine, M. J. (1987). Methimazole, carbimazole, and congenital skin defects. *Ann. Intern. Med.* 106: 60–61.

Vogt, T., Stolz, W., and Landthaler, M. (1995). Aplasia cutis congenita after exposure to methimazole: A causal relationship? *Br. J. Dermatol.* 133: 994–996.

Warkany, J. (1971). *Congenital Malformations. Notes and Comments.* Year Book Medical Publishers, Chicago, p. 442.

Wilson, L. C. et al. (1998). Choanal atresia and hypothelia following methimazole exposure *in utero.* A second report. *Am. J. Med. Genet.* 75: 220–222.

Wing, D. A. et al. (1994). A comparison of propylthiouracil versus methimazole in the treatment of hyperthyroidism in pregnancy. *Am. J. Obstet. Gynecol.* 170: 90–95.

Zolcinski, A. and Heimrath, T. (1966). Fetal damage following treatment of the pregnant woman with a thyreostatic drug. *Zentralbl. Gynaekol.* 88: 218–219.

Zolcinski, A., Heimrath, T., and Rzucidlo, Z. (1964). Effect of thiamizole (methimazole) on fetal development in rabbits. *Ginekol. Pol.* 35: 593–596.

22 Ethylene Oxide

Alternate names: Dimethylene oxide, 1,2-Epoxyethane

CAS #: 75-21-8

SMILES: C1CO1

INTRODUCTION

Ethylene oxide is a colorless gas used in the production of ethylene glycol, acrylonitrile, and nonionic surfactants. It is also used as a fumigant for foodstuffs and textiles, as a sterilizing agent for surgical instruments, and as an agricultural fungicide (*The Merck Index,* 2001). It is readily absorbed after dermal or inhalational exposure (Friedman and Polifka, 2000). The permissible occupational exposure limit is 1 ppm (8 h time-weighted average) (ACGIH, 2005). It has several trade names — Anproline®, Oxidoethane®, and Oxirane®, among others — and it is often referred to by its chemical name.

DEVELOPMENTAL TOXICOLOGY

ANIMALS

In animal studies, ethylene oxide displays developmental toxicity attributes in mice and rats when exposure is through the inhalational route. In the mouse, the chemical caused malformations, reduced fetal weight, and embryolethality when a regimen of 1200 ppm for single intervals ranging from 1 up to 25 h after mating was employed (Rutledge and Generoso, 1989). The mechanism of this early effect could involve a nonmutational imprinting process that causes changes in gene expression (Katoh et al., 1989). In rats, the chemical was not teratogenic, at least by the inhalational route of exposure, but it was maternally toxic and reduced fetal body weight and increased fetal death over the range of 100 to 1200 ppm given over a 10-day period during organogenesis (Snellings et al., 1979; Saillenfait et al., 1996). Dosages of 9 to 36 mg/kg/day by the intravenous route given to rabbit does for 4 or 9 days during organogenesis elicited embryotoxicity in their young (Kimmel et al., 1982).

HUMANS

In the human, developmental toxicity apparently has been limited to spontaneous abortion, as shown in Table 1. The evidence is not strong, but negative evidence has not been forthcoming to dispel the association. However, several reports have been critical of the methodology and conclusions

TABLE 1
Death/Spontaneous Abortion Recorded in Women Exposed to Ethylene Oxide during Pregnancy

Population (exposure)	Number of Pregnancies	Measure	Ref.
Chemical factory workers (0.55 ppm)	95	Increased incidence over unexposed factory workers	Yakubova et al., 1976
Hospital staff engaged in sterilizing materials (0.1–0.5 ppm)	146	Increased incidence over unexposed hospital staff (17 versus 6%)	Hemminki et al., 1982, 1983
Dental assistants (exposure data not given)	32	Weak statistical association (RR = 2.5, 95% CI 1.0–6.1)	Rowland et al., 1996

Note: RR is the relative risk; CI is the confidence interval.

made by the cited investigators (Austin, 1983; Gordon and Meinhardt, 1983; Olsen et al., 1997). No other developmental toxicity was apparent from analysis of the limited published studies.

One group of experts places the magnitude for spontaneous abortion as minimal to small (Friedman and Polifka, 2000).

CHEMISTRY

Ethylene oxide is a small molecule that is slightly hydrophobic. It has a low polar surface area. The calculated physicochemical and topological properties for ethylene oxide are listed below.

PHYSICOCHEMICAL PROPERTIES

Parameter	Value
Molecular weight	44.053 g/mol
Molecular volume	41.88 A^3
Density	1.109 g/cm^3
Surface area	56.09 A^2
LogP	0.154
HLB	21.540
Solubility parameter	19.095 $J^{(0.5)}/cm^{(1.5)}$
Dispersion	15.797 $J^{(0.5)}/cm^{(1.5)}$
Polarity	7.613 $J^{(0.5)}/cm^{(1.5)}$
Hydrogen bonding	7.556 $J^{(0.5)}/cm^{(1.5)}$
H bond acceptor	0.12
H bond donor	0.00
Percent hydrophilic surface	100.00
MR	10.879
Water solubility	3.131 log (mol/M^3)
Hydrophilic surface area	56.09 A^2
Polar surface area	12.53 A^2
HOMO	−11.411 eV
LUMO	2.516 eV
Dipole	1.991 debye

TOPOLOGICAL PROPERTIES (UNITLESS)

Parameter	Value
x0	2.121
x1	1.500
x2	1.061
xp3	0.000
xp4	0.000
xp5	0.000
xp6	0.000
xp7	0.000
xp8	0.000
xp9	0.000
xp10	0.000
xv0	1.822
xv1	1.077
xv2	0.612
xvp3	0.000
xvp4	0.000
xvp5	0.000
xvp6	0.000
xvp7	0.000
xvp8	0.000
xvp9	0.000
xvp10	0.000
k0	0.829
k1	1.333
k2	0.222
k3	0.000
ka1	1.298
ka2	0.206
ka3	0.000

REFERENCES

ACGIH (American Conference of Government Industrial Hygienists). (2001). *TLVs® and BEIs®. Threshold Limit Values for Chemical Substances and Physical Agents and Biological Exposure Indices*, ACGIH, Cincinnati, OH, p. 29.

Austin, S. G. (1983). Spontaneous abortions in hospital sterilizing staff. *Br. Med. J.* 286: 1976.

Friedman, J. M. and Polifka, J. E. (2000). *Teratogenic Effects of Drugs. A Resource for Clinicians (TERIS)*, Second Edition, Johns Hopkins University Press, Baltimore.

Gordon, J. E. and Meinhardt, T. J. (1983). Spontaneous abortions in hospital sterilizing staff. *Br. Med. J.* 186: 1976.

Hemminki, K. et al. (1982). Spontaneous abortions in hospital staff engaged in sterilizing instruments with chemical agents. *Br. Med. J.* 285: 1461–1462.

Hemminki, K., Mutanen, P., and Niemi, M.-L. (1983). Spontaneous abortion in hospital workers who used chemical sterilizing equipment during pregnancy. *Br. Med. J.* 286: 1976–1977.

Katoh, M. et al. (1989). Fetal anomalies produced subsequent to treatment of zygotes with ethylene oxide or ethyl methanesulfonate are not likely due to the usual genetic causes. *Mutat. Res.* 210: 337–344.

Kimmel, C. A. et al. (1982). Fetal development in New Zealand white (NZW) rabbits treated iv with ethylene oxide during pregnancy. *Toxicologist* 2: 70.

Olsen, G., Lucas, L., and Teta, J. (1997). Ethylene oxide exposure and risk of spontaneous abortion, preterm birth, and postterm birth. *Epidemiology* 8: 465–466.

Rowland, A. S. et al. (1996). Ethylene oxide exposure may increase the risk of spontaneous abortion, preterm birth, and postterm birth. *Epidemiology* 7: 363–368.

Rutledge, J. C. and Generoso, W. M. (1989). Fetal pathology produced by ethylene oxide treatment of the murine zygote. *Teratology* 39: 563–572.

Saillenfait, A. M. et al. (1996). Developmental toxicity of inhaled ethylene oxide in rats following short-duration exposure. *Fundam. Appl. Toxicol.* 34: 223–227.

Snellings, W. M. et al. (1979). Teratology and reproduction studies with rats exposed to 10, 33 or 100 ppm of ethylene oxide (EO). *Toxicol. Appl. Pharmacol.* 48: A84.

The Merck Index. An Encyclopedia of Chemicals, Drugs, and Biologicals, Thirteenth ed. (2001). Wiley, Hoboken, NJ.

Yakubova, Z. N. et al. (1976). Gynecological disorders in workers engaged in ethylene oxide production. *Kazan. Med. Zh.* 57: 558–560.

23 Tetracycline

Chemical name: [4S-(4α,4aα,5aα,6β,12aα)]-4-(Dimethylamino)-1,4,4a,5,5a,6-11,12a-octahydro-3,6,10,12,12a-pentahydroxy-6-methyl-1,11-dioxo-2-naphacenecarboxamide

Alternate name: Deschlorobiomycin

CAS #: 60-54-8

SMILES:
C12C(C(=C3C(C1)C(c4c(C3=O)c(ccc4)O)(C)O)O)(C(C(=C(C2N(C)C)O)C(N)=O)=O)O

INTRODUCTION

Tetracycline is a broad-spectrum antibiotic used in the treatment of both Gram-negative and Gram-positive organisms and infections due to mycoplasma, chlamydia, and rickettsia — for acne, chronic bronchitis, and treatment of gonorrhea and syphilis. It is prepared from cultures of certain streptomyces species. The drug inhibits bacterial protein synthesis by binding with the 30S and possibly the 50S ribosomal subunits of susceptible bacteria (Lacy et al., 2004). Tetracycline is one of a number of agents in the class, all of which have similar antimicrobial spectra. This one specifically is known as Sumycin®, Achromycin®, and by other trade names, and is a prescription drug. It has a pregnancy category of D, due to the warning on the package label stating that the use of drugs in the tetracycline class during tooth development (last half of pregnancy, infancy, and childhood to the age of 8 yr) may cause permanent discoloration of the teeth (*PDR*, 2004; also see below).

DEVELOPMENTAL TOXICOLOGY

ANIMALS

In animals, studies of oral administration, the pertinent route of administration in humans, are limited. In mice, 5 mg of tetracycline given over gestation caused only questionable abortion (Mela and Filippi, 1957), and in rats, dietary dosing of up to 200 mg throughout most of gestation did not result in any developmental toxicity (Hurley and Tuchmann-Duplessis, 1963). It should be mentioned that tetracycline did not elicit consistent teratogenic effects in any species by parenteral routes as shown in published studies. It is important to note, however, that in the pregnant rat, tetracycline injected at human therapeutic doses inhibited the calcification of fetal (calvarial) bones and biosynthesis of collagen in fetal bone and skin (Halme and Aer, 1968), effects analogous to the bony defects in humans (see below).

HUMANS

In humans, as confirmed by statements on the package label, tetracycline causes fluorescent deposition of a yellow or gray-brown stain in calcifying teeth and bones in fetuses, infants, and children over a long time interval (Cohlan et al., 1961; Davies et al., 1962; Rendle-Short, 1962; Totterman and Saxen, 1969; Glorieux et al., 1991). The effect is not a teratological finding in the traditional sense: There is no effect on development of the enamel or the likelihood of caries formation according to several groups of investigators (Genot et al., 1970; Rebich et al., 1985). It may be accompanied by hypoplasia of the tooth enamel (Witkop and Wolf, 1963) and is apparently only of cosmetic importance. Generally, only the deciduous teeth are involved, although the crowns of the permanent teeth may be stained (Baden, 1970). However, the effect is developmentally toxic and is included here for that reason. No other class of developmental toxicity is apparently affected. The effect occurs following treatment from 4 months of gestation to 8 yr of age, and it is said to occur from as little as 1 g/day in the third trimester (Cohlan, 1977). The usual therapeutic dose is 1 to 2 g/day. The original observations of tooth staining were made by Schwachman and Schuster almost 50 yr ago, in 1956. Several early reviews on the subject were published (Witkop and Wolf, 1963; Toaff and Ravid, 1968; Baden, 1970).

The magnitude of teratogenic risk for staining of dentition and bones is considered by one group of experts to be high, with documentation of the effect considered excellent (Friedman and Polifka, 2000). A reasonable estimate on incidence would be that there are virtually thousands of examples of the effect. No other drug of the tetracycline class has apparently elicited the staining of the dentition. Clearly, prenatal treatment in the second and third trimesters of pregnancy is contraindicated.

CHEMISTRY

Tetracycline is a large polar molecule. It is highly hydrophilic and is capable of participating in hydrogen bonding interactions both as an acceptor and donor. The calculated physicochemical and topological properties of tetracycline are as follows.

PHYSICOCHEMICAL PROPERTIES

Parameter	Value
Molecular weight	444.441 g/mol
Molecular volume	371.28 A^3
Density	1.213 g/cm^3
Surface area	454.26 A^2
LogP	−7.067
HLB	11.432
Solubility parameter	34.825 $J^{(0.5)}/cm^{(1.5)}$
Dispersion	25.142 $J^{(0.5)}/cm^{(1.5)}$
Polarity	8.321 $J^{(0.5)}/cm^{(1.5)}$
Hydrogen bonding	22.614 $J^{(0.5)}/cm^{(1.5)}$
H bond acceptor	3.25
H bond donor	1.87
Percent hydrophilic surface	56.05
MR	112.880

Continued.

Parameter	Value
Water solubility	3.934 log (mol/M^3)
Hydrophilic surface area	254.63 A^2
Polar surface area	191.10 A^2
HOMO	−9.208 eV
LUMO	−0.916 eV
Dipole	5.442 debye

TOPOLOGICAL PROPERTIES (UNITLESS)

Parameter	Value
x0	23.911
x1	14.772
x2	15.244
xp3	13.529
xp4	12.206
xp5	9.796
xp6	7.416
xp7	5.416
xp8	3.777
xp9	2.670
xp10	1.837
xv0	17.657
xv1	9.970
xv2	8.962
xvp3	7.022
xvp4	5.509
xvp5	4.179
xvp6	2.864
xvp7	1.929
xvp8	1.182
xvp9	0.764
xvp10	0.436
k0	47.563
k1	25.104
k2	8.294
k3	3.370
ka1	22.618
ka2	6.962
ka3	2.725

REFERENCES

Baden, E. (1970). Environmental pathology of the teeth. In *Thomas Oral Pathology*, 6th ed., R. J. Gorlin and H. M. Goldman, Eds., C. V. Mosby Co., St. Louis, pp. 189–191.

Cohlan, S. Q. (1977). Tetracycline staining of teeth. *Teratology* 15: 127–130.

Cohlan, S. Q., Bevelander, G., and Brass, S. (1961). Effect of tetracycline on bone growth in the premature infant. *Antimicrob. Agents Chemother.* 340: 347.

Davies, P. A., Little, K., and Aherne, W. (1962). Tetracycline and yellow teeth. *Lancet* 1: 743.

Friedman, J. M. and Polifka, J. E. (2000). *Teratogenic Effects of Drugs. A Resource for Clinicians (TERIS)* Second ed., Johns Hopkins University Press, Baltimore, MD.

Genot, M. T. et al. (1970). Effect of administration of tetracycline in pregnancy on the primary dentition of the offspring. *J. Oral Med.* 25: 75–79.

Glorieux, F. H. et al. (1991). Dynamic histomorphometric evaluation of human fetal bone formation. *Bone* 12: 377–381.

Halme, J. and Aer, J. (1968). Effect of tetracycline on synthesis of collagen and incoroporation of 45 calcium into bone in foetal and pregnant rats. *Biochem. Pharmacol.* 17: 1479–1484.

Hurley, L. S. and Tuchmann-Duplessis, H. (1963). Influence de la tetracycline sur la developpement pre- et post-natal du rat. *C. R. Acad. Sci. (Paris)* 257: 302–304.

Lacy, C. F. et al. (2004). *Drug Information Handbook (Pocket), 2004–2005*, Lexi-Comp., Inc., Hudson, OH.

Mela, V. and Filippi, B. (1957). Malformazioni congenite mandibolari da presunti stati carenzioli indotti con l'uso di antibiotico: Ia tetraciclina. *Minerva Stomatol.* 6: 307–316.

PDR® (Physicians' Desk Reference®). (2004). Medical Economics Co., Inc., Montvale, NJ.

Rebich, T., Kumar, J., and Brustman, B. (1985). Dental caries and tetracycline-stained dentition in an American Indian population. *J. Dent. Res.* 64: 462–464.

Rendle-Short, T. J. (1962). Tetracycline in teeth and bone. *Lancet* 1: 1188.

Schwachman, H. and Schuster, A. (1956). The tetracyclines: Applied pharmacology. *Pediatr. Clin. North Am.* 2: 295.

Toaff, R. and Ravid, R. (1968). Tooth discolouration due to tetracyclines. In *Drug-Induced Diseases* Vol. 3, L. Meyler and H. M. Peck, Eds., Excerpta Medica Foundation, New York, pp. 117–133.

Totterman, L. E. and Saxen, L. (1969). Incorporation of tetracycline into human foetal bones after maternal drug administration. *Acta Obstet. Gynecol. Scand.* 48: 542–549.

Witkop, C. J. and Wolf, R. O. (1963). Hypoplasia and intense staining of enamel following tetracycline therapy. *JAMA* 185: 1008–1011.

24 Caffeine

Chemical name: 1,3,7-Trimethylxanthine

Alternate names: Coffeine, guaranine, methyltheobromine, thein

CAS #: 58-08-2

SMILES: c12c(N(C(N(C1=O)C)=O)C)ncn2C

INTRODUCTION

Caffeine is a central nervous system stimulant. Chemically, it is of the methylated xanthine class, and it occurs naturally in more than 60 angiosperm plant genera. It constitutes 1 to 2% (dry weight) of roasted coffee beans, 3.5% of fresh tea leaves, and about 2% of mate leaves (Spiller, 1984). Caffeine is present in many commonly consumed beverages and candies, and in many over-the-counter (OTC) and prescription medicines, usually in combination with other chemicals, as cold and allergy tablets, headache medicines, diuretics, and stimulants. As such, it is one of the most widely used drugs in the world: The per capita consumption from all sources is estimated at 200 mg/day (3 to 7 mg/kg/day; see Barone and Roberts, 1996). More than 3 billion pounds of coffee beans are used each year in the United States, and more than 2 million pounds of caffeine are added to soft drinks each year (Weathersbee et al., 1977; press, 2005). Caffeine contents of representative products are shown in Table 1. An important consideration in the present discussion is that caffeine is consumed by a high proportion of pregnant women during their gestations: Average intakes are said to approximate 1 mg/kg/day (Barone and Roberts, 1996). As many as 26% of pregnant women consumed more than 400 mg caffeine/day in one study (Larroque et al., 1993). Caffeine readily crosses the placenta and enters fetal tissues. Up until 1980, caffeine was listed by the U.S. Food and Drug Administration (FDA) as "generally recognized as safe" (GRAS).

DEVELOPMENTAL TOXICOLOGY

ANIMALS

Caffeine is developmentally toxic in several species of laboratory animals. It is teratogenic in mice, rats, and rabbits. By the oral route (the usual route of administration), caffeine induced cleft palate, facial/skull, and digital defects in mice at doses of 50 to 300 mg/kg/day when administered prior to and through gestation or for 14 days in gestation (Knoche and Konig, 1964; Elmazar et al.,

---

TABLE 1
Caffeine Content of Representative Products

Product	Measure	Average Content (mg)
Coffee		
Ground roasted	5 oz cup	83
Instant	5 oz cup	66
Decaffeinated	5 oz cup	3
Tea		
Leaf or bag	5 oz cup	42
Instant	5 oz cup	28
Hot chocolate (cocoa)	5 oz cup	4
Colas		
Regular	12 oz container	35
Decaffeinated	6 oz container	Trace
Chocolate		
Milk	1 oz	6
Sweet	1 oz	20
Baking	1 oz	60
Medicines	Capsule or tablet	15–200

Source: From Schardein, J. L., *Chemically Induced Birth Defects*,
Third ed., Marcel Dekker, New York, 2000, compiled from various
sources.

1982). Fetal resorption was also recorded. In the rat, oral doses (by gavage, in drinking water or diet) over the range of 80 to 330 mg/kg/day either prior to and throughout gestation or throughout gestation alone elicited digit defects (ectrodactyly), low birth weights, and resorption (Fujii and Nishimura, 1972; Collins et al., 1981). Caffeine administered to rats under several different regimens also produced subtle behavioral changes in the offspring postnatally (Sobotka et al., 1979; Peruzzi et al., 1985). A definitive study in this species (rat) conducted under contemporaneous standards at doses over the range of 6 to 125 mg/kg/day on gestation days 0 through 19 demonstrated no observed effect level (NOEL), as both maternal and developmental effects were observed, however minor, at the lowest dose (Collins et al., 1981). The effect level for frank teratogenesis (reversible digit malformations) was 40 mg/kg/day, and most importantly, no selective toxicity to the fetus occurred, there being no hazard to development, at least in this species, at doses at least threefold greater than the average daily intake in humans (see below). In the rabbit, oral doses of 100 mg/kg/day delivered on gestation days 1 through 15 produced digit defects but no other developmental toxicity (Bertrand et al., 1970). At lower doses of 10 to 35 mg/kg administered orally in the drinking water to cynomolgus monkeys before, during, and after gestation resulted in decreased maternal weight and fetal birth weights (reversible) and increased stillbirths/miscarriages (Gilbert et al., 1988).

HUMANS

In the human, a number of studies have assessed the developmental toxicity potential of caffeine. Analysis of the contribution caffeine makes to adverse developmental effects is tenuous, because studies do not factor in alcohol consumption or cigarette smoking, both of which are active toxicants in their own right, and both are associated with caffeine consumption (Soyka, 1981; Larroque et al., 1993). With respect to teratogenicity, at least 12 studies evaluating caffeine from various sources

---

TABLE 2

Reports of Congenital Malformations of Infants Whose Mothers Consumed Caffeine during Pregnancy

Reports Associated with Malformations	Reports Not Associated with Malformations
Fedrick, 1974	Nelson and Forfar, 1971
Borlee et al., 1978	Heinonen et al., 1977
Jacobson et al., 1981	Kurppa et al., 1982
Furuhashi et al., 1985	Rosenberg et al., 1982; Linn et al., 1982; Kurppa et al., 1983;
	Narod et al., 1991; Tikkanen and Heinonen, 1991

Source: Modified from Schardein, J. L., *Chemically Induced Birth Defects*, Third ed., Marcel Dekker, New York, 2000, and Christian, M. S. and Brent, R. L., *Teratology*, 64, 51–78, 2001. With permission.

---

in intakes up to 10 to 30 mg/kg/day were conducted, as shown in Table 2. None showed convincing, consistent evidence of malformation induction by the chemical (Schardein, 2000; Christian and Brent, 2001). However, excess caffeine consumption has been associated with other classes of developmental toxicity. Birth weight, the most extensively studied of these endpoints, has been associated with a decrease in 11 published studies, as shown in Table 3. While it is known that birth weight is complicated by a number of demographic, medical, social, and behavioral characteristics, it is concluded in many of these reports that caffeine exerts a small but measurable effect on fetal growth. One study reported that heavy caffeine use was associated with a 105 g reduction in birth weight (Martin and Bracken, 1987). Similar conclusions were made by others (Watkinson and Fried, 1985; Fenster et al., 1991; Peacock et al., 1991). An increased risk of small-for-date infants was associated with excessive daily caffeine intake by another group of investigators (Fortier et al., 1993). While cigarette smoking was an associated factor with reduced birth weight, a subgroup among nonsmokers in the same study approached significance for decreased birth weight among women consuming large quantities of caffeine (Larroque et al., 1993). In at least one recent major study, researchers found no increased risk for intrauterine growth retardation (IUGR) from even high intakes of caffeine (Mills et al., 1993). Interestingly, low birth weights were also recorded in animal studies, in both rats and primates.

It was also suggested that excessive caffeine consumption is associated with miscarriage/spontaneous abortion, but the data are much more inconclusive than that for reduced birth weight (Table 4). These reports were limited to high doses, on the order of 48 to 162 mg caffeine per day, according to one study (Infante-Rivard et al., 1993). Other studies have not supported the consistent association of caffeine consumption with an increased risk of spontaneous abortion, as pointed out in the review by Christian and Brent (2001). Further, in the most convincing study published thus far, Klebanoff et al. (1999) conducted thorough studies on caffeine consumption and its potential association with spontaneous abortion by quantitating serum caffeine and paraxanthine levels (a metabolite of caffeine) from 487 women (compared to 2087 women as controls) as markers of caffeine intake during pregnancy. Based on an insignificant odds ratio for spontaneous abortion with paraxanthine concentrations of ≥1845 ng/ml (which was >95% of the women who had a spontaneous abortion), their results suggested that moderate consumption of caffeine was unlikely to increase the risk of spontaneous abortion.

In sum, it would appear that when consumed in excess, caffeine may have the potential to injure the embryo, as concluded by Christian and Brent (2001) and suggested earlier by Schardein (2000). The data suggested by studies on birth weights as provided in this discussion are an example of that property. These data are supported by animal studies as well. Used in moderation, caffeine

TABLE 3
Reports of Growth Retardation/Decreased Birth Weights of Infants Whose Mothers Consumed Caffeine during Pregnancy

Source	Quantity	Number of Subjects	Conclusion	Ref.
Cola or coffee	6 to 8 cups/day	5200[a]	Low birth weights	Mau and Netter, 1974
Coffee	1 to >7 cups/day	1500[a]	Low birth weights	van den Berg, 1977
Coffee	>8 mg/kg/day	12,205[a]	Low birth weights suggestive but not statistically significant (RR = 1.17, 95% CI 0.85–1.61) for >4 cups)	Linn et al., 1982
Coffee	Various	5093[a]	Significantly decreased birth weights with increasing consumption	Kuzma and Sokol, 1982
Coffee	>300 mg/day	286[a]	Decreased birth weight (P<0.05)	Watkinson and Fried, 1985
Coffee	Up to >5 cups/day	9921	Decreased birth weight >5 cups compared to <5 cups (3081 versus 3163 g)	Furuhashi et al., 1985
Coffee	1 to >300 mg/day	3891[a]	Decreased birth weight only among heavy consumption (105 g reduction) (RR = 2.3, 95% CI 1.1–5.2) for 151–300 mg; (RR = 4.6, 95% CI 2.0–10.5) for 300 mg	Martin and Bracken, 1987
Coffee	>3 mg/kg/day	48[a]	Average decrease of 121 g birth weight for women consuming >450 ml/day	Munoz et al., 1988
Beverages	>6 mg/kg/day	9564[a]	Decreased birth weight for consumption of >300 mg/day (RR = 2.79, 95% CI 0.89–8.69)	Caan and Goldhaber, 1989
Coffee	Up to> 400 mg/day	1513[a]	Decreased birth weight for consumption of >400 mg/day (3556 g versus 3664 and 3609 g for 0–200 and 200–400 mg/day)	Brooke et al., 1989
Beverages	>6 mg/kg/day	1230	Heavy consumption associated with fetal growth retardation	Fenster et al., 1991

Note: RR is the relative risk; CI is the confidence interval.

[a] Adjusted for one or more factors: smoking, alcohol, reproduction and education history, maternal and gestational age, weight, height, parity, ethnicity, psychological stress, or sex of the baby.

Source: Modified from Schardein, J. L., *Chemically Induced Birth Defects*, Third ed., Marcel Dekker, New York, 2000, and Christian, M. S. and Brent, R. L., *Teratology*, 64, 51–78, 2001. With permission.

consumption in pregnant women apparently does not pose a consistent, measurable risk to the human fetus with respect to congenital malformation, spontaneous abortion, or functional changes. In the case of effects on fetal growth and birth weight, excessively high consumption may exert an adverse effect on this class of developmental toxicity. Intake of less than 300 mg/day, the equivalent of three to four cups of coffee per day, has been suggested as a safe level (Berger, 1988). One group of experts (Friedman and Polifka, 2000) assigned no teratogenic risk to caffeine, zero to minimal risk for spontaneous abortion, and do not mention risk assessments for lowered birth weights.

Reviews of the developmental toxicity of caffeine and its history in this regard are available (Morris and Weinstein, 1981; Oser and Ford, 1981; Curatolo and Robertson, 1983; Ferguson, 1985;

TABLE 4
Reports of Spontaneous Abortion/Miscarriage of Infants Whose Mothers Consumed Caffeine during Pregnancy

Reports Associated with Death	Reports Not Associated with Death
Weathersbee et al., 1977	Fenster et al., 1991
Watkinson and Fried, 1985	Armstrong et al., 1992
Furuhashi et al., 1985	Mills et al., 1993
Srisuphan and Bracken, 1986	Kline et al., 1994
Infante-Rivard et al., 1993	
Parazzini et al., 1998	
Cnattingius et al., 2000[a]	
Wen et al., 2001[a]	

[a] Only in specific regimens.

Nash and Persaud, 1988; Berger, 1988; Al-Hachim, 1989; Nolen, 1989; Shiono and Klebanoff, 1993; Christian and Brent, 2001).

CHEMISTRY

Caffeine is a near-average-sized compound that is slightly hydrophobic. It has an average polar surface area compared to the other human developmental toxicants. Caffeine can act as a hydrogen bond acceptor. The calculated physicochemical and topological properties are listed below.

PHYSICOCHEMICAL PROPERTIES

Parameter	Value
Molecular weight	194.193 g/mol
Molecular volume	163.93 A^3
Density	1.176 g/cm^3
Surface area	205.60 A^2
LogP	0.677
HLB	11.103
Solubility parameter	30.118 J$^{(0.5)}$/cm$^{(1.5)}$
Dispersion	22.631 J$^{(0.5)}$/cm$^{(1.5)}$
Polarity	14.932 J$^{(0.5)}$/cm$^{(1.5)}$
Hydrogen bonding	13.113 J$^{(0.5)}$/cm$^{(1.5)}$
H bond acceptor	0.60
H bond donor	0.02
Percent hydrophilic surface	54.62
MR	53.058
Water solubility	1.110 log (mol/M^3)
Hydrophilic surface area	112.30 A^2
Polar surface area	68.14 A^2
HOMO	−9.174 eV
LUMO	−0.159 eV
Dipole	3.676 debye

TOPOLOGICAL PROPERTIES (UNTILESS)

Parameter	Value
x0	10.456
x1	6.537
x2	6.232
xp3	5.877
xp4	4.482
xp5	3.124
xp6	1.970
xp7	1.054
xp8	0.506
xp9	0.213
xp10	0.043
xv0	8.183
xv1	4.108
xv2	3.233
xvp3	2.316
xvp4	1.471
xvp5	0.867
xvp6	0.416
xvp7	0.174
xvp8	0.067
xvp9	0.020
xvp10	0.003
k0	16.046
k1	10.516
k2	3.539
k3	1.454
ka1	9.195
ka2	2.809
ka3	1.088

REFERENCES

Al-Hachim, G. M. (1989). Teratogenicity of caffeine: A review. *Eur. J. Obstet. Gynecol. Reprod. Biol.* 31: 237–247.

Armstrong, B. G., McDonald, A. D., and Slone, M. (1992). Cigarette, alcohol, and coffee consumption and spontaneous abortion. *Am. J. Public Health* 82: 85–87.

Barone, J. J. and Roberts, H. R. (1996). Caffeine consumption. *Food Chem. Toxicol.* 34: 119–129.

Berger, A. (1988). Effects of caffeine consumption on pregnancy outcome — a review. *J. Reprod. Med.* 33: 945–956.

Bertrand, M., Girod, J., and Rigaud, M. F. (1970). Ectrodactylie provoquee par la caffeine chez les rongeurs. Role des facteurs specifique et genetiques. *C. R. Soc. Biol. (Paris)* 164: 1488–1489.

Borlee, I., Lechat, M. F., Bouckaert, A., and Misson, C. (1978). Coffee, risk factor during pregnancy? *Louvain Med.* 97: 279–284.

Brooke, O. G. et al. (1989). Effects on birth weight of smoking, alcohol, caffeine, socioeconomic factors and psychological stress. *Br. Med. J.* 298: 795–801.

Caan, B. J. and Goldhaber, M. K. (1989). Caffeinated beverages and low birthweight — a case control study. *Am. J. Public Health* 79: 1299–1300.

Christian, M. S. and Brent, R. L. (2001). Teratogen update: Evaluation of the reproductive and developmental risks of caffeine. *Teratology* 64: 51–78.

Cnattingius, S. et al. (2000). Caffeine intake and the risk of first trimester spontaneous abortion. *N. Engl. J. Med.* 343: 1839–1845.

Collins, T. F. X. et al. (1981). A study of the teratogenic potential of caffeine given by oral intubation to rats. *Regul. Toxicol. Pharmacol.* 1: 355–378.

Curatolo, P. W. and Robertson, D. (1983). The health consequences of caffeine. *Ann. Intern. Med.* 98 (Part 1): 641–653.

Elmazar, M. M. A. et al. (1982). Studies on the teratogenic effects of different oral preparations of caffeine in mice. *Toxicology* 23: 57–72.

Fedrick, J. (1974). Anencephalus and maternal tea drinking: Evidence for a possible association. *Proc. R. Soc. Med.* 67: 356–360.

Fenster, L. et al. (1991). Caffeine consumption during pregnancy and fetal growth. *Am. J. Public Health* 81: 458–461.

Ferguson, A. (1985). Should pregnant women avoid caffeine? *Hum. Toxicol.* 4: 3–5.

Fortier, I., Marcoux, S., and Beaulac-Baillergeron, L. (1993). Relation of caffeine intake during pregnancy to intrauterine growth retardation and preterm birth. *Am. J. Epidemiol.* 137: 931–940.

Friedman, J. M. and Polifka, J. E. (2000). *Teratogenic Effects of Drugs. A Resource for Clinicians (TERIS)*, Second ed., Johns Hopkins University Press, Baltimore, MD.

Fujii, T. and Nishimura, H. (1972). Adverse effects of prolonged administration of caffeine on rat fetus. *Toxicol. Appl. Pharmacol.* 22: 449–457.

Furuhashi, N. et al. (1985). Effects of caffeine ingestion during pregnancy. *Gynecol. Obstet. Invest.* 19: 187–191.

Gilbert, S. G. et al. (1988). Adverse pregnancy outcome in the monkey (*Macaca fascicularis*) after chronic caffeine exposure. *J. Pharmacol. Exp. Ther.* 245: 1048–1053.

Heinonen, O. P., Slone, D., and Shapiro, S. (1977). *Birth Defects and Drugs in Pregnancy*, Publishing Sciences Group, Littleton, MA.

Infante-Rivard, C. et al. (1993). Fetal loss associated with caffeine intake before and during pregnancy. *JAMA* 270: 2940–2943.

Jacobson, M. F., Goldman, A. S., and Syme, R. H. (1981). Coffee and birth defects. *Lancet* 1: 1415–1416.

Klebanoff, M. A. et al. (1999). Maternal serum paraxanthine, a caffeine metabolite, and the risk of spontaneous abortion. *N. Engl. J. Med.* 341: 1639–1644.

Kline, J. et al. (1994). Fetal loss and caffeine intake. *JAMA* 272: 27–28.

Knoche, C. and Konig, J. (1964). Zur prenatalen toxizitat von Diphenylpyralin-8-chlortheophyllinat unter-berucksichtigung von erfahrungen mit Thalidomid und Caffein. *Arzneimittelforschung* 14: 415–424.

Kurppa, K. et al. (1982). Coffee consumption during pregnancy. *N. Engl. J. Med.* 306: 1548.

Kurppa, K. et al. (1983). Coffee consumption during pregnancy and selected congenital malformations: A nationwide case-control study. *Am. J. Public Health* 73: 1397–1399.

Kuzma, J. W. and Sokol, R. J. (1982). Maternal drinking behavior and decreased intrauterine growth. *Alcohol. Clin. Exp. Res.* 6: 396–402.

Larroque, B. et al. (1993). Effects on birth weight of alcohol and caffeine consumption during pregnancy. *Am. J. Epidemiol.* 137: 941–950.

Linn, S. et al. (1982). No association between coffee consumption and adverse outcomes of pregnancy. *N. Engl. J. Med.* 306: 141–144.

Martin, T. R. and Bracken, M. B. (1987). The association between low birth weight and caffeine consumption during pregnancy. *Am. J. Epidemiol.* 126: 813–821.

Mau, G. and Netter, P. (1974). [The effects of paternal cigarette smoking on perinatal mortality and the incidence of malformations]. *Dtsch. Med. Wochenschr.* 99: 1113–1118.

Mills, J. L. et al. (1993). Moderate caffeine use and the risk of spontaneous abortion and intrauterine growth retardation. *JAMA* 269: 593–597.

Morris, M. B. and Weinstein, L. (1981). Caffeine and the fetus — is trouble brewing? *Am. J. Obstet. Gynecol.* 140: 607–610.

Munoz, L. et al. (1988). Coffee consumption as a factor in iron deficiency anemia among pregnant women and their infants in Costa Rica. *Am. J. Clin. Nutr.* 48: 645–651.

Narod, S. A., Desanjose, S., and Victoria, C. (1991). Coffee during pregnancy. A reproductive hazard. *Am. J. Obstet. Gynecol.* 164: 1109–1114.

Nash, J. and Persaud, T. V. N. (1988). Reproductive and teratological risks of caffeine. *Anat. Anz.* 167: 265–270.

Nelson, M. M. and Forfar, J. O. (1971). Associations between drugs administered during pregnancy and congenital malformations of the fetus. *Br. Med. J.* 1: 523–527.

Nolen, G. A. (1989). The developmental toxicology of caffeine. In *Issues and Reviews in Teratology*, Vol. 4, H. Kalter, Ed., Plenum Press, New York, pp. 305–350.

Oser, B. L. and Ford, R. A. (1981). Caffeine: An update. *Drug Chem. Toxicol.* 4: 311–330.

Parazzini, F. et al. (1998). Coffee consumption and risk of hospitalized miscarriage before 12 weeks of gestation. *Hum. Reprod.* 13: 2286–2291.

Peacock, J. L., Bland, M. J., and Anderson, H. R. (1991). Effects on birthweight of alcohol and caffeine consumption in smoking women. *J. Epidemiol. Comm. Health* 45: 159–163.

Peruzzi, G. et al. (1985). Perinatal caffeine treatment: Behavioral and biochemical effects in rats before weaning. *Neurobehav. Toxicol. Teratol.* 7: 453–460.

Rosenberg, L. et al. (1982). Selected birth defects in relation to caffeine-containing beverages. *JAMA* 247: 1429–1432.

Schardein, J. L. (2000). *Chemically Induced Birth Defects*, Third ed., Marcel Dekker, New York, pp. 726, 727.

Shiono, P. and Klebanoff, M. A. (1993). Invited commentary: Caffeine and birth outcomes. *Am. J. Epidemiol.* 137: 951–954.

Sobotka, T. J., Spaid, S. L., and Brodie, R. E. (1979). Neurobehavioral teratology of caffeine exposure in rats. *Neurotoxicology* 1: 403–416.

Soyka, L. F. (1981). Caffeine ingestion during pregnancy: *In utero* exposure and possible effects. *Semin. Perinatol.* 5: 305–309.

Spiller, G. A. (1984). The chemical components of coffee. In *Progress in Clinical and Biological Research, Vol. 158. The Methylxanthine Beverages and Foods: Chemistry, Consumption and Health Effects*, G. A. Spiller, Ed., Alan R. Liss, New York, pp. 91–147.

Srisuphan, W. and Bracken, M. B. (1986). Caffeine consumption during pregnancy and association with late spontaneous abortion. *Am. J. Obstet. Gynecol.* 154: 14–20.

Tikkanen, J. and Heinonen, O. P. (1991). Maternal exposure to chemical and physical factors during pregnancy and cardiovascular malformations in the offspring. *Teratology* 43: 591–600.

van den Berg, B. J. (1977). Epidemiological observations of prematurity: Effects of tobacco, coffee and alcohol. In *The Epidemiology of Prematurity*, D. M. Reed and F. J. Stainley, Eds., Urban and Schwarzenberg, Baltimore, MD, pp. 157–176.

Watkinson, B. and Fried, P. A. (1985). Maternal caffeine use before, during and after pregnancy and effects upon offspring. *Neurobehav. Toxicol. Teratol.* 7: 9.

Weathersbee, P. S., Olsen, L. K., and Lodge, J. R. (1977). Caffeine and pregnancy. A retrospective survey. *Postgrad. Med.* 62: 64–69.

Wen, W. et al. (2001). The association of maternal caffeine consumption and nausea with spontaneous abortion. *Epidemiology* 12: 38–42.

25 Thalidomide

Chemical name: α-Phthalimidoglutarimide

CAS #: 50-35-1

SMILES: C1(C(NC(CC1)=O)=O)N2C(c3ccccc3C2=O)=O

INTRODUCTION

Thalidomide is a widely known agent that when introduced into the marketplace in Europe almost 50 yr ago, was promoted as a sedative/hypnotic. It was useful for treating the nausea and vomiting of pregnancy and was said to be effective against influenza. Following removal from the market globally in 1962, it was reintroduced in July 1998 by the biotechnology firm Celgene (for the first time in this country) as an immunomodulator, for therapeutic use in the treatment of erythema nodosum leprosum (ENL; a serious inflammatory condition of Hansen's disease) and in orphan status for treating Crohn's disease and a few other indications. Its mode of action is unclear (Lacy et al., 2004). Thalidomide is a prescription drug with the current trade name Thalomid®. It was known by as many as 70 or so trade names in the 46 countries where it was licensed to be used (Schardein, 2000). It has a pregnancy category of X. The package label for the drug has a "black box" warning for severe, life-threatening birth defects, the reason for its earlier removal from the market. Due to this property (see below), thalidomide is approved for marketing only under a special restricted distribution program approved by the U.S. Food and Drug Administration (FDA) termed S.T.E.P.S. ("System for Thalidomide Education and Prescribing Safety"). Under this program, only prescribers and pharmacists registered with the program are allowed to prescribe and dispense the drug, and patients must be advised of or agree to comply with the requirements of the program in order to receive the drug (*PDR*, 2002). A postmarketing surveillance scheme was also to be put into place (Yang et al., 1997).

DEVELOPMENTAL TOXICOLOGY IN ANIMALS

Because thalidomide has such an infamous history with regard to teratogenicity, a representation of testing responses from all of the 18 animal species in which it has been evaluated is presented in Table 1 (see Schardein [2000] for further details). It is evident that only certain breeds of the rabbit and eight of nine primate species show concordant malformations to that of the human (see below), and while most species demonstrate teratogenic effects of some type, per se, many also

TABLE 1
Representative Developmental Toxicity Studies Conducted in Animals with Thalidomide

Species	Results: Strain, Responses, Regimen	Ref.
Mouse	A, Swiss (some sources only): Nonconcordant malformations, growth retardation and embryolethality at 31 mg/kg for 9 days or 50 mg/kg for 15 days during organogenesis; other strains negative	Giroud et al., 1962; DiPaolo, 1963
Rat	Sprague-Dawley, Wistar (some sources only): Nonconcordant malformations, growth retardation and embryolethality at ~33 mg/kg for 11 days or 500 mg/kg for 3 days during organogenesis; other strains negative	King and Kendrick, 1962; Bignami et al., 1962
Rabbit	Most breeds tested (New Zealand White, Himalayan, Californian, Dutch belted (DB), chinchilla, common, hybrid, crossbred, Fauve de Bourgogne, Danish; concordant (limb) malformations, embryolethality at 25 mg/kg for 8 days or 250 mg/kg for 5 days in organogenesis; remaining breeds negative	Staples and Holtkamp, 1963; Lechat et al., 1964
Hamster	Syrian inbred: Low-frequency nonconcordant malformations at 0.75 mg/kg through gestation	Homburger et al., 1965
Guinea pig	Strain unspecified: No malformations at 1000 to 5000 mg/kg for 5 to 60 days following copulation over four generations	Arbab-Zadeh, 1966
Cat	Breed unspecified: Possibly concordant (limb) malformations at 500 mg/kg for 13 days during organogenesis	Somers, 1963
Dog	Beagle, mongrels: Possibly concordant (limb) malformations in beagles at 100 mg/kg for 17 days in gestation; 30 mg/kg for 13 days in gestation to mongrels also positive but nonconcordant; malformations and death observed	Weidman et al., 1963; Delatour et al., 1965
Armadillo	NA: One resultant embryo possibly concordant (limb) malformation at 100 mg/kg for 30 days in gestation	Marin-Padilla and Benirschke, 1963
Ferret	NA: Nonconcordant malformations at ? for 21 days in gestation	Steffek and Verrusio, 1972
Pig	NA: Nonconcordant malformations at 15 mg/kg for 4 days in gestation	Palludan, 1966
Cynomolgus monkey	NA: Concordant (limb) malformations, abortion at 10 mg/kg for 11 days in gestation	Delahunt and Lassen, 1964
Rhesus monkey	NA: Concordant (limb) malformations, growth retardation at 10–12 mg/kg for 1 to 3 days in gestation	Wilson and Gavan, 1967; Theisen et al., 1979
Stump-tailed monkey	NA: Concordant (limb, visceral) malformations at 5 mg/kg for 3 days in gestation	Vondruska et al., 1971
Bonnet monkey	NA: Concordant (limb, visceral) malformations at 5 mg/kg for 1 to 4 days in gestation	Hendrickx and Newman, 1973
Japanese monkey	NA: Concordant (limb) malformations at 20 mg/kg for 3 days in gestation	Tanimura et al., 1971
Baboon	NA: Concordant (limb) malformations, abortion, embryolethality at 5 mg/kg for 15–33 days in gestation	Hendrickx et al., 1966
Marmoset	NA: Concordant (limb) malformations, abortion at 45 mg/kg for 5 days in gestation	Poswillo et al., 1972
Green monkey	NA: Concordant (limb, visceral, skeletal) malformations, abortion, embryolethality at 10 mg/kg for up to 24 days in gestation	Hendrickx and Sawyer, 1978
Bushbaby	NA: No developmental toxicity at 20 mg/kg for 15 days in gestation	Wilson and Fradkin, 1969

have not, at least under the circumstances tested. It is also clear from the data in Table 1 that the most sensitive species to thalidomide are three species of primates — the baboon, the stump-tailed monkey, and the bonnet monkey — all of which showed developmental toxicity/teratogenicity at a dose of 5 mg/kg/day; the mouse, rat, and dog appeared to be the least sensitive of those tested, only eliciting malformations (nonconcordant at that) at a dose level of ~30 mg/kg. Two species, the guinea pig and the bushbaby (a primate), were not responsive, at least under the experimental regimens utilized. It is interesting too, that thalidomide also causes skeletal defects in fish and sea urchin embryos (unpublished data), as well as heart malformations in chick embryos (Gilani, 1973).

DEVELOPMENTAL TOXICOLOGY IN HUMANS

In the human, thalidomide is a well-known and prototypic teratogen, being responsible for about 8000 infants born in the early 1960s with prominent deformities of the limbs and other organs. It was one of the first drugs to be clearly shown to be a human teratogen. The history of this remarkable drug has been told in a number of published forums, especially by Lear (1962), Pfeiffer and Kosenow (1962), Mellin and Katzenstein (1962), Taussig (1962a, 1962b, 1962c, 1963), Smithells (1965), Kelsey (1965, 1988), Sjostrom and Nilsson (1972), McFadyen (1976), Insight Team (1979), Fine (1972), Rosenberg and Glueck (1973), Quibell (1981), Newman (1985, 1986), Stromland and Miller (1993), Green (1996), Miller and Stromland (1999), Schardein (2000), Stephens and Brynner (2001), among others. Much attention has been paid to this drug in the past: In the period 1963 to 1985, over 800 scientific papers were published on the subject of thalidomide embryopathy as listed in *Index Medicus* (Stephens, 1988).

PRE-TRAGEDY HISTORY

Thalidomide was first synthesized in 1953 by Chemie Grunenthal in Germany, and clinical trials proceeded in 1954. Test marketed in the Hamburg area in November 1956, it was placed on the market in West Germany as a whole under the patented trade name Contergan on October 1, 1957. By 1961, approximately 1 million tablets were being sold daily in West Germany (Lenz, 1965).

The licensee in the United States, Wm. S. Merrell Co., filed a New Drug Application (NDA) on thalidomide (code name MER32, proposed trade name Kevadon) in the United States on September 12, 1960, but the drug was never licensed for sale in this country, because an FDA reviewer (Dr. Frances Kelsey) squelched approval 1 day before becoming automatically effective, due to what she believed was insufficient detail about animal and clinical studies performed and limited information regarding the stability of the drug. It was determined later that neither developmental nor reproductive toxicity studies were performed pre-NDA; this oversight was changed later by legislation (see below). Further, there were 1600 clinical reports of nerve damage post-marketing (Stephens and Brynner, 2001). However, some 624 women in the United States took the drug during pregnancy following distribution to 1267 physicians for investigational use (Curran, 1971). Later, in 1962, Dr. Kelsey was awarded the Distinguished Federal Civilian Service Medal from President John F. Kennedy for her efforts in preventing the tragedy from occurring in the United States. The first case history presented of the defect phocomelia, though not recognized at the time as being drug related, was by a scientist, Weidenbach, in December 1959 at a meeting in Germany (unpublished). Two additional cases with the defect were reported at another meeting, again in Germany in September 1960 by Kosenow and Pfeiffer (Kosenow and Pfeiffer, 1960). The first scientific publication on the increasing incidence of the defect in Germany was written by Wiedemann (1961), in which he reported on three recent cases seen by him as a syndrome.

The first reports attributing thalidomide to the birth defects appeared almost simultaneously by two physicians — Dr. William McBride in Australia on December 16, 1961, based on six cases (McBride, 1961), and by Dr. Widukind Lenz in Germany on December 29, 1961, on knowledge of 41 cases (Lenz, 1961). The U.S. licensee Merrell was first informed of the teratogenic concerns of thalidomide in late November of 1961 (Green, 1996); they withdrew the NDA application on March 8, 1962. The drug was removed from the market on November 26, 1961, in Germany and on December 21, 1961, in England following association to the limb defect (phocomelia). The epidemic of cases of malformation subsided by August, 1962, 9 months after withdrawal from most countries, confirming the drug's involvement.

THE TRAGEDY UNFOLDS

Malformations

The first known case of embryopathy (absent ears) was a girl born December 26, 1956, in Stolberg, the site of the Chemie Grunenthal plant, where her father worked; he had brought the drug home from the plant. Drug-induced defects were primarily phocomelia of the arms (80%) and malformed ears (20%; see also Lenz, 1964). The malformations were so unusual and unexpected that even teratology pioneers were disbelievers of the event early on (Fraser, 1988; Warkany, 1988). The drug was said to increase dysmelia by 80-fold (Lenz, 1971). The pattern of malformations of the limbs is shown in Table 2. These were always bilateral and usually grossly symmetrical (Lenz, 1971; Smithells, 1973; Sugiura et al., 1979; Newman, 1986). The evolution sequence of limb involvement was thumb → radius → humerus → ulna (Smithells, 1973). Put another way, the malformations in affected German subjects were as follows: arms only (53%), arms + legs (25%), ears only (11%), arms + ears (6%), arms, legs + ears (2%), and legs only (1%)(Lenz, 1964). In Japan, the incidences were somewhat different: arms only (70%), arms + legs (14%), arms, legs + ears (5%), ears only (5%), arms + ears (3%), and other organs (3%)(Kajii, 1965). Oddly enough, defects were concordant in only four of eight twins examined (Schmidt and Salzano, 1980), and malformations in identical twins were not identical (Stephens and Brynner, 2001). There were malformations comprising the thalidomide syndrome other than the limb and ear defects already mentioned. These are tabulated in Table 3. Initially, the defects were confused with other syndromes: Goldenhar, Mobius, Wilder-vanck, Duane, and LADD (Smithells and Newman, 1992). The limb reduction defects (and cardiac malformations) were also mimicked by the Holt–Oram syndrome (Brent and Holmes, 1988). A number of other rare birth defects cited from a number of references associated with thalidomide treatment include scoliosis, disc lesions, dysgenesis of sacrum, absence or poor development of muscles, epileptic electroencephalogram (EEG) discharges, abnormalities of internal genitalia,

TABLE 2
Pattern of Limb Malformations Induced by Thalidomide in 154 Children

Pattern	Percent (%) Incidence
Upper limb amelia or phocomelia with normal legs	38
Upper limb amelia or phocomelia with less severe leg defects	11
Forearm defects with normal legs	11
Four-limb phocomelia	9
Lower limb phocomelia or femoral hypoplasia with less severe upper limb defects	6
Forearm defects with less severe leg defects	3
Lower limb defects with normal upper limbs	1
Others (thumbs abnormal in 88%)	5

Source: Modified after Smithells, R. W., *Br. Med. J.*, 1, 269–272, 1973.

TABLE 3
Malformations Comprising the Thalidomide Syndrome

System	Malformations
Limbs[a]	Thumb aplasia
	Hip dislocation
	Femora hypoplasia
	Girdle hypoplasia
Ears	Anotia
	Microtia
	Other abnormalities
Eyes	Microphthalmia, coloboma
	Refractive errors
	Cataracts, squint, pupillary abnormalities
Face	Hypoplastic nasal bridge
	Expanded nasal tip, choanal atresia
Central nervous system	Facial nerve paralysis
	Deafness
	Marcus Gunn or jaw-winking phenomenon
	Crocodile-tear syndrome
	Convulsive disorders?
Respiratory system	Laryngeal and tracheal abnormalities
	Abnormal lobulation of lungs
Heart and blood vessels	Capillary hemangioma extending from dorsum of the nose to the philtrum in the midline
	Congenital heart disease (conotruncal malformations)
Abdominal and visceral	Inguinal hernia
	Cryptorchidism
	Intestinal atresias
	Absent gallbladder and appendix
	Abnormal kidney position
	Horseshoe kidney
	Double ureter
	Vaginal atresia
	Anal atresia, anal stenosis

[a] Other than those tabulated in Table 2.

Source: From Brent, R. L. and Holmes, L. B., *Teratology*, 38, 241–251, 1988 (from many sources). With permission.

obesity in second and third decades of life, and problems secondary to profound deafness (Folb and Dukes, 1990). A number of defects, other than those of the limb, were found among survivors, as determined by one investigator; these have been tabulated in Table 4. The deciduous teeth of thalidomide-exposed children were apparently not abnormally developed (Stahl, 1968). Newman (1985, 1986) and Miller and Stromland (1999) described many of the constellation of defects induced by the drug.

Birth defects were eventually reported from 31 countries, as shown in Table 5. The total number of reported cases of thalidomide embryopathy is not known with certainty but has ranged from estimates made in the press or literature from 5850 (Lenz, 1988) to 12,000 (press, 1998). More reliable estimates are on the order of 7000 to 8000 (*Look*, 5/28/68; *Newsweek*, 2/3/75; Insight Team, 1979; Schardein, 2000), to account for underrepresentation of early deaths and stillborns and incomplete ascertainment of survivors. The timetable for induction of defects with thalidomide was

TABLE 4
Malformations Other Than Limb
Observed in 200 Surviving Children

Malformation	Number
Loss of hearing	57
Abducens paralysis	40
Facial paralysis	28
Anotia	26
Cryptorchidism	20
Renal malformations	17
Microtia	15
Congenital heart disease	13
Inguinal hernia	11
Anal stenosis or atresia	5
Pyloric stenosis	4
Duodenal stenosis or atresia	2

Source: Modified after Ruffing, L., *Birth Defects*, 13, 287–300, 1977.

established (Table 6). The critical period then, includes days 20 to 36 following conception or days 34 to 50 following the first day of the last menses (Nowack, 1965; Lenz, 1965, 1968; Kreipe, 1967). The embryopathy was diagnosed by ultrasound as early as the 17th week of gestation (Gollop et al., 1987).

The recommended therapeutic dose of thalidomide was 16 mg/day (about 0.32 mg/kg/day). Thalidomide embryopathy was induced by as little as one 50 or 100 mg capsule (Taussig, 1963; Lenz, 1964). A blood level of 0.9 µg/ml was sufficient to induce defects according to investigators (Beckman and Kampf, 1961), and the drug was so nontoxic systemically that willful medication of over 14 g of thalidomide was unsuccessful in a suicide attempt (Neuhaus and Ibe, 1960). The risk of teratogenesis from thalidomide has ranged from 2% (Burley, 1962) to 20 or 25% (Ellenhorn, 1964; Tuchmann-Duplessis, 1965) to 100% (Lenz, 1962, 1966); a more likely risk estimate would be in the 10 to 50% range for the drug (Newman, 1985). One group of experts places the teratogenic risk of thalidomide as high, maybe as much as 50% or greater (Friedman and Polifka, 2000). There are known thalidomide-resistant pregnancies (Mellin and Katzenstein, 1962; Jones and Williamson, 1962; Smithells, 1962; Petersen, 1962; Kohler et al., 1962; Pembrey et al., 1970; Kajii et al., 1973). It is recorded that one woman delivered two babies with malformations after taking thalidomide (Insight Team, 1979).

Growth Retardation

This class of developmental toxicity has not been associated with thalidomide embryopathy, except for a report that there was poor linear growth among 202 thalidomide children in late childhood who had limb deformities: They were shorter than normal children but grew at a normal pace in later years (Brook et al., 1977). This report was not further corroborated.

Death

Thalidomide may induce abortion at high doses, according to Lenz (1988). In fact, he stated that 40% of exposed fetal cases died in the first year. This was corroborated in a statement made earlier by Smithells (1973) that there is increased mortality at birth and in the first year of life. Another

TABLE 5
Extent of Reported Cases of Thalidomide Embryopathy[a]

Country	Dates Marketed, Where Known	Approximate Number of Cases
Argentina	To 3/62	?
Australia	/60–12/61	26–39
Austria	/60–/62	19
Belgium	3/59–6/62	35
Brazil	3/59–6/62	204
Canada	4/61–3/62	115–122+
Denmark	10/59–12/61	15–20
Egypt	?	3+
England/Wales	4/58–12/61	435–456
Finland	9/59–12/61	50
France	Not marketed	1+
Ireland	5/59–1/62	36
Israel	Few weeks	2
Italy	/60–9/62	86
Japan	1/58–9/62	309–1000
Kenya	?	1
Lebanon	?	7
Mexico	– /61	4
Netherlands	1/59–11/61	26+
New Zealand	?	8
Norway	11/59–12/61	11
Peru	?	1
Portugal	8/60–12/61	8
Scotland	?	56
Spain	5/61–5/62	5
Sweden	1/59–12/61	158
Switzerland	9/58–12/61	12
Taiwan	/58–9/62	38
Uganda	?	3
United States	Not marketed	17[b]
West Germany	11/56–11/61	2600–4734

[a] See Schardein, J. L., *Chemically Induced Birth Defects*, Third ed., Marcel Dekker, New York, 2000 for further details.
[b] Seven cases from foreign sources (Lenz, 1966).

study recorded an incidence of 9% for miscarriages occurring among 70 drug-exposed pregnancies (Maouris and Hirsch, 1988). Warkany (1971) attributed a 45% mortality rate to cardiac, gastrointestinal, or renal malformations among offspring exposed prenatally to thalidomide.

Functional Deficit

Functional changes in the central nervous system were reported (Holmes et al., 1972; Ruffing, 1977). A recent review indicated that autism occurred 30 times more often among thalidomide-exposed children than in the normal population, and that approximately 5% of survivors of thalidomide exposure are autistic (Stromland et al., 1994). Mental retardation secondary to sensory deprivation, as from hearing impairment and deafness (Rosendal, 1963; Zetterstrom, 1966; Takemori et al., 1976), visual dysfunction (Cullen, 1964; Cant, 1966; Gilkes and Strode, 1963;

TABLE 6
Timetable of Thalidomide-Induced Malformations

Malformation	Days in Gestation
Cranial nerve palsy	35–37
Duplication of thumbs, abnormal ears (anotia)	35–38
Duplication of vagina	35–39
Eye defects	35–42
Thumb aplasia	35–43
Heart and vessel abnormalities	36–45
Thumb hypoplasia	38–40
Amelia (arms)	38–43
Ectopic kidneys and hydronephrosis	38–43
Phocomelia (arms)	38–47 or 49
Hip dislocation	38–48
Microtia	39–43
Duodenal atresia	40–45 or 47
Phocomelia (legs)	40 or 42–47
Pyloric stenosis	40–47
Anal atresia	41–43
Amelia (legs)	41–45
Duodenal stenosis	41–48
Gallbladder atresia	42–43
Choanal atresia	43–46
Respiratory defects	43–46
Urogenital defects	45–47
Triphalangism of thumbs	46–50
Rectal stenosis	49–50

Source: Compiled from Schardein, J. L. *Chemically Induced Birth Defects,* Third ed., Marcel Dekker, New York, 2000 and data from other sources.

Zetterstrom, 1966; Rafuse et al., 1967; Miller and Stromland, 1991), and epilepsy and severe learning disorders (Stephenson, 1976; Newman, 1977, 1985; Stromland et al., 1994) have all been recorded as functional deficits induced by the drug. In one report, 4 of 56 children aged 7 to 10 years with thalidomide-induced limb malformations had subnormal intelligence, a proportion greater than would be expected (McFie and Robertson, 1973).

AFTERWARD

The history of the thalidomide disaster would be incomplete without discussing what we have learned from the event. First, after 45 years, it is still not understood why thalidomide is such an active toxicant. Chemically, none of the elements of the molecule are teratogenic in animals (Smith et al., 1965), and neither are the 12 major hydrolysis products (Fabro et al., 1965) or the racemeic R-enantiomers (Blaschke et al., 1979). Further, the urinary metabolite is different in reactive than in nonreactive species (Fabro, 1981). The teratogenic property is not related to different rates of metabolism in the various species (Schumacher and Gillette, 1966). Moreover, a number of the almost 100 thalidomide-related chemicals or analogs have not been teratogenic in either rabbits or primates (Giacone and Schmidt, 1970; Jonsson et al., 1972; Wuest, 1973), and only two agents have shown similar teratogenic activity in thalidomide exposure. These are methyl-4-phthalimidoglutaramate (WU-385; see also Wuest et al., 1968; McNulty and Wuest,

1969) and 2(2,6-dioxopiperiden-2-yl) phthalimidine (EM_{12}; see also Schumacher et al., 1972). The latter is considered even more potent than thalidomide. Unfortunately, these responses have not added to our understanding of the chemical nature of thalidomide. However, we have learned this much from its chemical structure: The linkage point between the two rings comprising thalidomide seems to be essential for teratogenicity (Helm and Frankus, 1982). An intact ring, receptor reaction, reactive imide group, and relative ring stability appear to be necessary for teratogenicity (Ackermann, 1981). Moreover, while at least two dozen different mechanisms have been proposed over the years, none provided a full explanation of just how thalidomide acts in producing limb malformations (see reviews: Keberle et al., 1965; Helm et al., 1981; Stephens, 1988; Stephens et al., 2000). These proposed mechanisms can be placed into hypothetical categories of thalidomide action as follows: DNA synthesis or transcription, synthesis or function of growth factors, integrins, angiogenesis, chondrogenesis, or cell injury or apoptosis. The net result is that we still do not know the pathogenesis of thalidomide embryopathy. One of the most recent and plausible explainations of how thalidomide works on the limbs is via an angiogenesis pathway (Stephens and Fillmore, 2000). It might happen in this manner: Growth factors (FCF-2 and IGF-1) attach to receptors on limb bud mesenchymal cells and initiate some second messenger system (perhaps Sp-1), which activates α-v and β-3 integrin subunit genes. The resulting integrin proteins stimulate angiogenesis in the developing limb bud. Several steps in the pathway depend on the activation of genes with primarily GC Sp-1 binding site promoters (GGGCGG). Thalidomide specifically binds to these promotor sites and inhibits the transcription of those genes. Inhibition interferes with normal angiogenesis, which results in truncation of the limb. It remains to be seen if this mechanism can be shown to be operative. However, it was recently determined that the embryonic target tissue is the neural crest, the precursor of sensory and autonomic nerves, and therefore, the biochemical lesion should be sought in the neural crest, not the limb buds (McCredie and Willert, 2000).

It is not unexpected that litigation has followed in the wake of thalidomide for almost 20 years. The first tort case against a pharmaceutical company involving a litogen dealt with thalidomide. The trial in Alsdorf, West Germany, ended in December 1969, after 283 days (Curran, 1971). Thirty million dollars was paid by the manufacturer Grunenthal to 2000 survivors, plus $27 million was added by the German government (Insight Team, 1979). In addition, settlements were made to individuals in some of the countries affected by the drug. Included were Ontario and Quebec in Canada; England and Wales (plus reported suits filed against Distillers [a corporation who took over Grunenthal later]); Japan; and Sweden, amounts totaling about $120 million. In Belgium, a mother poisoned her malformed week-old daughter and was charged with infanticide; she was acquitted in court. In the United States, the licensee (Merrell) settled all 13 cases brought against it; only one (McCarrick case) was not, as the company did not consider her defects to be due to the drug (the case was later settled). The total settlements in North America may have reached $50 million. The legal aftermath outside this country due to thalidomide litigation was discussed by Teff and Munro (1976). What was the fate of the West German manufacturer Chemie Grunenthal who was responsible for the tragedy? Reorganized under Distillers, it was eventually taken over by Guinness, and all records of the event were destroyed, according to press reports. What of the survivors? About 5000 exist today, and many, if not most, lead active, "normal" lives within the bounds of their limitations. Obstetrical problems were one of these (Chamberlain, 1989). An interesting Web site exists for Canadian survivors (www. thalidomide.ca/action/index.html). While it has been alleged that birth defects have occurred among children of adults purported to have thalidomide embryopathy (Clementi et al., 1997; Neumann et al., 1998), this may not be the case among the so-called "generational affected" individuals (Smithells, 1998).

One major positive event resulting from the thalidomide tragedy was a tightening of laws regulating drug safety testing. Public Law 87-781, the Kefauver–Harris Drug Amendments of 1962, was the direct effect of thalidomide that led to stricter testing requirements of drug safety. It would be less likely today that a new drug could be introduced into the market with the toxicity that thalidomide displayed. The U.S. Teratology Society published a recent position paper on thalidomide (U.S. Teratology Society, 2000).

New Beginnings

Beginning as long ago as 1965, the use of thalidomide was initiated in Brazil, Argentina, and Venezuela for treating leprosy and other immunopathological conditions in those countries. Prior to its official reintroduction, it was manufactured in Brazil and Argentina and was available either through pharmacies (Brazil) or governmental health agencies (Argentina; see Castilla et al., 1996). The situation was updated more recently (Miller and Stromland, 1999; Ances, 2002). Since 1965, some 34 cases of embryopathy were identified from 10 (Chile excluded) South American countries (Gollop et al., 1987; Cutler, 1994; Rocha, 1994; Jones, 1994; Castilla et al., 1996; Castilla, 1997). By mid-1999, some 15,000 prescriptions had been written, and by the year 2000, some 20,000 prescriptions were filed (Stephens and Brynner, 2001). Dosages taken are in the range of 400 to 1600 mg/day, much greater than before. But by 2002, some 86 cases of embryopathy, identical to that described half a century earlier, had been recorded in Brazil (Mamiya, 2003). The full story is not yet complete.

CHEMISTRY

Thalidomide is a hydrophilic compound that is near average size and of larger polarity compared to the other human developmental toxicants. It can participate in hydrogen bonding interactions, primarily as a hydrogen bond acceptor. The calculated physicochemical and topological properties of thalidomide are as follows.

Physicochemical Properties

Parameter	Value
Molecular weight	258.233 g/mol
Molecular volume	206.37 A^3
Density	1.259 g/cm^3
Surface area	240.86 A^2
LogP	−3.932
HLB	17.695
Solubility parameter	27.096 J$^{(0.5)}$/cm$^{(1.5)}$
Dispersion	23.047 J$^{(0.5)}$/cm$^{(1.5)}$
Polarity	10.370 J$^{(0.5)}$/cm$^{(1.5)}$
Hydrogen bonding	9.771 J$^{(0.5)}$/cm$^{(1.5)}$
H bond acceptor	1.15
H bond donor	0.29
Percent hydrophilic surface	83.28
MR	69.089
Water solubility	0.585 log (mol/M^3)
Hydrophilic surface area	200.59 A^2
Polar surface area	97.88 A^2
HOMO	−10.717 eV
LUMO	−1.448 eV
Dipole	5.641 debye

Topological Properties (Unitless)

Parameter	Value
x0	13.568
x1	9.092
x2	8.467
xp3	7.432
xp4	6.584
xp5	5.155
xp6	3.216
xp7	2.278
xp8	1.437
xp9	0.874
xp10	0.483
xv0	9.881
xv1	5.900
xv2	4.493
xvp3	3.347
xvp4	2.419
xvp5	1.626
xvp6	0.855
xvp7	0.499
xvp8	0.282
xvp9	0.142
xvp10	0.069
k0	21.286
k1	13.959
k2	5.413
k3	2.380
ka1	11.885
ka2	4.183
ka3	1.728

REFERENCES

Ackermann, H. (1981). Assessment of teratogenic action of phthalimide derivatives. *Anwend. Pflanzenschutzm. Mittein Steuer. Biol. Prozesse* 187: 197–202.

Ances, B. M. (2002). New concerns about thalidomide. *Obstet. Gynecol.* 99: 125–128.

Arbab-Zadeh, A. (1966). Toxische und teratogene wirkungen des Thalidomid. *Dtsch. Z. Gesamte Gerichtl. Med.* 57: 285–290.

Beckman, R. and Kampf, H. H. (1961). Zur quantitativen Bestimmung und zum qualitativen Nachweis von *N*-phthalyl-glutaminsauremid (thalidomide). *Arzneimittelforschung* 11: 45–47.

Bignami, G. et al. (1962). Drugs and congenital abnormalities. *Lancet* 2: 1333.

Blaschke, G. et al. (1979). [Chromatographic separation of racemic thalidomide and teratogenic activity of its enantiomers]. *Arzneimittelforschung* 29: 1640–1642.

Brent, R. L. and Holmes, L. B. (1988). Clinical and basic science lessons from the thalidomide tragedy: What have we learned about the causes of limb defects? *Teratology* 38: 241–251.

Brook, C. G. D., Jarvis, S. N., and Newman, C. G. H. (1977). Linear growth of children with limb deformities following exposure to thalidomide *in utero*. *Acta Paediatr. Scand.* 66: 673–675.

Burley, D. M. (1962). Thalidomide and congenital abnormalities. *Lancet* 1: 271.

Cant, J. S. (1966). Minor ocular abnormalities associated with thalidomide. *Lancet* 1: 1134.

Castilla, E. E. (1997). Thalidomide, a current teratogen in South America. *Teratology* 55: 160.

Castilla, E. E. et al. (1996). Thalidomide, a current teratogen in South America. *Teratology* 54: 273–277.

Chamberlain, G. (1989). The obstetrical problems of the thalidomide children. *Br. Med. J.* 298: 6.

Clementi, M., Turolla, L., and Tenconi, R. (1997). Two cases of congenital anomalies in children of thalidomide victims. *Teratology* 56: 397.

Cullen, J. F. (1964). Ocular defects in thalidomide babies. *Br. J. Ophthalmol.* 48: 151–153.

Curran, W. J. (1971). The thalidomide tragedy in Germany: The end of a historic medicolegal trial. *N. Engl. J. Med.* 284: 481–482.

Cutler, J. (1994). Thalidomide revisited. *Lancet* 343: 795–796.

Delahunt, C. S. and Lassen, L. J. (1964). Thalidomide syndrome in monkeys. *Science* 146: 1300–1305.

Delatour, P., Dams, R., and Favre-Tissot, M. (1965). Thalidomide: Embryopathies chez le chien. *Therapie* 20: 573–589.

DiPaolo, J. A. (1963). Congenital malformations in strain A mice: Its experimental production by thalidomide. *JAMA* 183: 139–141.

Ellenhorn, M. J. (1964). The FDA and the prevention of drug embryopathy. *J. New Drugs* 4: 12–20.

Fabro, S. (1981). Biochemical basis of thalidomide teratogenicity. In *The Biochemical Basis of Chemical Teratogenesis*, M. R. Juchau, Ed., Elsevier/North Holland, New York, pp. 159–178.

Fabro, S. et al. (1965). The metabolism of thalidomide: Some biological effects of thalidomide and its metabolites. *Br. J. Pharmacol.* 25: 352–362.

Fine, R. A. (1972). *The Great Drug Deception. The Shocking Story of MER 29 and the Folks Who Gave You Thalidomide.* Stein & Day, New York.

Folb, P. I. and Dukes, M. N. (1990). *Drug Safety in Pregnancy*, Elsevier, Amsterdam, p. 5.

Fraser, F. C. (1988). Thalidomide perspective: What did we learn? *Teratology* 38: 201–202.

Friedman, J. M. and Polifka, J. E. (2000). *Teratogenic Effects of Drugs. A Resource for Clinicians (TERIS)*, Second ed., Johns Hopkins University Press, Baltimore, MD.

Giacone, J. and Schmidt, H. L. (1970). Internal malformations produced by thalidomide and related compounds in rhesus monkeys. *Anat. Rec.* 166: 306.

Gilani, S. H. (1973). Cardiovascular malformations in the chick embryo induced by thalidomide. *Toxicol. Appl. Pharmacol.* 25: 77–83.

Gilkes, M. J. and Strode, M. (1963). Ocular anomalies in association with developmental limb abnormalities of drug origin. *Lancet* 1: 1026–1027.

Giroud, A., Tuchmann-Duplessis, H., and Mercier-Parot, L. (1962). [Influence of thalidomide on fetal development]. *Bull. Acad. Nat. Med. (Paris)* 146: 343–345.

Gollop, T. R., Eigier, A., and GuiduglioNeto, J. (1987). Prenatal diagnosis of thalidomide syndrome. *Prenat. Diagn.* 7: 295–298.

Green, M. D. (1996). *Bendectin and Birth Defects. The Challenges of Mass Tort Substances Litigation*, University of Pennsylvania Press, Philadelphia.

Helm, F. C. and Frankus, E. (1982). Chemical structure and teratogenic activity of thalidomide-related compounds. *Teratology* 25: 47A.

Helm, F. C. et al. (1981). Comparative teratological investigations of compounds structurally and pharmacologically related to thalidomide. *Arzneimittelforschung* 31: 941–949.

Hendrickx, A. and Newman, L. (1973). Appendicular skeletal and visceral malformations induced by thalidomide in bonnet monkeys. *Teratology* 7: 151–160.

Hendrickx, A. G. and Sawyer, R. H. (1978). Developmental staging and thalidomide teratogenicity in the green monkey (*Cercopithecus aethiops*). *Teratology* 18: 393–404.

Hendrickx, A. G., Axelrod, L. R., and Clayborn, L. D. (1966). "Thalidomide" syndrome in baboons. *Nature* 210: 958–959.

Holmes, L. B. et al. (1972). *Mental Retardation. An Atlas of Diseases with Associated Physical Abnormalities*, Macmillan, New York, pp. 132–133.

Homburger, F. et al. (1965). Susceptibility of certain inbred strains of hamsters to teratogenic effects of thalidomide. *Toxicol. Appl. Pharmacol.* 7: 686–698.

Insight Team (of the *Sunday Times of London*). (1979). *Suffer the Children: The Story of Thalidomide*, Viking Press, New York.

Jones, E. E. and Williamson, D. A. (1962). Thalidomide and congenital anomalies. *Lancet* 1: 222.

Jones, G. R. (1994). Thalidomide: 35 years on and still deforming. *Lancet* 343: 1041.

Jonsson, N. A., Mikiver, L., and Selberg, U. (1972). Chemical structure and teratogenic properties. 2. Synthesis and teratogenic activity in rabbits of some derivatives of phthalimide, isoindoline-1-one, 1,2-ben-zisothiazoline-3-one-1,1-dioxide and 4(3*H*)-quinazolinone. *Acta Pharm. Suec.* 9: 431–446.

Kajii, T. (1965). Thalidomide experience in Japan. *Ann. Paediatr.* 295: 341–354.

Kajii, T., Kida, M., and Takahashi, K. (1973). The effect of thalidomide intake during 115 human pregnancies. *Teratology* 8: 163–166.

Keberle, H. et al. (1965). Theories on the mechanism of action of thalidomide. In *Embryopathic Activity of Drugs*, J. M. Robson, F. M. Sullivan, and R. L. Smith, Eds., Little, Brown, Boston, pp. 210–226.

Kelsey, F. O. (1965). Problems raised for the FDA by the occurrence of thalidomide embryopathy in Germany, 1960–1961. *Am. J. Public Health* 55: 703–707.

Kelsey, F. O. (1988). Thalidomide update: Regulatory aspects. *Teratology* 38: 221–226.

King, C. T. G. and Kendrick, F. J. (1962). Teratogenic effects of thalidomide in the Sprague-Dawley rat. *Nature* 2: 1116.

Kohler, H. G., Ockenfels, H., and Dunn, P. M. (1962). Thalidomide and congenital abnormalities. *Lancet* 1: 326.

Kosenow, W. and Pfeiffer, R. A. (1960). Micromelia, haemangioma und duodenal stenosis exhibit. *German Pediatric Society Meeting Abstracts*, Kassel.

Lacy, C. F. et al. (2004). *Drug Information Handbook (Pocket), 2004–2005*, Lexi-Comp., Inc., Hudson, OH.

Lear, J. (1962). The unfinished story of thalidomide. *SR* (September 1), pp. 35–40.

Lechat, P. et al. (1964). Resultats negatifs d'une Recherche des effect Teratogenes eventuels du *N*-Methyl-phthalimide. *Therapie* 19: 1393–1403.

Lenz, W. (1961). Kindliche Missbildungen nach Medikament-Einnahme wahrend der Graviditat? *Dtsch. Med. Wochenschr.* 86: 2555–2556.

Lenz, W. (1962). Thalidomide and congenital abnormalities. *Lancet* 1: 271.

Lenz, W. (1964). Chemicals and malformations in man. In *Congenital Malformations, 22nd International Conference, New York, 1963*, M. Fishbein, Ed., International Medical Congress, New York, pp. 263–276.

Lenz, W. (1966). Malformations caused by drugs in pregnancy. *Am. J. Dis. Child.* 112: 99–106.

Lenz, W. (1971). How can the teratogenic action of a factor be established in man? *South. Med. J.* 64 (Suppl. 1): 41–50.

Lenz, W. (1988). A short history of thalidomide embryopathy. *Teratology* 38: 203–215.

Mamiya, K. (2003). Thalidomide: The victim's point of view. *Cong. Anom.* 43: 211–212.

Maouris, P. G. and Hirsch, P. J. (1988). Pregnancy in women with thalidomide-induced disabilities. Case report and a questionnaire study. *Br. J. Obstet. Gynecol.* 95: 717–719.

Marin-Padilla, M. and Benirschke, K. (1963). Thalidomide induced alterations in the blastocyst and placenta of the armadillo, *Dasypus novemcinctus mexicanus*, including a choriocarcinoma. *Am. J. Pathol.* 43: 999–1016.

McBride, W. G. (1961). Thalidomide and congenital abnormalities. *Lancet* 2: 1358.

McCredie, J. and Willert, H. G. (2000). Thalidomide revisited. *Reprod. Toxicol.* 14: 566–567.

McFadyen, R. E. (1976). Thalidomide in America. A brush with tragedy. *Clio Med.* 11: 79–93.

McFie, J. and Robertson, J. (1973). Psychological test results of children with thalidomide deformities. *Dev. Med. Child Neurol.* 15: 719–727.

McNulty, W. P. and Wuest, H. M. (1969). Thalidomide, teratogeny, and structure: Teratogenic action of WU 385 on the rhesus. *Teratology* 2: 265.

Mellin, G. W. and Katzenstein, M. (1962). The saga of thalidomide. *N. Engl. J. Med.* 267: 1184 passim 1244.

Miller, M. T. and Stromland, K. (1991). Ocular motility in thalidomide embryopathy. *J. Pediatr. Ophthalmol. Strab.* 28: 47–54.

Miller, M. T. and Stromland, K. (1999). Teratogen update: Thalidomide: A review, with focus on ocular findings and new potential uses. *Teratology* 60: 306–321.

Neuhaus, G. and Ibe, K. (1960). Clinical observations on a suicide attempt with 144 tablets of Contergan Forte (*N*-phthalyl-glutarimide). *Med. Klin.* 55: 544–545.

Neumann, L., Pelz, J., and Kunze, J. (1998). Unilateral terminal aphalangia in father and daughter — exogenous or genetic cause. *Am. J. Med. Genet.* 78: 366–370.

Newman, C. G. H. (1977). Clinical observations on the thalidomide syndrome. *Proc. R. Soc. Med.* 70: 225–227.

Newman, C. G. H. (1985). Teratogen update: Clinical aspects of thalidomide embryopathy — a continuing preoccupation. *Teratology* 32: 133–144.

Newman, C. G. H. (1986). The thalidomide syndrome: Risks of exposure and spectrum of malformations. *Clin. Perinatol.* 13: 555–573.

Palludan, B. (1966). Swine in teratological research. In *Swine in Biomedical Research*, L. K. Bustad and R. O. McClellan, Eds., Battelle Memorial Institute, Columbus, OH, pp. 51–78.

PDR® (*Physicians' Desk Reference®*). (2002). Medical Economics Co., Inc., Montvale, NJ.

Pembrey, M. E., Clarke, C. A., and Frais, M. M. (1970). Normal child after maternal thalidomide ingestion in critical period of pregnancy. *Lancet* 1: 275–277.

Petersen, C. E. (1962). Thalidomid und Missbildungen beitrage zur Frage der Atiologie emes gehauft aufge-tretenen Fehlbildungfskomplexes. *Med. Welt.* 14: 753–756.

Pfeiffer, R. A. and Kosenow, W. (1962). Zur Frage einer exogenen Verursachung von schweren Extremitaten-missbildungen. *Munch. Med. Wochenschr.* 104: 68–74.

Poswillo, D. E., Hamilton, W. J., and Sopher, D. (1972). The marmoset as an animal model for teratological research. *Nature* 239: 460–462.

Quibell, E. P. (1981). The thalidomide embryopathy. *Practitioner* 225: 721–726.

Rafuse, E. V., Arstikaitis, M., and Brent, H. P. (1967). Ocular findings in thalidomide children. *Can. J. Ophthalmol.* 2: 222–225.

Rocha, J. (1994). Thalidomide given to women in Brazil. *Br. Med. J.* 308: 1061.

Rosenberg, M. and Glueck, B. C. (1973). First-cause no harm: An early warning system for iatrogenic disease. *Prev. Med.* 2: 82–87.

Rosendal, T. (1963). Thalidomide and aplasia-hypoplasia of the otic labyrinth. *Lancet* 1: 724–725.

Ruffing, L. (1977). Evaluation of thalidomide children. *Birth Defects* 13: 287–300.

Schardein, J. L. (2000). *Chemically Induced Birth Defects*, Third ed., Marcel Dekker, New York, pp. 89–119.

Schmidt, M. and Salzano, F. M. (1980). Dissimilar effects of thalidomide in dizygotic twins. *Acta Genet. Med. Gemellol. (Roma)* 29: 295–297.

Schumacher, H. and Gillette, J. (1966). Embryotoxic effects of thalidomide. *Fed. Proc.* 25: 353.

Schumacher, H. J. et al. (1972). The teratogenic activity of a thalidomide analogue, EM_{12} in rabbits, rats, and monkeys. *Teratology* 5: 233–240.

Sjostrom, H. and Nilsson, R. (1972). *Thalidomide and the Power of the Drug Companies*, Penguin Books, London.

Smith, R. L. et al. (1965). Studies on the relationship between the chemical structure and embryotoxic activity of thalidomide and related compounds. In *Embryopathic Activity of Drugs*, J. M. Robson, F. M. Sullivan, and R. L. Smith, Eds., Little, Brown, Boston, pp. 194–209.

Smithells, D. (1998). Dominant gene probably caused some defects ascribed to thalidomide. *Br. Med. J.* 316: 149.

Smithells, R. W. (1962). Thalidomide and malformations in Liverpool. *Lancet* 1: 1270–1273.

Smithells, R. W. (1965). The thalidomide legacy. *Proc. R. Soc. Med.* 58: 491–492.

Smithells, R. W. (1973). Defects and disabilities of thalidomide children. *Br. Med. J.* 1: 269–272.

Smithells, R. W. and Newman, C. G. H. (1992). Recognition of the thalidomide defects. *J. Med. Genet.* 29: 716–723.

Somers, G. F. (1963). The foetal toxicity of thalidomide. *Proc. Eur. Soc. Study Drug Toxicol.* 1: 49–58.

Stahl, A. (1968). Clinical, orthodontic, and radiological findings in the jaws and face of children with dysmelia associated with thalidomide embryopathy. *Inter. Dental J.* 18: 631–638.

Staples, R. E. and Holtkamp, D. E. (1963). Effects of parental thalidomide treatment on gestation and fetal development. *Exp. Mole. Pathol.* Suppl. 2: 81–106.

Steffek, A. J. and Verrusio, A. C. (1972). Experimentally induced oral–facial malformations in the ferret (*Mustela putorius* Juro). *Teratology* 5: 268.

Stephens, T. D. (1988). Proposed mechanisms of action for thalidomide embryopathy. *Teratology* 38: 229–239.

Stephens, T. and Brynner, R. (2001). *Dark Remedy. The Impact of Thalidomide and Its Revival as a Vital Medicine*, Perseus, Cambridge, MA.

Stephens, T. D. and Fillmore, B. J. (2000). Hypothesis: Thalidomide embryopathy-proposed mechanisms of action. *Teratology* 61: 189–195.

Stephens, T. D., Bunde, C. J. W., and Fillmore, B. J. (2000). Mechanism of action in thalidomide teratogenesis. *Biochem. Pharmacol.* 59: 1489–1499.

Stephenson, J. B. F. (1976). Epilepsy: Neurological complication of thalidomide embryopathy. *Dev. Med. Child Neurol.* 18: 189–197.

Stromland, K. and Miller, M. T. (1993). Thalidomide embryopathy: Revisited 27 years later. *Acta Ophthalmol.* 71: 238–245.

Stromland, K. et al. (1994). Autism in thalidomide embryopathy: A population study. *Dev. Med. Child Neurol.* 36: 351–356.

Sugiura, Y. et al. (1979). Thalidomide dysmelia in Japan. *Cong. Anom.* 19: 1–19.

Takemori, S., Tanoka, Y., and Suzuki, J.-I. (1976). Thalidomide anomalies of the ear. *Arch Otolaryngol.* 102: 425–427.

Tanimura, T., Tanaka, O., and Nishimura, H. (1971). Effects of thalidomide and quinine dihydrochloride on Japanese and rhesus monkey embryos. *Teratology* 4: 247.

Taussig, H. B. (1962a). A study of the German outbreak of phocomelia. The thalidomide syndrome. *JAMA* 180: 1106–1114.

Taussig, H. B. (1962b). The thalidomide syndrome. *Sci. Am.* 207: 29–35.

Taussig, H. B. (1962c). Thalidomide and phocomelia. *Pediatrics* 30: 654–659.

Taussig, H. B. (1963). The evils of camouflage as illustrated by thalidomide. *N. Engl. J. Med.* 269: 92–94.

Teff, H. and Munro, C. R. (1976). *Thalidomide. The Legal Aftermath*, Lexington Books (Saxon House), Lanham, MD.

Theisen, C. T. et al. (1979). Unusual muscle abnormalities associated with thalidomide treatment in a rhesus monkey: A case report. *Teratology* 19: 313–320.

Tuchmann-Duplessis, H. (1965). Design and interpretation of teratogenic tests. In *Embryopathic Activity of Drugs*, F. M. Robson, R. M. Sullivan, and R. L. Smith, Eds., Little, Brown, Boston, pp. 56–93.

U.S. Teratology Society. (2000). Teratology Society Public Affairs Committee Position Paper. Thalidomide. *Teratology* 62: 172–173.

Vondruska, J. F., Fancher, O. E., and Calandra, J. C. (1971). An investigation into the teratogenic potential of captan, folpet, and difolatan in nonhuman primates. *Toxicol. Appl. Pharmacol.* 18: 619–624.

Warkany, J. (1971). *Congenital Malformations. Notes and Comments*, Year Book Medical Publishers, Chicago.

Warkany, J. (1988). Why I doubted that thalidomide was the cause of the epidemic of limb defects of 1959 to 1961. *Teratology* 38: 217–219.

Weidman, W. H., Young, H. H., and Zollman, P. E. (1963). The effect of thalidomide on the unborn puppy. *Proc. Staff Meet. Mayo Clin.* 38: 518–522.

Wiedemann, H.-R. (1961). Hinweis, auf eme derzeitige Haufung hypo und aplastischer Fehlbildungen der Gleidmassen. *Med. Welt* 37: 1863–1866.

Wilson, J. G. and Fradkin, R. (1969). Teratogeny in nonhuman primates, with notes on breeding procedures in *Macaca mulatta*. *Ann. NY Acad. Sci.* 162: 267–277.

Wilson, J. G. and Gavan, J. A. (1967). Congenital malformations in nonhuman primates: Spontaneous and experimentally induced. *Anat. Rec.* 158: 99–110.

Wuest, H. M. (1973). Experimental teratology and the thalidomide problem. *Teratology* 8: 242.

Wuest, H. M., Fox, R. R., and Crary, D. D. (1968). Relationship between teratogeny and structure in the thalidomide field. *Experientia* 24: 993–994.

Yang, Q. et al. (1997). The return of thalidomide: Are birth defects surveillance systems ready? *Am. J. Med. Genet.* 73: 251–258.

Zetterstrom, B. (1966). Ocular malformations caused by thalidomide. *Acta Ophthalmol.* 44: 391–395.

26 Primidone

Chemical name: 5-Ethyldihydro-5-phenyl-4,6(1*H*,5*H*)-pyrimidinedione

Alternate names: Primaclone, desoxyphenobarbital

CAS # 125-33-7

SMILES: C1(c2ccccc2)(C(NCNC1=O)=O)CC

INTRODUCTION

Primidone is a barbiturate-type anticonvulsant used therapeutically for over 50 years in the management of focal, psychomotor, and grand mal seizures. It acts mechanistically by decreasing neuron excitability, thereby raising seizure thresholds (Lacy et al., 2004). One of its two active metabolites is the structural analog drug, phenobarbital, also a probable developmental toxicant. Primidone is available by prescription under the trade name Mysoline® and several other trade names, and it has a pregnancy category classification of D. The package label carries a use in pregnancy warning that recent reports suggest an association between the use of anticonvulsant drugs by women with epilepsy and an elevated incidence of birth defects in children born to these women (among them, primidone; see below; also see *PDR*, 2004).

DEVELOPMENTAL TOXICOLOGY

ANIMALS

Animal studies with primidone are limited. Oral doses by either dietary or gavage administration to mice over the range of 100 to 250 (gavage) or 500 to 2500 mg/kg/day for 5 or 11 days during organogenesis induced low incidences of cleft palate but no other developmental toxicity (Sullivan and McElhatton, 1975). Rats given a gavage dose of 120 mg/kg/day on gestation days 8 to 20 had maternal toxicity and evidenced embryolethality, and the resulting pups showed decreased activity among females, and a specific learning deficit (Pizzi et al., 1998).

TABLE 1
Case Reports of Primidone Embryopathy in Humans

Number of Cases	Ref.
1[a]	Lowe, 1973
2	Seip, 1976
2[a]	Rudd and Freedom, 1979
1	Shih et al., 1979
2[a]	Myrhe and Williams, 1981
1	Thomas and Buchanan, 1981
1	Nau et al., 1981
4	Rating et al., 1982
1	Ohta et al., 1982
1	Krauss et al., 1984
10	Hoyme et al., 1986; Hoyme, 1990

[a] Monotherapy.

HUMANS

Reports of developmental toxicity in humans have centered on case reports on a specific embryopathy as shown in 27 cases tabulated in Table 1. It is an accepted fact that anticonvulsants are difficult to interpret with respect to toxicity due to confounding factors of multiple drug therapy and that epilepsy itself may result in malformations, among other factors. Nevertheless, a syndrome of minor dysmorphic features, not yet completely delineated, was recorded in studies tabulated above that include facial dysmorphism, microcephaly, poor somatic development, short stature, and cardiac defects. Hirsutism, hypoplastic nails, and alveolar prominence were also noted, with various descriptions fitting both the Noonan syndrome (Burn and Baraitser, 1982) and the syndrome of defects produced by another anticonvulsant drug, phenytoin (Seip, 1976). In addition to the case reports of embryopathy shown above, a number of studies also reported the drug associated with increased abnormalities of varied types and retarded growth (Fedrick, 1973; Martinez and Snyder, 1973; Nakane et al., 1980; Neri et al., 1983; Majewski and Steger, 1984; Battino et al., 1992; Olafsson et al., 1998). Neither mortality nor functional impairments have been associated with the retarded growth and abnormalities. A number of additional studies recorded malformations resulting from primidone in combination with other anticonvulsants; the combination of valproic acid, carbamazepine, and primidone is considered by some the most risky of all regimens (Murasaki et al., 1988). In a small number of reports, researchers did not find primidone-induced malformations with or without combined drug therapy (Annegers et al., 1974; Kaneko et al., 1988, 1993; Samren et al., 1997; Canger et al., 1999). The embryopathy was associated with intake to therapeutic levels of the drug, 125 to 1500 mg/day up to a maximum of 2 g/day, and treatment was limited, where recorded, to the first trimester. The magnitude of teratogenic risk has been placed at small to moderate by one group of experts (Friedman and Polifka, 2000). A number of review articles on monotherapy and combination therapy of primidone with other anticonvulsants were cited in another reference (Schardein, 2000).

CHEMISTRY

Primidone is an average-sized molecule that is slightly hydrophilic. It is of average polarity in comparison to the other human developmental toxicants. Primidone can engage in hydrogen bonding both as an acceptor and donor. The calculated physicochemical and topological properties for this compound are listed below.

PHYSICOCHEMICAL PROPERTIES

Parameter	Value
Molecular weight	218.255 g/mol
Molecular volume	199.16 A^3
Density	1.0148 g/cm^3
Surface area	248.63 A^2
LogP	−0.616
HLB	11.685
Solubility parameter	23.142 J$^{(0.5)}$/cm$^{(1.5)}$
Dispersion	20.662 J$^{(0.5)}$/cm$^{(1.5)}$
Polarity	6.844 J$^{(0.5)}$/cm$^{(1.5)}$
Hydrogen bonding	7.862 J$^{(0.5)}$/cm$^{(1.5)}$
H bond acceptor	1.07
H bond donor	0.52
Percent hydrophilic surface	57.15
MR	63.183
Water solubility	1.058 log (mol/M^3)
Hydrophilic surface area	142.10 A^2
Polar surface area	64.52 A^2
HOMO	−9.419 eV
LUMO	0.328 eV
Dipole	3.439 debye

TOPOLOGICAL PROPERTIES (UNITLESS)

Parameter	Value
x0	11.596
x1	7.714
x2	6.534
xp3	6.087
xp4	5.184
xp5	3.368
xp6	1.900
xp7	1.143
xp8	0.688
xp9	0.234
xp10	0.124
xv0	9.118
xv1	5.337
xv2	3.814
xvp3	3.070
xvp4	2.120
xvp5	1.220
xvp6	0.552
xvp7	0.286
xvp8	0.126
xvp9	0.036
xvp10	0.014
k0	16.256
k1	12.457

Continued.

Parameter	Value
k2	5.104
k3	2.080
ka1	10.980
ka2	4.155
ka3	1.595

REFERENCES

Annegers, J. F. et al. (1974). Do anticonvulsants have a teratogenic effect? *Arch. Neurol.* 31: 364–373.

Battino, D. et al. (1992). Intrauterine growth in the offspring of epileptic mothers. *Acta Neurol. Scand.* 86: 555–557.

Burn, J. and Baraitser, M. (1982). Primidone teratology or Noonan syndrome. *J. Pediatr.* 100: 836.

Canger, R. et al. (1999). Malformations in offspring of women with epilepsy: A prospective study. *Epilepsia* 40: 1231–1236.

Fedrick, J. (1973). Epilepsy and pregnancy: A report from the Oxford Record Linkage Study. *Br. Med. J.* 2: 442–448.

Friedman, J. M. and Polifka, J. E. (2000). *Teratogenic Effects of Drugs. A Resource for Clinicians (TERIS)*, Second ed., Johns Hopkins University Press, Baltimore, MD.

Hoyme, H. E. et al. (1986). Fetal primidone effects. *Teratology* 33: 76C.

Hoyme, H. E. (1990). Teratogenically induced fetal anomalies. *Clin. Perinatol.* 17: 547–565.

Kaneko, S. et al. (1988). Teratogenicity of antiepileptic drugs: Analysis of possible risk factors. *Epilepsia* 29: 459–467.

Kaneko, S. et al. (1993). Teratogenicity of antiepileptic drugs and drug specific malformations. *Jpn. J. Psychiatr. Neurol.* 47: 306–308.

Krauss, C. M. et al. (1984). Four siblings with similar malformations after exposure to phenytoin and primidone. *J. Pediatr.* 105: 750–755.

Lacy, C. F. et al. (2004). *Drug Information Handbook (Pocket), 2004–2005*, Lexi-Comp. Inc., Hudson, OH.

Lowe, C. R. (1973). Congenital malformations among infants born to epileptic women. *Lancet* 1: 9–10.

Majewski, F. and Steger, M. (1984). Fetal head growth retardation associated with maternal phenobarbital/primidone and/or phenytoin therapy. *Eur. J. Pediatr.* 141: 188–189.

Martinez, G. and Snyder, R. D. (1973). Transplacental passage of primidone. *Neurology* 23: 381–383.

Murasaki, O. et al. (1988). Reexamination of the teratological effect of antiepileptic drugs. *Jpn. J. Psychiatr. Neurol.* 42: 592–593.

Myhre, S. A. and Williams, R. (1981). Teratogenic effects associated with maternal primidone therapy. *J. Pediatr.* 99: 160–162.

Nakane, Y. et al. (1980). Multiinstitutional study on the teratogenicity and fetal toxicity of antiepileptic drugs: A report of the Collaborative Study Group in Japan. *Epilepsia* 21: 663–680.

Nau, H. et al. (1981). Valproic acid and its metabolites, placental transfer, neonatal pharmacokinetics, transfer via mothers milk and clinical status in neonates of epileptic mothers. *J. Pharmacol. Exp. Therap.* 219: 768–777.

Neri, A. et al. (1983). Neonatal outcome in infants of epileptic mothers. *Eur. J. Obstet. Gynecol. Reprod. Biol.* 16: 263–268.

Ohta, S. et al. (1982). A case of primidone embryopathy associated with barbiturate withdrawal syndrome. *Teratology* 26: 36A.

Olafsson, E. et al. (1998). Pregnancies of women with epilepsy: A population-based study in Iceland. *Epilepsia* 39: 887–892.

PDR® (*Physicians' Desk Reference®*). (2004). Medical Economics Co., Inc., Montvale, NJ.

Pizzi, W. J., Newman, A. S., and Shansky, A. (1998). Primidone-induced embryolethality and DRL deficits in surviving offspring. *Neurotoxicol. Teratol.* 20: 3–7.

Rating, D. et al. (1982). Teratogenic and pharmacokinetic studies of primidone during pregnancy and in the offspring of epileptic women. *Acta Paediatr. Scand.* 71: 301–311.

Rudd, N. L. and Freedom, R. M. (1979). A possible primidone embryopathy. *J. Pediatr.* 94: 835–837.

Samren, E. B. et al. (1997). Maternal use of antiepileptic drugs and the risk of major congenital malformations: A joint European prospective study of human teratogenesis associated with maternal epilepsy. *Epilepsia* 38: 981–990.

Schardein, J. L. (2000). *Chemically Induced Birth Defects*, Third ed., Marcel Dekker, New York, p. 213.

Seip, M. (1976). Growth retardation, dysmorphic facies and minor malformations following massive exposure to phenobarbitone *in utero. Acta Paediatr. Scand.* 65: 617–621.

Shih, L. Y., Diamond, N., and Kushnick, T. (1979). Primidone induced teratology — clinical observations. *Teratology* 18: 47A.

Sullivan, F. M. and McElhatton, P. R. (1975). Teratogenic activity of the antiepileptic drugs phenobarbital, phenytoin, and primidone in mice. *Toxicol. Appl. Pharmacol* 34: 271–282.

Thomas, D. and Buchanan, N. (1981). Teratogenic effects of anticonvulsants. *J. Pediatr.* 99: 163.

27 Fluconazole

Chemical name: 2,4-Difluoro-α,α′-bis(1*H*-1,2,4-triazol-1-ylmethyl)benzyl alcohol

CAS #: 86386-73-4

SMILES: c1cc(c(cc1F)F)C(Cn2cncn2)(Cn3ncnc3)O

INTRODUCTION

Fluconazole is a synthetic triazole chemical used therapeutically as an antifungal drug, given orally for the treatment of vaginal candidiasis, or given parenterally at higher doses for other mycotic infections or for other drug-resistant organisms. It acts by interfering with cytochrome P450 activity, decreasing ergosterol synthesis, and inhibiting cell membrane formation (Lacy et al., 2004). The drug is available by prescription as Diflucan® or as several other trade names. It is widely used, ranking 65th among the most frequently prescribed drugs (www.rxlist.com.top200.htm). Fluconazole has a pregnancy category of C (in this case meaning that animal studies show adverse effects and no controlled human studies are available). However, the package label states with reference to use in pregnancy that "there have been reports of multiple congenital abnormalities in infants whose mothers were being treated for 3 or more months with high doses (400 to 800 mg/day) fluconazole therapy for coccidiodomycosis (an unindicated use). The relationship between fluconazole use and these events is unclear" (*PDR*, 2005; see below).

DEVELOPMENTAL TOXICOLOGY

ANIMALS

Only the rat and rabbit have been investigated in the laboratory for developmental toxicity potential in animals, but the results are unpublished. Indicated on the package label is that fluconazole in the rat at oral doses in the range of 80 to 320 mg/kg/day during organogenesis resulted in cleft palate, wavy ribs, and abnormal craniofacial ossification, all teratogenic responses; embryolethality was also recorded, and developmental variations were observed at lower doses of 25 mg/kg/day and higher. Also indicated on the package label is that in the rabbit, oral doses of 5 to 75 mg/kg/day during organogenesis resulted in abortion at the highest dose and maternal toxicity over the entire

TABLE 1
Developmental Toxicity Profile of Fluconazole in Humans

Case Number	Malformations	Growth Retardation	Death	Functional Deficit	Ref.
1	Multiple: skull, palate, skeleton		✔		Lee et al., 1992
2	Multiple: head, face, ears, jaw, heart, skeleton, vessels	✔			Pursley et al., 1996
3	Multiple: head, face, heart, palate, ears, skeleton		✔		Pursley et al., 1996
4	Multiple: skull, face, ears, skeleton, digits	✔			Aleck and Bartley, 1997
5	Brain, heart		✔		Sanchez and Moya, 1998
6	Multiple: face, heart, skeleton				Lopez-Rangel and Van Allen, 2004
7	Multiple: skull, eyes, skeleton				Briggs et al., 2005 (FDA case)
8–10	Cleft palate				Briggs et al., 2005 (FDA cases)
11	Limbs, digits		✔		Briggs et al., 2005 (FDA case)
12	Brain				Briggs et al., 2005 (FDA case)
13	Body wall				Briggs et al., 2005 (FDA case)
14	Ears (deafness)				Briggs et al., 2005 (FDA case)

range of doses but no congenital malformations. The highest dose levels in the two species exceeded the human dose level by 5- to 20-fold.

HUMANS

In the human, developmental toxicity was recorded in a number of studies, as tabulated in Table 1. However, in only six studies has a consistent pattern of multiple malformations been produced (cases 1 through 4, 6, 7). Defects observed in common in these cases included brachycephaly and abnormal calvarial development, abnormal facies, cleft palate, bowing of femurs and thinning of ribs and long bones, arthrogryposis, and congenital heart disease, as summarized by others (Friedman and Polifka, 2000). The malformations resembled those described in Antley-Bixler syndrome, an autosomal recessive disease (Briggs et al., 2005). The remaining cases tabulated (5, 8 to 14) had no consistent pattern and are not considered fluconazole-related. In the six positive cases, drug treatment extended over the first trimester, ranging from prior to conception and throughout pregnancy; the shortest interval was somewhat longer than 5 weeks. The doses eliciting the malformations in common were 400 mg/day (cases 1, 3, and 6), 800 mg/day (cases 2 and 7), and 800 to 1200 mg/day (case 4). The usual therapeutic doses of fluconazole recommended range from 150 to 800 mg/day depending upon the seriousness of the infection. The effective doses were thus at the upper levels of the usual dose range or were supratherapeutic. In contrast to these positive reports, a number of studies comprising over 900 pregnancies found no increase in congenital malformations or a specific syndrome of defects, as described here, from fluconazole treatment in the first trimester (Inman et al., 1994; Mastroiacovo et al., 1996; Kremery et al., 1996; Campomori and Bonati, 1997; Wilton et al., 1998; Jick, 1999; Sorensen et al., 1999). However, doses recorded in these studies, with two exceptions, were in the low range of 50 to 150 mg/day. Two reports (Kremery et al., 1996; Campomori and Bonati, 1997) cited higher doses, in the 600 to 1000 mg/day range, but without apparent fluconazole-induced malformations. King et al. (1998) came to the conclusion that fluconazole was not an active teratogen in the human. However, the consensus among several reviewers that the rare pattern of an identifiable phenotype in infants of mothers treated during pregnancy, including during the critical first trimester, at doses of 400 mg/day or greater strongly suggests a causal relationship to the drug (Friedman and Polifka, 2000; Schardein, 2000; Briggs et al., 2005). Growth retardation and postnatal death occurred in two of the six affected cases and therefore cannot be excluded as insignificant features. Several published reviews on fluconazole treatment during pregnancy have appeared (Wiesinger et al., 1996; King et al., 1998).

CHEMISTRY

Fluconazole is an average-sized polar molecule. It is slightly hydrophilic and can participate in hydrogen bonding primarily as an acceptor. The calculated physicochemical and topological properties are shown in the following.

PHYSICOCHEMICAL PROPERTIES

Parameter	Value
Molecular weight	306.275 g/mol
Molecular volume	239.95 A^3
Density	1.331 g/cm^3
Surface area	287.65 A^2
LogP	−0.131
HLB	13.833
Solubility parameter	31.088 $J^{(0.5)}/cm^{(1.5)}$
Dispersion	24.149 $J^{(0.5)}/cm^{(1.5)}$
Polarity	11.868 $J^{(0.5)}/cm^{(1.5)}$
Hydrogen bonding	15.570 $J^{(0.5)}/cm^{(1.5)}$
H bond acceptor	1.00
H bond donor	0.37
Percent hydrophilic surface	66.49
MR	77.230
Water solubility	−1.645 log (mol/M^3)
Hydrophilic surface area	191.26 A^2
Polar surface area	81.65 A^2
HOMO	−9.921 eV
LUMO	−0.776 eV
Dipole	4.114 debye

TOPOLOGICAL PROPERTIES (UNITLESS)

Parameter	Value
x0	15.579
x1	10.566
x2	9.865
xp3	7.786
xp4	6.835
xp5	4.082
xp6	3.179
xp7	1.968
xp8	1.292
xp9	0.816
xp10	0.431
xv0	11.342
xv1	6.395

Continued.

Parameter	Value
xv2	4.842
xvp3	3.213
xvp4	2.252
xvp5	1.272
xvp6	0.758
xvp7	0.407
xvp8	0.208
xvp9	0.100
xvp10	0.040
k0	25.921
k1	16.844
k2	7.266
k3	4.110
ka1	14.572
ka2	5.794
ka3	3.135

REFERENCES

Aleck, K. A. and Bartley, D. L. (1997). Multiple malformation syndrome following fluconazole use in pregnancy: Report of an additional patient. *Am. J. Med. Genet.* 72: 2253–2256.

Briggs, G. G., Freeman, R. K., and Yaffe, S. J. (2005). *Drugs in Pregnancy and Lactation. A Reference Guide to Fetal and Neonatal Risk*, Seventh ed., Lippincott Williams & Wilkins, Philadelphia.

Campomori, A. and Bonati, M. (1997). Fluconazole treatment for vulvo-vaginal candidiasis during pregnancy. *Ann. Pharmacother.* 31: 118–119.

Friedman, J. M. and Polifka, J. E. (2000). *Teratogenic Effects of Drugs. A Resource for Clinicians (TERIS)*, Second ed., Johns Hopkins University Press, Baltimore, MD.

Inman, W., Pearce, G., and Wilton, L. (1994). Safety of fluconazole in the treatment of vaginal candidiasis. A prescription-event monitoring study, with special reference to the outcome of pregnancy. *Eur. J. Clin. Pharmacol.* 46: 115–118.

Jick, S. S. (1999). Pregnancy outcomes after maternal exposure to fluconazole. *Pharmacotherapy* 19: 221–222.

King, C. T. et al. (1998). Antifungal therapy during pregnancy. *Clin. Infect. Dis.* 27: 1151–1160.

Kremery, V., Huttova, M., and Masar, O. (1996). Teratogenicity of fluconazole. *Pediatr. Infect. Dis.* 15: 841.

Lacy, C. F. et al. (2004). *Drug Information Handbook (Pocket), 2004–2005*, Lexi-Comp, Inc., Hudson, OH.

Lee, B. E. et al. (1992). Congenital malformations in an infant born to a woman treated with fluconazole. *Pediatr. Infect. Dis.* 11: 1062–1064.

Lopez-Rangel, E. and Van Allen, M. I. (2004). Prenatal exposure to fluconazole. A identifiable dysmorphic phenotype. *Birth Defects Res. (A)* 70: 261.

Mastroiacovo, P. et al. (1996). Prospective assessment of pregnancy outcomes after first trimester exposure to fluconazole. *Am. J. Obstet. Gynecol.* 175: 1645–1650.

PDR® (*Physicians' Desk Reference®*). (2005). Medical Economics Co., Inc., Montvale, NJ.

Pursley, T. J. et al. (1996). Fluconazole-induced congenital anomalies in three infants. *Clin. Infect. Dis.* 22: 336–340.

Sanchez, J. M. and Moya, G. (1998). Fluconazole teratogenicity. *Prenat. Diagn.* 18: 862–863.

Schardein, J. L. (2000). *Chemically Induced Birth Defects*, Third ed., Marcel Dekker, New York, pp. 399–400.

Sorenson, H. T. et al. (1999). Risk of malformations and other outcomes in children exposed to fluconazole *in utero. Br. J. Clin. Pharmacol.* 48: 234–238.

Wiesinger, E. E. C. et al. (1996). Fluconazole in *Candida albicans* sepsis during pregnancy: Case report and review of the literature. *Infection* 24: 263–266.

Wilton, L. V. et al. (1998). The outcomes of pregnancy in women exposed to newly marketed drugs in general practice in England. *Br. J. Obstet. Gynaecol.* 105: 882–889.

28 Ergotamine

Chemical name: 12′-Hydroxy-2′-methyl-5′α-(phenylmethyl)ergotaman-3′,6′-18-trione

CAS #: 113-15-5

SMILES: [nH]1cc2CC3C(c4c2c1ccc4)=CC(CN3C)C(NC5(C(N6C(O5)
(C7N(C(C6Cc8ccccc8)=O)CCC7)O)=O)C)=O

INTRODUCTION

Ergotamine is an ergot alkaloid occurring naturally in the plant *Claviceps purpurea*. It has vaso-constrictive action and is used therapeutically in the treatment of vascular headaches, such as migraine. Ergotamine is distinguished from the other three main classes of ergot alkaloids, all of which also have strong oxytocic activity, and some which also have therapeutic value. Mechanistically, ergotamine has partial agonist and/or antagonist activity against tryptaminergic, dopaminergic, and α-adrenergic receptors, producing depression of central vasomotor centers (Lacy et al., 2004). It crosses the placenta. It is available as a prescription drug under the trade names Ergomar® and Wigraine®, among other names. Ergotamine occurs in a combined form with caffeine as Cafergot®. The drug has a pregnancy category of X. This contraindication presumably is due to its oxytocic properties in constricting the uterine vessels and/or increasing myometrial tone leading to reduced placental blood flow and its attendant toxicity. It is stated that this may contribute to fetal growth retardation (observed in animals; see *PDR*, 2002).

TABLE 1
Developmental Toxicity Profile of Ergotamine in Humans

Case Number	Malformations	Growth Retardation	Death	Functional Deficit	Ref.
1	Heart				Anon., 1971
2	Multiple: abdomen, renal, gastrointestinal, genital		✔		Peeden et al., 1979
3	Multiple: brain, skeleton, digits, muscle (cerebro-arthrodigital syndrome)	✔			Spranger et al., 1980
4	Gastrointestinal	✔	✔		Graham et al., 1983
5	None	✔			Graham et al., 1983
6–9	None		✔		Graham et al., 1983
10	None		✔		Au et al., 1985
11	Multiple: brain, limbs, skeleton	✔		✔	Hughes and Goldstein, 1988
12–15	Brain				Czeizel, 1989
16	Limbs			✔	Verloes et al., 1990
17	(Fetal stress syndrome)	✔			de Groot et al., 1993
18	Brain				Barkovich et al., 1995
19	Mobius sequence				Smets et al., 2004
20–28	Various (heart/digits, genital cited)				Briggs et al., 2005

DEVELOPMENTAL TOXICOLOGY

ANIMALS

Ergotamine given orally to laboratory animals has not shown teratogenic potential. Doses of up to 300 mg/kg/day given to mice during organogenesis caused a reduction in fetal weight and retarded ossification (Grauwiler and Schon, 1973). Lower doses of up to 100 mg/kg/day given to rats during organogenesis caused the same fetal toxicity as in mice, and increased mortality (Grauwiler and Schon, 1973). In rabbits given up to 30 mg/kg/day during organogenesis, no developmental toxicity was elicited (Grauwiler and Schon, 1973). The drug was maternally toxic in all three species.

HUMANS

In the human, a variety of cases recording developmental toxicity apparently resultant from treatment with ergotamine were published (Table 1). Of the 28 cases described in the literature, at least 22 depicted a diversity of congenital defects, consistent with a disruptive vascular mechanism, due presumably to the known vasospasmic action of the drug (Raymond, 1995). Higher doses of the drug than the usual recommended doses of two to six 1- to 2-mg tablets/day and idiosyncratic response are other factors related to this teratogenic activity (Briggs et al., 2005). The combination of ergotamine with other drugs, particularly caffeine, as Cafergot, may enhance the described effect. Treatment during pregnancy was apparently not confined to a characteristic time frame. In the cited cases, retarded fetal growth was observed in five cases, death in seven, and functional deficits in two (paraplegia in both); thus, these classes must also be considered as being contributory to the developmental toxicity profile of ergotamine. In contrast, in two studies, researchers found no increased incidence of congenital malformations and concluded that ergotamine was probably not teratogenic (Heinonen et al., 1977; Wainscott et al., 1978). One group of experts determined the magnitude of teratogenic risk to be minimal (Friedman and Polifka, 2000). Several reviews were published on ergotamine and its developmental toxicity potential (deGroot et al., 1993; Raymond, 1995).

CHEMISTRY

Ergotamine is one of the largest human developmental toxicants. This polar compound is slightly hydrophobic. It is capable of participating in hydrogen bonding. The calculated physicochemical and topological properties of ergotamine are as follows.

PHYSICOCHEMICAL PROPERTIES

Parameter	Value
Molecular weight	581.672 g/mol
Molecular volume	502.80 A^3
Density	1.234 g/cm^3
Surface area	577.34 A^2
LogP	0.511
HLB	11.806
Solubility parameter	25.956 J$^{(0.5)}$/cm$^{(1.5)}$
Dispersion	22.581 J$^{(0.5)}$/cm$^{(1.5)}$
Polarity	5.406 J$^{(0.5)}$/cm$^{(1.5)}$
Hydrogen bonding	11.601 J$^{(0.5)}$/cm$^{(1.5)}$
H bond acceptor	2.01
H bond donor	0.83
Percent hydrophilic surface	57.68
MR	161.020
Water solubility	−4.704 log (mol/M^3)
Hydrophilic surface area	332.99 A^2
Polar surface area	127.69 A^2
HOMO	−8.028 eV
LUMO	−0.317 eV
Dipole	6.005 debye

TOPOLOGICAL PROPERTIES (UNITLESS)

Parameter	Value
x0	29.673
x1	20.676
x2	20.338
xp3	18.595
xp4	17.028
xp5	14.492
xp6	10.745
xp7	8.578
xp8	6.681
xp9	4.770
xp10	3.377
xv0	24.247
xv1	15.118
xv2	12.784
xvp3	10.132
xvp4	8.123
xvp5	6.020
xvp6	4.011
xvp7	2.820

Continued.

Parameter	Value
xvp8	1.902
xvp9	1.184
xvp10	0.719
k0	69.035
k1	30.341
k2	11.313
k3	4.826
ka1	27.217
ka2	9.573
ka3	3.950

REFERENCES

Anon. (1971). New Zealand Committee on Adverse Drug Reactions. Sixth Annual Report. *N. Z. Med. J.* 74: 184–191.

Au, K. L., Woo, J. S. K., and Wong, V. C. W. (1985). Intrauterine death from ergotamine overdosage. *Eur. J. Obstet. Gynecol. Reprod. Biol.* 19: 313–315.

Barkovich, A. J., Rowley, H., and Bollen, A. (1995). Correlation of prenatal events with the development of polymicrogyria. *Am. J. Neuroradiol.* 16: 822–827.

Briggs, G. G., Freeman, R. K., and Yaffe, S. J. (2005). *Drugs in Pregnancy and Lactation. A Reference Guide to Fetal and Neonatal Risk*, Seventh ed., Lippincott Williams & Wilkins, Philadelphia.

Czeizel, A. (1989). Teratogenicity of ergotamine. *J. Med. Genet.* 26: 69–71.

de Groot, A. N. J. A. et al. (1993). Ergotamine-induced fetal stress: Review of side effects of ergot alkaloids during pregnancy. *Eur. J. Obstet. Gynecol. Reprod. Biol.* 51: 73–77.

Friedman, J. M. and Polifka, J. E. (2000). *Teratogenic Effects of Drugs. A Resource for Clinicians (TERIS)*, Second ed., Johns Hopkins University Press, Baltimore, MD.

Graham, J. M., Marin-Padilla, M., and Hoefnagel, D. (1983). Jejunal atresia associated with Cafergot ingestion during pregnancy. *Clin. Pediatr. (Phila.)* 22: 226–228.

Grauwiler, J. and Schon, H. (1973). Teratological experiments with ergotamine in mice, rats, and rabbits. *Teratology* 7: 227–236.

Heinonen, O. P., Slone, D., and Shapiro, S. (1977). *Birth Defects and Drugs in Pregnancy*, Publishing Sciences Group, Littleton, MA.

Hughes, H. E. and Goldstein, D. A. (1988). Birth defects following maternal exposure to ergotamine, beta blockers, and caffeine. *J. Med. Genet.* 25: 396–399.

Lacy, C. F. et al. (2004). *Drug Information Handbook (Pocket), 2004–2005*, Lexi-Comp, Inc., Hudson, OH.

PDR® (Physicians' Desk Reference®). (2002). Medical Economics Co., Inc., Montvale, NJ.

Peeden, J. N., Wilroy, R. S., and Soper, R. G. (1979). Prune perineum. *Teratology* 20: 233–236.

Raymond, G. V. (1995). Teratogen update: Ergot and ergotamine. *Teratology* 51: 344–347.

Smets, K., Zecic, A., and Willems, J. (2004). Ergotamine as a possible cause of Mobius sequence: Additional clinical observation. *J. Child Neurol.* 19: 398.

Spranger, J. W. et al. (1980). Cerebroarthrodigital syndrome: A newly recognized formal genesis syndrome in three patients with apparent arthromyodysplasia and sacral agenesis, brain malformation, and digital hypoplasia. *Am. J. Med. Genet.* 5: 13–24.

Verloes, A. et al. (1990). Paraplegia and arthrogryposis multiplex of the lower extremities after intrauterine exposure to ergotamine. *J. Med. Genet.* 27: 213–214.

Wainscott, G. et al. (1978). The outcome of pregnancy in women suffering from migraine. *Postgrad. Med. J.* 54: 98–102.

29 Propylthiouracil

Chemical name: 6-Propyl-2-thiouracil, 2,3-Dihydro-6-propyl-2-thioxo-4(1-*H*)pyrimidinone

Alternate name: PTU

CAS # 51-52-5

SMILES: C1(=CC(NC(N1)=S)=O)CCC

INTRODUCTION

Propylthiouracil is a thiocarbamide chemical derivative used in the palliative treatment of hyperthyroidism and in the management of thyrotoxic crises. Hyperthyroidism is said to complicate some 3% of pregnancies (Burrow, 1985), and the drug acts against this condition by inhibiting the synthesis of thyroid hormones by blocking the oxidation of iodine in the thyroid gland (Lacy et al., 2004). It is known by its generic name, as well as by the trade name Propyl-Thyracil®, among other names. Propylthiouracil has a pregnancy category of D. This is because of the warning on the package label that states that the drug can cause fetal harm when administered to a pregnant woman. Because the drug readily crosses placental membranes, it can induce goiter and even cretinism in the developing fetus. If the drug is used during pregnancy, or if the patient becomes pregnant while taking the drug, the patient should be warned of the potential hazard to the fetus (*PDR*, 2005; see below).

DEVELOPMENTAL TOXICOLOGY

ANIMALS

In a number of species (tests were conducted many years ago), propylthiouracil administered orally to laboratory animals caused thyroid lesions, including enlargement. The lesions were induced in rats (Jost, 1957a, 1957b), guinea pigs (Webster and Young, 1948), and rabbits (Krementz et al., 1957). In mice, loss of hearing but no thyroid abnormalities was produced from prenatal treatment (Deol, 1973).

TABLE 1
Reports of Thyroid Alterations Attributed to Propylthiouracil in Humans

Astwood and VanderLaan, 1946	Mestman et al., 1974
Bain, 1947	Refetoff et al., 1974
French and Van Wyck, 1947	Worley and Crosby, 1974
Lahey and Bartels, 1947	Mujtaba and Burrow, 1975
Reveno, 1948	Hayek and Brooks, 1975
Eisenberg, 1950	Ibbertson et al., 1975
Seligman and Pescovitz, 1950	Serup and Petersen, 1977
Astwood, 1951	Burrow et al., 1978
Hepner, 1952	Serup, 1978
Saye et al., 1952	Sugrue and Drury, 1980
Pearlman, 1954	Weiner et al., 1980
Aaron et al., 1955	Cheron et al., 1981
Waldinger et al., 1955	Check et al., 1982
Bongiovanni et al., 1956	Kock and Merkus, 1983
Branch and Tuthill, 1957	Ramsay et al., 1983
Man et al., 1958	Burrow, 1985
Becker and Sudduth, 1959	Becks and Burrow, 1991
Greenman et al., 1962	Belfar et al., 1991
Herbst and Selenkow, 1963	Wing et al., 1994
Reveno and Rosenbaum, 1964	Soliman et al., 1994
Burrow, 1965	van Loon et al., 1995
Herbst and Selenkow, 1965	Nicolini et al., 1996
Martin and Matus, 1966	Bruner and Dellinger, 1997
Burrow et al., 1968	Momotani et al., 1997
Hollingsworth and Austin, 1969	Ochoa-Maya et al., 1999
Hollingsworth and Austin, 1971	Gallagher et al., 2001
Ayromlooi, 1972	

HUMANS

In humans, propylthiouracil treatment during pregnancy causes suppression of thyroid function in the fetus, and goiter may result as compensation for the induced hypothyroidism. This constitutes a functional deficit with respect to class of developmental toxicity affected. One review estimated that up to 12% of infants born to women treated with the drug during pregnancy develop transient neonatal hypothyroidism (Briggs et al., 2005). Over 60 cases are tabulated in reports presented in Table 1 as being representative of the resultant thyroid findings due to increased levels of fetal pituitary thyrotropin (Refetoff et al., 1974). Many cases tabulated in Table 1 resulted from combination therapy with other antithyroid medication, especially iodides. They include small fetal goiters and, occasionally, cretinism. No airway obstruction is usually observed. The defects are concordant with the induced ones in animals, lending credence to their significance.

The previous reports published generally presented no adverse effects on growth and development, including subsequent intellectual development (Seligman and Peskovitz, 1950; Burrow et al., 1968, 1978; Eisenstein et al., 1992). Abortion and death were infrequent findings. Malformations of nonthyroid organ systems were infrequently reported and are not significant as a class of developmental toxicity to be included here.

The timetable of treatment resulting in thyroid effects ranged throughout gestation. However, treatment does not result in fetal goiter until the 11th or 12th week due to lack of hormone production earlier (Burr, 1981). In most instances, the enlarged thyroid gland regresses spontaneously in the postnatal period. Dosages producing the alterations were within the range of therapeutic doses of 150 to 900 mg/day orally.

One group of experts places the risk of goiter induction as small to moderate (Friedman and Polifka, 2000). Several reviews pertinent to this discussion were published (Hepner, 1952; Klevit, 1969; Burr, 1981; Davis et al., 1989; Becks and Burrow, 1991; Diav-Citrin and Ornoy, 2002).

CHEMISTRY

Propylthiouracil is a hydrophobic compound that is smaller than the other human developmental toxicants within this compilation. It is of average polarity. Propylthiouracil can participate as both a donor and acceptor during hydrogen bonding interactions. The calculated physicochemical and topological properties are shown below.

PHYSICOCHEMICAL PROPERTIES

Parameter	Value
Molecular weight	170.235 g/mol
Molecular volume	144.53 A^3
Density	1.532 g/cm^3
Surface area	184.06 A^2
LogP	2.679
HLB	12.546
Solubility parameter	25.268 $J^{(0.5)}/cm^{(1.5)}$
Dispersion	22.102 $J^{(0.5)}/cm^{(1.5)}$
Polarity	8.249 $J^{(0.5)}/cm^{(1.5)}$
Hydrogen bonding	9.049 $J^{(0.5)}/cm^{(1.5)}$
H bond acceptor	0.76
H bond donor	0.53
Percent hydrophilic surface	60.90
MR	48.926
Water solubility	1.990 log (mol/M^3)
Hydrophilic surface area	112.08 A^2
Polar surface area	51.81 A^2
HOMO	−9.087 eV
LUMO	−0.911 eV
Dipole	5.835 debye

TOPOLOGICAL PROPERTIES (UNITLESS)

Parameter	Value
x0	8.268
x1	5.220
x2	4.572
xp3	2.973
xp4	2.758
xp5	1.567
xp6	1.099
xp7	0.260
xp8	0.164
xp9	0.000

Continued.

Parameter	Value
xp10	0.000
xv0	7.125
xv1	3.954
xv2	2.771
xvp3	1.540
xvp4	1.155
xvp5	0.580
xvp6	0.350
xvp7	0.080
xvp8	0.046
xvp9	0.000
xvp10	0.000
k0	11.455
k1	9.091
k2	4.133
k3	2.844
ka1	8.516
ka2	3.708
ka3	2.494

REFERENCES

Aaron, H. H., Schneierson, S. J., and Siegel, E. (1955). Goiter in newborn infant due to mothers ingestion of propylthiouracil. *JAMA* 159: 848–850.

Astwood, E. B. (1951). The use of antithyroid drugs during pregnancy. *J. Clin. Endocrinol. Metab.* 11: 1045–1056.

Astwood, E. B. and VanderLaan, W. P. (1946). Treatment of hyperthyroidism with propylthiouracil. *Ann. Intern. Med.* 25: 813–821.

Ayromlooi, J. (1972). Congenital goiter due to maternal ingestion of iodides. *Obstet. Gynecol.* 39: 818–822.

Bain, L. (1947). Propylthiouracil in pregnancy: Report of a case. *South. Med. J.* 40: 1020–1021.

Becker, W. F. and Sudduth, P. G. (1959). Hyperthyroidism and pregnancy. *Ann. Surg.* 149: 867–872.

Becks, G. P. and Burrow, G. N. (1991). Thyroid disease and pregnancy. *Med. Clin. North Am.* 75: 121–150.

Belfar, H. L. et al. (1991). Sonographic findings in maternal hyperthyroidism. Fetal hyperthyroidism/fetal goiter. *J. Ultrasound Med.* 10: 281–284.

Bongiovanni, A. M. et al. (1956). Sporadic goiter of the newborn. *J. Clin. Endocrinol. Metab.* 16: 146–152.

Branch, L. K. and Tuthill, S. W. (1957). Goiters in twins resulting from propylthiouracil given during pregnancy. *Ann. Intern. Med.* 46: 145–148.

Briggs, G. G., Freeman, R. K., and Yaffe, S. J. (2005). *Drugs in Pregnancy and Lactation. A Reference Guide to Fetal and Neonatal Risk*, Seventh ed., Lippincott Williams & Wilkins, Philadelphia.

Bruner, J. P. and Dellinger, E. H. (1997). Antenatal diagnosis and treatment of fetal hypothyroidism: A report of two cases. *Fetal Diagn. Ther.* 12: 200–204.

Burr, W. A. (1981). Thyroid disease. *Clin. Obstet. Gynecol.* 8: 341–351.

Burrow, G. N. (1965). Neonatal goiter after maternal propylthiouracil therapy. *J. Clin. Endocrinol. Metab.* 25: 403–408.

Burrow, G. N. (1985). The management of thyrotoxicosis in pregnancy. *N. Engl. J. Med.* 313: 562–565.

Burrow, G. N. et al. (1968). Children exposed *in utero* to propylthiouracil. Subsequent intellectual and physical development. *Am. J. Dis. Child.* 116: 161–165.

Burrow, G. N., Klatskin, E. H., and Genel, M. (1978). Intellectual development in children whose mothers received propylthiouracil during pregnancy. *Yale J. Biol. Med.* 51: 151–156.

Check, J. H. et al. (1982). Prenatal treatment of thyrotoxicosis to prevent intrauterine growth retardation. *Obstet. Gynecol.* 60: 122–124.

Cheron, R. G. et al. (1981). Neonatal thyroid function after propylthiouracil therapy for maternal Graves disease. *N. Engl. J. Med.* 304: 525–528.

Davis, L. E. et al. (1989). Thyrotoxicosis complicating pregnancy. *Am. J. Obstet. Gynecol.* 160: 63–70.

Deol, M. S. (1973). Congenital deafness and hypothyroidism. *Lancet* 2: 105–106.

Diav-Citrin, O. and Ornoy, A. (2002). Teratogen update: Antithyroid drugs — methimazole, carbimazole, and propylthiouracil. *Teratology* 65: 38–44.

Eisenberg, L. (1950). Thyrotoxicosis complicating pregnancy. *NY State J. Med.* 50: 1618–1619.

Eisenstein, Z. et al. (1992). Intellectual capacity of subjects exposed to methimazole or propylthiouracil *in utero. Eur. J. Pediatr.* 151: 558–559.

French, F. S. and Van Wyck, J. J. (1947). Fetal hypothyroidism. *J. Pediatr.* 64: 589–600.

Friedman, J. M and Polifka, J. E. (2000). *Teratogenic Effects of Drugs. A Resource for Clinicians (TERIS)*, Second ed., Johns Hopkins University Press, Baltimore, MD.

Gallagher, M. P. et al. (2001). Neonatal thyroid enlargement associated with propylthiouracil therapy of Graves' disease during pregnancy. *J. Pediatr.* 139: 896–900.

Greenman, G. W. et al. (1962). Thyroid dysfunction in pregnancy. *N. Engl. J. Med.* 267: 426–431.

Hayek, A. and Brooks, M. (1975). Neonatal hyperthyroidism following intrauterine hypothyroidism. *J. Pediatr.* 87: 446–448.

Hepner, W. R. (1952). Thiourea derivatives and the fetus. A review and report of a case. *Am. J. Obstet. Gynecol.* 63: 869–874.

Herbst, A. L. and Selenkow, H. A. (1963). Combined antithyroid–thyroid therapy of hyperthyroidism in pregnancy. *Obstet. Gynecol.* 21: 543–550.

Herbst, A. L. and Selenkow, H. A. (1965). Hyperthyroidism during pregnancy. *N. Engl. J. Med.* 273: 627–633.

Hollingsworth, D. R. and Austin, E. (1969). Observations following I131 for Graves disease during first trimester of pregnancy. *South. Med. J.* 62: 1555–1556.

Hollingsworth, D. R. and Austin, E. (1971). Thyroxine derivatives in amniotic fluid. *J. Pediatr.* 79: 923–929.

Ibbertson, H. K., Seddon, R. J., and Craxson, M. S. (1975). Fetal hypothyroidism complicating medical treatment of thyrotoxicosis in pregnancy. *Clin. Endocrinol.* 4: 521–523.

Jost, A. (1957a). Action du propylthiouracile sur la thyroide de foetus de rat intacts ou decapites. *C. R. Soc. Biol. (Paris)* 151: 1295–1298.

Jost, A. (1957b). Le probleme des interrelations thyreo-hypophysaires chez le foetus et l'action du propylthiouracile sur la thyroide foetale du rat. *Rev. Suisse Zool.* 64: 821–834.

Klevit, H. D. (1969). Iatrogenic thyroid disease. In *Endocrine and Genetic Diseases of Childhood*, L. I. Gardner, Ed., W.B. Saunders, Philadelphia, pp. 243–252.

Kock, H. C. L. V. and Merkus, J. M. W. M. (1983). Graves' disease during pregnancy. *Eur. J. Obstet. Gynecol. Reprod. Biol.* 14: 323–330.

Krementz, E. T., Hooper, R. G., and Kempson, R. L. (1957). The effect on the rabbit fetus of the maternal administration of propylthiouracil. *Surgery* 41: 619–631.

Lacy, C. F. et al. (2004). *Drug Information Handbook (Pocket), 2004–2005*, Lexi-Comp, Inc., Hudson, OH.

Lahey, F. H. and Bartels, E. C. (1947). The use of thiouracil, thiobarbital and propylthiouracil in patients with hyperthyroidism. *Ann. Surg.* 125: 572–581.

Man, E. B., Shaver, B. A., and Cooke, R. E. (1958). Studies of children born to women with thyroid disease. *Am. J. Obstet. Gynecol.* 75: 728–741.

Martin, M. M. and Matus, R. N. (1966). Neonatal exophthalmos with maternal thyrotoxicosis. *Am. J. Dis. Child.* 111: 545–547.

Mestman, J. H., Manning, P. R., and Hodgman, J. (1974). Hyperthyroidism and pregnancy. *Arch. Intern. Med.* 134: 434–439.

Momotani, N. et al. (1997). Effects of propylthiouracil and methimazole on fetal thyroid status in mothers with Graves' disease. *J. Clin. Endocrinol. Metab.* 82: 3633–3636.

Mujtaba, Q. and Burrow, G. N. (1975). Treatment of hyperthyroidism in pregnancy with propylthiouracil and methimazole. *Obstet. Gynecol.* 46: 282–286.

Nicolini, U. et al. (1996). Prenatal treatment of fetal hypothyroidism: Is there more than one option? *Prenat. Diagn.* 16: 443–448.

Ochoa-Maya, M. R. et al. (1999). Resolution of fetal goiter after discontinuation of propylthiouracil in a pregnant woman with Graves' hyperthyroidism. *Thyroid* 9: 1111–1114.

PDR® (*Physicians' Desk Reference®*). (2005). Medical Economics Co., Inc., Montvale, NJ.

Pearlman, L. N. (1954). Goitre in a premature infant. *Can. Med. Assoc. J.* 70: 317–319.

Ramsay, I., Kaur, S., and Krassas, G. (1983). Thyrotoxicosis in pregnancy: Results of treatment by antithyroid drugs combined with T4. *Clin. Endocrinol.* 18: 73–85.

Refetoff, S. et al. (1974). Neonatal hypothyroidism and goiter of each of two sets of twins due to maternal therapy with antithyroid drugs. *J. Pediatr.* 85: 240–244.

Reveno, W. S. (1948). Propylthiouracil in the treatment of toxic goiter. *J. Clin. Endocrinol. Metab.* 8: 866–874.

Reveno, W. S. and Rosenbaum, H. (1964). Observations on the use of antithyroid drugs. *Ann. Intern. Med.* 60: 982–989.

Saye, E. B. et al. (1952). Congenital thyroid hyperplasia in twins. Report of a case following administration of thiouracil and iodine to mother during pregnancy. *JAMA* 149: 1399.

Seligman, B. and Pescovitz, H. (1950). Suffocative goiter in newborn infant. *NY State J. Med.* 50: 1845–1847.

Serup, J. (1978). Maternal propylthiouracil to manage fetal hyperthyroidism. *Lancet* 2: 896.

Serup, J. and Petersen, S. (1977). Hyperthyroidism during pregnancy treated with propylthiouracil. The significance of maternal and foetal parameters. *Acta Obstet. Gynecol. Scand.* 56: 463–466.

Soliman, S. et al. (1994). Color Doppler imaging of the thyroid gland in a fetus with congenital goiter: A case report. *Am. J. Perinatol.* 11: 21–23.

Sugrue, D. and Drury, M. I. (1980). Hyperthyroidism complicating pregnancy: Results of treatment by antithyroid drugs in 77 pregnancies. *Br. J. Obstet. Gynaecol.* 87: 970–975.

van Loon, A. J. et al. (1995). *In utero* diagnosis and treatment of fetal hypothyroidism, caused by maternal use of propylthiouracil. *Prenat. Diagn.* 15: 599–604.

Waldinger, C., Wermer, O. S., and Sobel, E. H. (1955). Thyroid function in infant with congenital goiter resulting from exposure to propylthiouracil. *J. Am. Med. Wom. Assoc.* 10: 196–197.

Webster, R. C. and Young, W. C. (1948). Thiouracil treatment of female guinea pig: Effect on gestation and offspring. *Anat. Rec.* 101: 722–723.

Weiner, S. et al. (1980). Antenatal diagnosis and treatment of a fetal goiter. *J. Reprod. Med.* 24: 39–42.

Wing, D. A. et al. (1994). A comparison of propylthiouracil versus methimazole in the treatment of hyperthyroidism in pregnancy. *Am. J. Obstet. Gynecol.* 170: 90–95.

Worley, R. J. and Crosby, W. M. (1974). Hyperthyroidism during pregnancy. *Am. J. Obstet. Gynecol.* 119: 150–155.

30 Medroxyprogesterone

Chemical name: (6α)-17-Hydroxy-6-methylpregn-4-ene-3,20-dione

Alternate names: Acetoxymethylprogesterone, methylacetoxyprogesterone

CAS #: 520-85-4

SMILES: C12C3C(C(CC3)(C(C)=O)O)(CCC1C4(C(C(C(C2)C)=CC(CC4)=O)C)C

INTRODUCTION

Medroxyprogesterone is a derivative of progestogen and is used either orally or intramuscularly as a contraceptive and in treating endometrial or renal carcinoma as well as secondary amenorrhea or abnormal uterine bleeding due to hormonal imbalance. It inhibits secretion of pituitary gonadotropin, which prevents follicular maturation and ovulation (Lacy et al., 2004). It is available commercially under the trade name Provera®, Depo-Provera® (injectable), among other names, and when combined as an oral contraceptive with the estrogen ethinyl estradiol, as Provest®. It has a pregnancy category of X, due to the package label warning that states that "Several reports suggest an association between intrauterine exposure to progestational drugs in the first trimester of pregnancy and genital abnormalities in male and female fetuses. There are insufficient data to quantify the risk to exposed female fetuses, but because some of these drugs induce mild virilization of the external genitalia of the female fetus and because of the increased association of hypospadias in the male fetus, it is prudent to avoid the use of these drugs during the first trimester of pregnancy" (*PDR*, 2005). (See below for more information.)

DEVELOPMENTAL TOXICOLOGY

ANIMALS

Laboratory studies in animals have produced pseudohermaphroditism in rats and primates. In rats, parenteral (subcutaneous) doses of 0.25 to 5 mg (Revesz et al., 1960) or oral doses of 10 mg/kg (Kawashima et al., 1977) late in gestation were effective in masculinizing female fetuses. In two species of primates (cynomolgus and baboon), a parenteral (intramuscular) dose of 300 mg/kg for a single day or for 19 days during organogenesis elicited masculinization (abnormal external genitalia) of female offspring and hypospadias in male offspring, as well as increased mortality (Prahalada and Hendrickx, 1982, 1983). No pseudohermaphroditism was produced in rabbit fetuses from prenatal treatment of up to 30 mg/kg subcutaneously for 3, 6, or 9 consecutive days during the days 7 to 15 gestation window (Andrew and Staples, 1977). However, cleft palate, reduced fetal body weight, and increased mortality with no maternal toxicity were produced. Similar results were obtained in mice at much higher doses by the same investigators. As will be apparent later, the masculinization observed in the female offspring and the hypospadias in the male offspring are concordant with that described in humans (see below).

HUMANS

In the human, as stated on the package label, genital ambiguity (masculinization in females and feminization in males as hypospadias) was reported, as shown in Table 1. Three cases in females and seven cases in males were recorded early on, and no contemporary reports apparently have been published, with the exception of a recent publication reporting that intake of progestins (presumably including medroxyprogesterone, but no drugs were mentioned) was associated with increased hypospadias risk. The odds ratios ranged from 3.1 to 5 (95% confidence interval [CI] ranges 1.8 to 10.0) depending on when the drug was taken (Carmichael et al., 2004). According to an U.S. Food and Drug Administration (FDA) publication, only 14 cases of fetal ambiguous genitalia due to progestogens, including medroxyprogesterone, were known to them 25 years ago (Dayan and Rosa, 1981).

The genital lesions are identical to those produced by androgens. They were first discovered almost half a century ago (Jones, 1957; Wilkins et al., 1958) and were described in detail by others more recently (Keith and Berger, 1977; Schardein, 2000). Basically, in females, there is phallic enlargement (clitoral hypertrophy), with or without labioscrotal fusion, and the labia are usually enlarged. In some cases, masculinization may have progressed to the degree that labioscrotal fusion resulted in the formation of a urogenital sinus. There is usually a normal vulva, endoscopic evidence of a cervix, and a palpable, though sometimes infantile, uterus. In males, hypospadias (feminization, incomplete masculinization, or ambiguous genitalia) occurs anywhere from a subcoronal location to a site at the base of the penile shaft. It was proposed that the progestogen interferes with the fusion of the urethral fold, leading to hypospadias. In both females and males, the lesions correlate with the time of exposure and the dose of the progestogen.

These genital malformations have been produced prior to the 12th week (or later) at doses, where provided, within the 2.5 to 10 mg (oral) or 400 to 1000 mg/week (parenteral) therapeutic

TABLE 1
Reports of Ambiguous Genitalia in Infants Associated with Medroxyprogesterone in Humans

Ref.	Male	Female
Eichner, 1963		✔
Burstein and Wasserman, 1964	✔	✔
		✔[a]
Goldman and Bongiovanni, 1967	✔	
Aarskog, 1970, 1979	✔	
	✔	
	✔	
Harlap et al., 1975	✔	
	✔	

[a] Manufacturer's case cited.

levels, and are much lower than those recorded in animals. No adverse effects on pubertal development, sexual maturation or sexually dimorphic behavior, or intellectual development were found in several studies among both females and males who had been exposed *in utero* (Jaffe et al., 1988, 1989, 1990; Pardthaisong et al., 1992). While increased incidence of perinatal death and of low birth weight were reported from a cohort study of 1431 infants whose mothers were exposed to medroxyprogesterone around the time of conception (Gray and Pardthiasong, 1991; Pardthiasong and Gray, 1991), such adverse developmental toxicity has not been corroborated in other studies reported thereafter. In a number of studies, researchers have not found any fetal effects, including malformations other than genital, from medroxyprogesterone treatment in pregnancy (Rawlings, 1962; Schwallie and Assenzo, 1973; Nash, 1975; Heinonen et al., 1977; Resseguie, 1985; Yovich et al., 1988). With respect to nongenital malformations, the FDA had, beginning in late 1978, warned via the package label for this drug as well as for other progestins, of the use of these drugs in pregnancy. This restriction was lifted in 1999, removing warnings from the package inserts for nongenital malformations for all progestational agents (Brent, 2000).

One group of experts placed the magnitude of teratogenic risk of the drug for virilization of female genitalia to be minimal (Friedman and Polifka, 2000). While the risk was not mentioned by these authors for lesions in males, the data reviewed here suggest a similar if not greater risk. Several pertinent reviews on this subject were published (Nash, 1975; Keith and Berger, 1977; Schardein, 1980; Schwallie, 1981; Wilson and Brent, 1981).

CHEMISTRY

Medroxyprogesterone is a large hydrophobic molecule with average polarity. It can participate in hydrogen bonding primarily as a hydrogen bond acceptor. The calculated physicochemical and topological properties of medroxyprogesterone are listed below.

PHYSICOCHEMICAL PROPERTIES

Parameter	Value
Molecular weight	344.494 g/mol
Molecular volume	339.54 A^3
Density	0.950 g/cm^3
Surface area	423.62 A^2
LogP	3.750
HLB	2.107
Solubility parameter	21.161 $J^{(0.5)}/cm^{(1.5)}$
Dispersion	18.734 $J^{(0.5)}/cm^{(1.5)}$
Polarity	3.886 $J^{(0.5)}/cm^{(1.5)}$
Hydrogen bonding	9.041 $J^{(0.5)}/cm^{(1.5)}$
H bond acceptor	0.93
H bond donor	0.27
Percent hydrophilic surface	15.51
MR	99.542
Water solubility	−2.199 log (mol/M^3)
Hydrophilic surface area	65.70 A^2
Polar surface area	60.69 A^2
HOMO	−10.043 eV
LUMO	−0.121 eV
Dipole	3.525 debye

TOPOLOGICAL PROPERTIES (UNITLESS)

Parameter	Value
x0	18.198
x1	11.632
x2	12.149
xp3	11.237
xp4	9.185
xp5	7.520
xp6	5.770
xp7	4.304
xp8	3.290
xp9	2.343
xp10	1.503
xv0	16.100
xv1	10.116
xv2	9.908
xvp3	9.013
xvp4	7.579
xvp5	5.856
xvp6	4.538
xvp7	3.205
xvp8	2.373
xvp9	1.521
xvp10	0.916
k0	34.948
k1	18.367
k2	5.747
k3	2.304
ka1	17.454
ka2	5.280
ka3	2.079

REFERENCES

Aarskog, D. (1970). Clinical and cytogenetic studies in hypospadias. *Acta Paediatr. Scand. Suppl.* 203: 7–62.

Aarskog, D. (1979). Maternal progestins as a possible cause of hypospadias. *N. Engl. J. Med.* 300: 75–78.

Andrew, F. D. and Staples, R. E. (1977). Prenatal toxicity of medroxyprogesterone acetate in rabbits, rats and mice. *Teratology* 15: 25–32.

Brent, R. L. (2000). Nongenital malformations and exposure to progestational drugs during pregnancy: The final chapter of an erroneous allegation. *Teratology* 61: 449.

Burstein, R. and Wasserman, H. C. (1964). The effect of Provera on the fetus. *Obstet. Gynecol.* 23: 931–934.

Carmichael, S. L. et al. (2004). Hypospadias and maternal intake of progestins and oral contraceptives. *Birth Defects Res. (A)* 70: 255.

Dayan, E. and Rosa, F. W. (1981). Fetal ambiguous genitalia associated with sex hormone use early in pregnancy. FDA, Division of Drug Experience, *ADR Highlights* 1–14.

Eichner, E. (1963). Clinical uses of 17α-hydroxy-6α-methylprogesterone acetate in gynecologic and obstetric practice. *Am. J. Obstet. Gynecol.* 86: 171–176.

Friedman, J. M. and Polifka, J. E. (2000). *Teratogenic Effects of Drugs. A Resource for Clinicians (TERIS)*, Second ed., Johns Hopkins University Press, Baltimore, MD.

Goldman, A. S. and Bongiovanni, A. M. (1967). Induced genital anomalies. *Ann. NY Acad. Sci.* 142: 755–767.

Gray, R. H. and Pardthiasong, T. (1991). *In utero* exposure to steroid contraceptives and survival during infancy. *Am. J. Epidemiol.* 134: 804–811.

Harlap, S., Prywes, R., and Davies, A. M. (1975). Birth defects and oestrogens and progesterones in pregnancy. *Lancet* 1: 682–683.

Heinonen, O. P., Slone, D., and Shapiro, S. (1977). *Birth Defects and Drugs in Pregnancy*, Publishing Sciences Group, Littleton, MA.

Jaffe, B. et al. (1988). Long-term effects of MPA on human progeny: Intellectual development. *Contraception* 37: 607–619.

Jaffe, B. et al. (1989). Aggression, physical activity levels and sex role identity in teenagers exposed *in utero* to MPA. *Contraception* 40: 351–363.

Jaffe, B. et al. (1990). Health, growth and sexual development of teenagers exposed *in utero* to medroxyprogesterone acetate. *Paediatr. Perinat. Epidemiol.* 4: 184–195.

Jones, H. W. (1957). Female hermaphroditism without virilization. *Obstet. Gynecol. Surv.* 12: 433–460.

Kawashima, K. et al. (1977). Virilizing activities of various steroids in female rat fetuses. *Endocrinol. Jpn.* 24: 77–81.

Keith, L. and Berger, G. S. (1977). The relationship between congenital defects and the use of exogenous progestational contraceptive hormones during pregnancy: A 20-year review. *Int. J. Gynaecol. Obstet.* 15: 115–124.

Lacy, C. F. et al. (2004). *Drug Information Handbook (Pocket), 2004–2005*, Lexi-Comp, Inc., Hudson, OH.

Nash, H. A. (1975). Depo-Provera: A review. *Contraception* 12: 377–393.

Pardthiasong, T. and Gray, R. H. (1991). *In utero* exposure to steroid contraceptives and outcome of pregnancy. *Am. J. Epidemiol.* 134: 795–803.

Pardthiasong, T., Yenchit, C., and Gray, R. (1992). The long-term growth and development of children exposed to depo-provera during pregnancy or lactation. *Contraception* 45: 313–324.

PDR® (*Physicians' Desk Reference®*). (2005). Medical Economics Co., Inc., Montvale, NJ.

Prahalada, S. and Hendrickx, A. G. (1982). Teratogenicity of medroxyprogesterone acetate (MPA) in cynomolgus monkeys. *Teratology* 25: 67A–68A.

Prahalada, S. and Hendrickx, A. G. (1983). Effect of medroxyprogesterone acetate (MPA) on the fetal development in baboons. *Teratology* 27: 69A–70A.

Rawlings, W. J. (1962). Progestogens and the foetus. *Br. Med. J.* 1: 336–337.

Resseguie, L. J. (1985). Congenital malformations among offspring exposed *in utero* to progestins, Olmstead Co., Minnesota, 1936–1974. *Fertil. Steril.* 43: 514–519.

Revesz, C., Chappel, C. I., and Gaudry, R. (1960). Masculinization of female fetuses in the rat by progestational compounds. *Endocrinology* 66: 140–144.

Schardein, J. L. (1980). Congenital abnormalities and hormones during pregnancy: A clinical review. *Teratology* 22: 251–270.

Schardein, J. L. (2000). *Chemically Induced Birth Defects*, Third ed., Marcel Dekker, NY, pp. 286–289, 298–299.

Schwallie, P. C. (1981). The effect of depot-medroxyprogesterone acetate on the fetus and nursing infant: A review. *Contraception* 23: 375–386.

Schwallie, P. C. and Assenzo, J. R. (1973). Contraceptive use — efficacy study utilizing Depo-Provera administered as an intramuscular injection once every 90 days. *Fertil. Steril.* 24: 331–339.

Wilkins, L. et al. (1958). Masculinization of female fetus associated with administration of oral and intramuscular progestins during gestation: Nonadrenal pseudohermaphroditism. *J. Clin. Endocrinol. Metab.* 18: 559–585.

Wilson, J. G. and Brent, R. L. (1981). Are female sex hormones teratogenic? *Am. J. Obstet. Gynecol.* 141: 567–580.

Yovich, J. L., Turner, S. R., and Draper, R. (1988). Medroxyprogesterone acetate therapy in early pregnancy has no apparent fetal effects. *Teratology* 38: 135–144.

31 Cocaine

Chemical name: Benzoylmethylecgonine, [1*R*-(*exo,exo*)]-3-(Benzoyloxy)-
8-methyl-8-azabicyclo[3.2.1]octane-2-carboxylic acid methyl ester

Alternate names: Benzoylmethylecgonine, ecgonine methyl ester, (street jargon — see below)

CAS #: 50-36-2

SMILES: C1(C2N(C(CC1OC(c3ccccc3)=O)CC2)C)C(OC)=O

INTRODUCTION

Cocaine is an alkaloid derived from the coca plant, *Erythroxylum coca*, indigenous to Peru and Bolivia, its properties recognized for at least 5000 years. It is limited therapeutically as a local (topical) anesthetic, but abuse of the chemical is a much greater concern based on its recreational use as a central nervous system stimulant providing euphoria. One estimate was that close to 11% of the U.S. population in 1988 were regular cocaine users, and 2 to 3% were believed to use the drug during pregnancy (Lindenberg et al., 1991). There is no evidence that these statistics have improved today. In fact, a national agency estimated that 1% of infants born and up to 4% of selected populations in the United States in 1995 were exposed to cocaine *in utero* (National Pregnancy and Health Survey, 1995). Mothers at highest risk are Black, single, separated or divorced, and have less than a secondary school education (Streissguth et al., 1991). The cost to society of this abuse was placed at $300 billion some time ago (Kandall, 1991).

Mechanistically, the drug blocks both the initiation and conduction of nerve impulses by decreasing the neuronal membrane's permeability to sodium ions, which results in inhibition of depolarization with resultant blockade of conduction; it also interferes with the uptake of norepinephrine by adrenergic nerve terminals, producing vasoconstriction (Lacy et al., 2004). For therapeutic uses (as generic name), cocaine has a pregnancy category of C. Pregnancy is a nonmedicinal use (as "base," "blow," "coke," "crack," "freebase," or "lady," as it is known by) and carries a risk factor of X (contraindicated in pregnancy). The usual therapeutic dose of 1 to 4% for topical application is undoubtedly greatly exceeded by the various routes of administration when abused.

DEVELOPMENTAL TOXICOLOGY

ANIMALS

In animals, cocaine has not been studied in the laboratory by topical application, the route for its therapeutic use in humans. However, its nonmedicinal use is by intranasal inhalation or intravenous injection in its abused regimen, and cocaine has shown developmental toxicity in several forms by one or more of these routes. In mice, subcutaneous injections of 60 mg/kg/day during organogenesis were teratogenic, producing eye and skeletal defects as well as increased mortality (Mahalik et al., 1980). In rats, developmental toxicity was observed as postnatal behavioral effects when given subcutaneously at 10 mg/kg/day for 15 days during gestation (Smith et al., 1989). Similar changes were observed upon oral administration of the drug to this species (Foss and Riley, 1988; Hutchings et al., 1989). In rabbits, 4 mg/kg of "crack" cocaine administered by intravenous injection during gestation elicited no developmental toxicity (Atlas and Wallach, 1991). Similarly, continuous subcutaneous administration of cocaine via a minipump during portions of the gestational period to monkeys caused no developmental toxicity (Howell et al., 2001). However, reduced density and number of cerebral cortical neurons are observed in the brains of primates exposed *in utero* to cocaine in other studies (Lidow and Song, 2001; Lidow, 2003).

HUMANS

In humans, cocaine produces adverse outcomes on development when abused during pregnancy. This drug has proven to be the most significant developmental toxicant of the past decade. As it affects all classes of developmental toxicity, these effects will be discussed separately. As pointed out by Friedman and Polifka (2000), interpretation of effects in pregnancy associated with the drug are problematic due to a number of confounding factors, not the least of which include study design, reporting, and documentation. It is widely known that there is a correlation between cocaine use and the use of other drugs of abuse, including heroin, methadone, methamphetamine, marijuana, tobacco, and alcohol (Frank et al., 1988). It follows that the lines between effects induced by cocaine or another of these drugs or their combination is tenuous at best. Illicitly obtained cocaine also varies greatly in purity, and it is commonly adulterated; thus, effects produced may vary from study to study depending on the chemical composition of cocaine, with resulting study conclusions quite variable. Then, too, there are literally hundreds of publications attesting to its toxicity, and no review of its effects can be considered definitive. Nonetheless, it is abundantly clear that cocaine induces a variety of adverse developmental effects in thousands of subjects when abused during pregnancy.

Malformations

The initial reports on malformations associated with cocaine use involved two questionable case reports in the 1970s, one with limb defects (Kushnick et al., 1972) and one with ocular malformations (Chan et al., 1978). These reports went largely unnoticed, because other drugs were taken in addition to cocaine, and because they represented single case reports some years apart. No further associations between cocaine use and adverse effects in pregnancy surfaced until 1987. Then, in a large number of reports, researchers reported that use of cocaine in pregnancy is associated with the induction of congenital malformation. This "fetal cocaine syndrome," as it was termed by some investigators, was characterized from 32 cases as a distinct phenotype consisting of neurologic irritability, large fontanels, prominent glabella, marked periorbital and eyelid edema, low nasal bridge with transverse crease, short nose, lateral soft tissue nasal buildup, and small toenails (Fries et al., 1993). Other severe malformations included cleft lip/cleft palate, atypical facial cleft, abnormal brain stem evoked potentials (BSER), interventricular hemorrhage, arthrogryposes, and genitourinary abnormalities. Inhibition of growth parameters was part of the syndrome (see below). Based on the descriptive effects, the syndrome is considered due to disruptive vasoconstrictive phenomena. The exact mechanism is unknown at present, but it was suggested by a number of investigators that the malformations produced by the drug may be related to vasoconstriction in the placenta and hypoxia in the fetus, with the resulting intermittent vascular disruption and ischemia causing fetal damage in the various organs. One group of investigators placed the congenital malformation rate (among 50 exposed subjects) as ranging from 4.5 to 10% (Neerhof et al., 1989). Significantly, the rat is a good model for the central nervous system abnormalities observed in the human (Webster et al., 1991). Cocaine animal models were discussed further by Hutchings and Dow-Edwards (1991). Increased malformations affected a number of different organs and are decribed as follows:

Malformations in general: Bingol et al., 1987; MacGregor et al., 1987; Little et al., 1989; Neerhof et al., 1989; Hoyme et al., 1988, 1990; Hannig and Phillips, 1991; Fries et al., 1993; Robin and Zackai, 1994; Delaney-Black et al., 1994; Hume et al., 1994; Burkett et al., 1998

Limbs: Hoyme et al., 1988; Vandenanker et al., 1991; Sarpong and Headings, 1992; Sheinbaum and Badell, 1992; Viscarello et al., 1992

Central nervous system: Chasnoff et al., 1986, 1987; Bingol et al., 1987; Oro and Dixon, 1987; Ferriero et al., 1988; Tenorio et al., 1988; Greenland et al., 1989; Kobori et al., 1989; Sims and Walther, 1989; Dixon and Bejar, 1989; Kramer et al., 1990; Kapur et al., 1991; Dominguez et al., 1991; Heier et al., 1991; Volpe, 1992; Dusick et al., 1993; Gieron-Korthals et al., 1994; Cohen et al., 1994; Suchet, 1994; Dogra et al., 1994; McLenan et al., 1994; Singer et al., 1993, 1994; Smit et al., 1994; King et al., 1995; Scafidi et al., 1996; Shaw et al., 1996; Behnke et al., 1998; Frank et al., 1998; Bellini et al., 2000; Harvey, 2004

Ocular: Isenberg et al., 1987; Ricci and Molle, 1987; Teske and Trese, 1987; Dominguez et al., 1991; Good et al., 1992; Stafford et al., 1994; Silva-Araujo and Tavares, 1996; Silva-Araujo et al., 1996; Heffelfinger et al., 1997; Block et al., 1997; Church et al., 1998

Cardiovascular: Ferriero et al., 1988; Little et al., 1989; Frassica et al., 1990; Shaw et al., 1991; Plessinger and Woods, 1991; Bulbul et al., 1994

Genitourinary: Chasnoff et al., 1985, 1987, 1988, 1989; Bingol et al., 1986; Ryan et al., 1987; MacGregor et al., 1987; Chavez et al., 1989; Rajegowda et al., 1991; Battin et al., 1995

Gastrointestinal: Hoyme et al., 1988; Telsey et al., 1988; Czyrko et al., 1991; Porat and Brodsky, 1991; Drongowski et al., 1991; Spinazzola et al., 1992; Sehgal et al., 1993; Lopez et al., 1995

Miscellaneous anomalies: Bingol et al., 1986; Rosenstein et al., 1990; Plessinger and Woods, 1991; Viscarello et al., 1992; Lezcano et al., 1994; Martinez et al., 1994; Esmer et al., 2000; Markov et al., 2003; Kashiwagi et al., 2003

Growth Retardation

Adverse effects from cocaine use on prenatal and postnatal growth and birth weight, including prematurity, are probably the most common developmental toxicity endpoints affected adversely by the drug. A number of publications attest to this effect (Chan et al., 1978; Madden et al., 1986; MacGregor et al., 1987; Ryan et al., 1987; Bingol et al., 1987; Landy and Hinson, 1988; Frank et al., 1988, 1996, 2001; Cherukuri et al., 1988; Chouteau et al., 1988; Zuckerman et al., 1989; Neerhof et al., 1989; Keith et al., 1989; Little et al., 1989; Chasnoff et al., 1989; Petitti and Coleman, 1990; Rosenak et al., 1990; Plessinger and Woods, 1991; Lester et al., 1991; Forman et al., 1993; Behnke and Eyler, 1993; Dusick et al., 1993; Sehgal et al., 1993; Holzman and Paneth, 1994; Singer et al., 1994, 2001; Hulse et al., 1997a; Ostrea et al., 1997; Andrews et al., 2000; Bandstra et al., 2001). Several studies have indicated this effect to be on the order of 27 to 36% incidence rates for intrauterine growth retardation or lowered birth weight, growth curves at less than the 25th percentile, some 40% premature births, and 17 to 43% born with microcephaly or small head circumference (Fulroth et al., 1989; Hadeed and Siegel, 1989; Burkett et al., 1990; Fries et al., 1993). The postnatal growth deficiencies have been shown to revert to control levels in time (Azuma and Chasnoff, 1993; Day et al., 1994; Griffith et al., 1994; Harsham et al., 1994; Hurt et al., 1995a; Richardson et al., 1996).

Death

Fetal perinatal and postnatal morbidity or mortality, and abortion and stillbirth appear to be associated with the syndrome (Madden et al., 1986; Chasnoff et al., 1987; Bingol et al., 1987; Oro and Dixon, 1987; Landy and Hinson, 1988; Critchley et al., 1988; Chouteau et al., 1988; Bauchner et al., 1988; Ferriero et al., 1988; Frank et al., 1988; Greenland et al., 1989; Little et al., 1989; Keith et al., 1989; Neerhof et al., 1989; Apple and Roe, 1990; Meeker and Reynolds, 1990; Morild and Stajic, 1990; Ostrea et al., 1997). Abruptio placentae is often associated with fetal death and is considered an induced effect of cocaine exposure (Acker et al., 1983; Chasnoff et al., 1985, 1987, 1989; Cregler and Mark, 1986; Bingol et al., 1987; Oro and Dixon, 1987; Landy and Hinson, 1988; Cherukuri et al., 1988; Townsend et al., 1988; Keith et al., 1989; Little et al., 1989; Collins et al., 1989; Neerhof et al., 1989; Slutsker, 1992; Holzman and Paneth, 1994; Frank et al., 1996; Hulse et al., 1997b; Addis et al., 2001).

Functional Deficit

Neurological deficits, neurophysiological dysfunction, and behavioral alterations termed "neuroteratology" by Scanlon (1991) have been reported as occurring in high numbers in infants exposed to cocaine prenatally and is a component of the cocaine developmental toxicity pattern (Chasnoff et al., 1985; Ryan et al., 1987; LeBlanc et al., 1987; Doberczak et al., 1988; Oro and Dixon, 1987; Cherukuri et al., 1988; Smith et al., 1989; Little et al., 1989; Neerhof et al., 1989; Van Dyke and Fox, 1990; Murphy and Hoff, 1990; Neuspiel and Hamel, 1991; Singer et al., 1991, 1993, 2000; Lester and Tronick, 1994; Bendersky et al., 1995; Lester et al., 1991, 1995, 1996, 1998; Needlman et al., 1995; Frank et al., 1996, 1998, 2001; Mayes, 1996; Chiriboga, 1998; Chiriboga et al., 1999; Mayes et al., 1998; Woods et al., 1995; Morrow et al., 2001; Behnke et al., 2002; M.W. Lewis et al., 2004; Arendt et al., 2004). Deficits in performance of standardized tests (Chasnoff et al., 1989; Hume et al., 1989; Delaney-Black et al., 2000; Myers et al., 2003; Noland et al., 2003), decrements in motor skills or cognition (Arendt et al., 1999; Singer et al., 2002; Messinger et al., 2004), and delayed language development (Nulman et al., 2001; Morrow et al., 2003, 2004; B.A. Lewis et al., 2004) are examples of problems encountered in neonates or children exposed prenatally to cocaine. In contrast to the large number of researchers who consider, as a result of their studies, cocaine to be responsible for significant developmental toxicity, in a number of studies, no such evidence was found either generally or for specific effects in their investigations (Gillogley et al., 1990; Handler et al., 1991; Lutiger et al., 1991; Martin et al., 1992; Torfs et al., 1994; Eyler et al., 1994; Hurt et al., 1995b; Kistin et al., 1996; Sprauve et al., 1997). The association between cocaine and the developmental toxicity reported was reviewed in a number of publications; representative of these are Chasnoff et al., 1989; Rosenak et al., 1990; Dow-Edwards, 1991, 1996; Scanlon, 1991; Lutiger et al., 1991; Lindenberg et al., 1991; Koren et al., 1989, 1992; Dow-Edwards et al., 1992, 1999; Slutsker, 1992; Needlman et al., 1995; Friedman and Polifka, 2000; Lester, 2000; Keller and Snyder-Keller, 2000; Addis et al., 2001; Briggs et al., 2002; Vidaeff and Mastrobattista, 2003. One group of experts considers the magnitude of teratogenic risk and pregnancy complications, including placental abruption, due to cocaine to be small to moderate (Friedman and Polifka, 2000). Koren and associates (1992) consider there to be an unrealistic high perception of teratogenic risk by cocaine and state that counseling is effective in preventing termination of many otherwise desired pregnancies.

CHEMISTRY

Cocaine is an average-sized human developmental toxicant. It is hydrophobic and of average polarity. Cocaine is a hydrogen bond acceptor. The calculated physicochemical and topological properties are as follows.

PHYSICOCHEMICAL PROPERTIES

Parameter	Value
Molecular weight	303.358 g/mol
Molecular volume	276.50 A^3
Density	1.209 g/cm^3
Surface area	336.44 A^2
LogP	1.682
HLB	6.219
Solubility parameter	21.823 $J^{(0.5)}/cm^{(1.5)}$
Dispersion	19.463 $J^{(0.5)}/cm^{(1.5)}$
Polarity	4.368 $J^{(0.5)}/cm^{(1.5)}$
Hydrogen bonding	8.851 $J^{(0.5)}/cm^{(1.5)}$
H bond acceptor	0.69
H bond donor	0.03
Percent hydrophilic surface	33.39
MR	83.419
Water solubility	−1.055 log (mol/M^3)
Hydrophilic surface area	112.33 A^2
Polar surface area	62.16 A^2
HOMO	−9.472 eV
LUMO	−0.170 eV
Dipole	3.086 debye

TOPOLOGICAL PROPERTIES (UNITLESS)

Parameter	Value
x0	15.690
x1	10.613
x2	9.430
xp3	8.364
xp4	6.838
xp5	5.371
xp6	3.641
xp7	2.565
xp8	1.546
xp9	0.943
xp10	0.600
xv0	12.898
xv1	7.673
xv2	6.058
xvp3	4.956
xvp4	3.767
xvp5	2.654
xvp6	1.658

Continued.

Parameter	Value
xvp7	0.936
xvp8	0.508
xvp9	0.239
xvp10	0.127
k0	28.329
k1	16.844
k2	7.266
k3	3.440
ka1	15.340
ka2	6.281
ka3	2.872

REFERENCES

Acker, D. et al. (1983). Abruptio placentae associated with cocaine use. *Am. J. Obstet. Gynecol.* 146: 220–221.

Addis, A. et al. (2001). Fetal effects of cocaine: An updated meta-analysis. *Reprod. Toxicol.* 15: 341–369.

Andrews, K., Francis, D. J., and Riese, M. L. (2000). Prenatal cocaine exposure and prematurity: Neurodevelopmental growth. *J. Dev. Behav. Pediatr.* 21: 262–270.

Apple, F. S. and Roe, S. J. (1990). Cocaine associated fetal death *in utero*. *J. Anal. Toxicol.* 14: 259–260.

Arendt, R. et al. (1999). Motor development of cocaine-exposed children at age two years. *Pediatrics* 103: 86–92.

Arendt, R. E. et al. (2004). Children prenatally exposed to cocaine: Developmental outcomes and environmental risks at seven years of age. *J. Dev. Behav. Pediatr.* 25: 83–90.

Atlas, S. J. and Wallach, E. E. (1991). Effects of intravenous cocaine on reproductive function in the mated rabbit. *Am. J. Obstet. Gynecol.* 165: 1785–1790.

Azuma, S. D. and Chasnoff, I. J. (1993). Outcome of children prenatally exposed to cocaine and other drugs: A path analysis of three-year data. *Pediatrics* 92: 396–402.

Bandstra, E. S. et al. (2001). Intrauterine growth of full-term infants: Impact of prenatal cocaine exposure. *Pediatrics* 108: 1309–1319.

Battin, M., Albersheim, S., and Newman, D. (1995). Congenital genitourinary tract abnormalities following cocaine exposure *in utero*. *Am. J. Perinatol.* 12: 425–428.

Bauchner, H. et al. (1988). Risk of sudden infant death syndrome among infants with *in utero* exposure to cocaine. *J. Pediatr.* 113: 831–834.

Behnke, M. and Eyler, F. D. (1993). The consequences of prenatal substance use for the developing fetus, newborn, and young child. *Int. J. Addict.* 28: 1341–1391.

Behnke, M. et al. (1998). Incidence and description of structural brain abnormalities in newborns exposed to cocaine. *J. Pediatr.* 132: 291–294.

Behnke, M. et al. (2002). Cocaine exposure and developmental outcome from birth to 6 months. *Neurotoxicol. Teratol.* 24: 283–295.

Bellini, C., Massocco, D., and Serra, G. (2000). Prenatal cocaine exposure and the expanding spectrum of brain malformations. *Arch. Int. Med.* 160: 2393.

Bendersky, M. et al. (1995). Measuring the effects of prenatal cocaine exposure. In *Mothers, Babies, and Cocaine: The Role of Toxins in Development*, M. Lewis and M. Bendersky, Eds., Lawrence Erlbaum, Hillsdale, NJ, pp. 163–178.

Bingol, N. et al. (1986). Prune belly syndrome associated with maternal cocaine use. *Am. J. Hum. Genet.* 39: A51.

Bingol, N. et al. (1987). Teratogenicity of cocaine in humans. *J. Pediatr.* 220: 93–96.

Block, S. S., Moore, B. D., and Scharre, J. E. (1997). Visual anomalies in young children exposed to cocaine. *Optom. Vis. Sci.* 74: 28–36.

Briggs, G. G., Freeman, R. K., and Yaffe, S. J. (2002). *Drugs in Pregnancy and Lactation. A Reference Guide to Fetal and Neonatal Risk*, Sixth ed., Lippincott Williams & Wilkins, Philadelphia.

Bulbul, Z. R., Rosenthal, D. N., and Kleinman, C. S. (1994). Myocardial infarction in the perinatal period secondary to maternal cocaine abuse. A case report and literature review. *Arch. Pediatr. Adolesc. Med.* 148: 1092–1096.

Burkett, G., Yasin, S., and Palow, D. (1990). Perinatal implications of cocaine exposure. *J. Reprod. Med.* 35: 35–42.

Burkett, G. et al. (1998). Prenatal care in cocaine-exposed pregnancies. *Obstet. Gynecol.* 92: 193–200.

Chan, C. C., Fishman, M., and Egbert, P. R. (1978). Multiple ocular anomalies associated with maternal LSD ingestion. *Arch. Ophthalmol.* 96: 282–284.

Chasnoff, I. J. et al. (1985). Cocaine use in pregnancy. *N. Engl. J. Med.* 313: 666–669.

Chasnoff, I. J. et al. (1986). Perinatal cerebral infarction and maternal cocaine use. *J. Pediatr.* 108: 456–459.

Chasnoff, I. J., Burns, K. A., and Burns, W. J. (1987). Cocaine use in pregnancy: Perinatal morbidity and mortality. *Neurobehav. Teratol.* 9: 291–293.

Chasnoff, I. J., Chisum, G. M., and Kaplan, W. R. (1988). Maternal cocaine use and genitourinary tract malformation. *Teratology* 37: 201–204.

Chasnoff, I. J. et al. (1989). Temporal patterns of cocaine use in pregnancy. *JAMA* 261: 1741–1744.

Chavez, G. F., Mulinare, J., and Cordero, J. (1989). Maternal cocaine use during early pregnancy as a risk factor for congenital urogenital anomalies. *JAMA* 262: 795–798.

Cherukuri, R. et al. (1988). A cohort study of alkaloidal cocaine ("crack") in pregnancy. *Obstet. Gynecol.* 72: 147–151.

Chiriboga, C. A. (1998). Neurological correlates of fetal cocaine exposure. *Ann. NY Acad. Sci.* 846: 109–125.

Chiriboga, C. A. et al. (1999). Dose-response effect of fetal cocaine exposure on newborn neurologic function. *Pediatrics* 103: 79–85.

Chouteau, M., Namerow, P. B., and Leppert, P. (1988). The effect of cocaine abuse on birth weight and gestational age. *Obstet. Gynecol.* 72: 351–354.

Church, M. W. et al. (1998). Effects of prenatal cocaine on hearing, vision, growth, and behavior. *Ann. NY Acad. Sci.* 846: 12–28.

Cohen, H. L. et al. (1994). Neurosonographic findings in full-term infants born to maternal cocaine abusers: Visualization of subependymal and periventricular cysts. *J. Clin. Ultrasound* 22: 327–333.

Collins, E., Hardwick, R. J., and Jeffery, H. (1989). Perinatal cocaine intoxication. *Med. J. Aust.* 150: 331–340.

Cregler, L. L. and Mark, J. (1986). Special report: Medical complications of cocaine use. *N. Engl. J. Med.* 315: 1495–1500.

Critchley, H. O. D. et al. (1988). Fetal death *in utero* and cocaine abuse: Case report. *Br. J. Obstet. Gynaecol.* 95: 195–196.

Czyrko, C. et al. (1991). Maternal cocaine abuse and necrotizing enterocolitis: Outcome and survival. *J. Pediatr. Surg.* 26: 414–421.

Day, N. L. et al. (1994). Alcohol, marijuana, and tobacco: Effects of prenatal exposure on offspring growth and morphology at age six. *Alcohol. Clin. Exp. Res.* 18: 786–794.

Delaney-Black, V., Covington, C., and Sokol, R. J. (1994). Maternal cocaine consumption: Birth outcome and child development. *Fetal Matern. Med. Rev.* 6: 119–134.

Delaney-Black, V. et al. (2000). Teacher-assessed behavior of children prenatally exposed to cocaine. *Pediatrics* 106: 782–791.

Dixon, S. D. and Bejar, R. (1989). Echoencephalographic findings associated with maternal cocaine and methamphetamine use: Incidence and clinical correlates. *J. Pediatr.* 115: 770–778.

Doberczak, T. M. et al. (1988). Neonatal neurologic and electroencephalographic effects of intrauterine cocaine exposure. *J. Pediatr.* 113: 354–358.

Dogra, V. S. et al. (1994). Neurosonographic abnormalities associated with maternal history of cocaine use in neonates of appropriate size for their gestational age. *AJNR Am. J. Neuroradiol.* 15: 697–702.

Dominguez, R. et al. (1991). Brain and ocular abnormalities in infants with *in utero* exposure to cocaine and other street drugs. *Am. J. Dis. Child.* 145: 688–695.

Dow-Edwards, D. L. (1991). Cocaine effects on fetal development. A comparison of clinical and animal research findings. *Neurotoxicol. Teratol.* 13: 347–352.

Dow-Edwards, D. (1996). Comparability of human and animal studies of developmental cocaine exposure. *NIDA Res. Monogr. Ser.* 164: 146–174.

Dow-Edwards, D., Chasnoff, I. J., and Griffith, D. R. (1992). Cocaine use during pregnancy: Neurobehavioral changes in the offspring. In *Perinatal Substance Abuse*, T. B. Sonderegger, Ed., Johns Hopkins University Press, Baltimore, MD, pp, 184–206.

Dow-Edwards, D. et al. (1999). Cocaine and development: Clinical, behavioral, and neurobiological perspectives — a symposium report. *Neurotoxicol. Teratol.* 21: 481–490.

Drongowski, R. A. et al. (1991). Contribution of demographic and environmental factors to the etiology of gastroschisis: A hypothesis. *Fetal Diagn. Ther.* 6: 14–27.

Dusick, A. M. et al. (1993). Risk of intracranial hemorrhage and other adverse outcomes after cocaine exposure in a cohort of 323 very low birth weight infants. *J. Pediatr.* 122: 438–445.

Esmer, M. C. et al. (2000). Cloverleaf skull and multiple congenital anomalies in a girl exposed to cocaine *in utero*: Case report and review of the literature. *Childs Nerv. Syst.* 16: 176–180.

Eyler, F. D. et al. (1994). Prenatal cocaine use: A comparison of neonates matched on maternal risk factors. *Neurotoxicol. Teratol.* 16: 81–87.

Ferriero, D. M., Partridge, J. C., and Wong, D. F. (1988). Congenital defects and stroke in cocaine-exposed neonates. *Ann. Neurol.* 24: 348–349.

Forman, R. et al. (1993). Maternal and neonatal characteristics following exposure to cocaine in Toronto. *Reprod. Toxicol.* 7: 619–622.

Foss, J. A. and Riley, E. P. (1988). Behavioral evaluation of animals exposed prenatally to cocaine. *Teratology* 37: 517.

Frank, D. A. et al. (1988). Cocaine use during pregnancy — prevalence and correlates. *Pediatrics* 82: 888–895.

Frank, D. A. et al. (1996). Maternal cocaine use: Impact on child health and development. *Curr. Probl. Pediatr.* 26: 52–70.

Frank, D. A., Augustyn, M., and Zuckerman, B. S. (1998). Neonatal neurobehavioral and neuroanatomic correlates of prenatal cocaine exposure: Problems of dose and confounding. *Ann. NY Acad. Sci.* 846: 40–50.

Frank, D. A. et al. (2001). Growth, development, and behavior in early childhood following prenatal cocaine exposure: A systematic review. *JAMA* 285: 1613–1625.

Frassica, J. J. et al. (1990). Cardiovascular abnormalities in infants prenatally exposed to cocaine. *Am. J. Cardiol.* 66: 525.

Friedman, J. M. and Polifka, J. E. (2000). *Teratogenic Effects of Drugs. A Resource for Clinicians (TERIS)*, Second ed., Johns Hopkins University Press, Baltimore, MD.

Fries, M. H. et al. (1993). Facial features of infants exposed prenatally to cocaine. *Teratology* 48: 413–420.

Fulroth, R., Phillips, V., and Durand, D. J. (1989). Perinatal outcome of infants exposed to cocaine and/or heroin *in utero*. *Am. J. Dis. Child.* 143: 905–910.

Gieron-Korthals, M. A., Helal, A., and Martinez, C. R. (1994). Expanding spectrum of cocaine induced central nervous system malformations. *Brain Dev.* 16: 253–256.

Gillogley, K. M. et al. (1990). The perinatal impact of cocaine, amphetamine, and opiate use detected by universal intrapartum screening. *Am. J. Obstet. Gynecol.* 163: 1535–1542.

Good, W. V. et al. (1992). Abnormalities of the visual system in infants exposed to cocaine. *Ophthalmology* 99: 341–346.

Greenland, V. C., Delke, I., and Minkoff, H. L. (1989). Vaginally administered cocaine overdose in a pregnant woman. *Obstet. Gynecol.* 74: 476–477.

Griffith, D. R., Azuma, S. D., and Chasnoff, I. J. (1994). Three-year outcome of children exposed prenatally to drugs. *J. Am. Acad. Child. Adolesc. Psychiatry* 33: 20–27.

Hadeed, A. J. and Siegel, S. R. (1989). Maternal cocaine use during pregnancy: Effect on the newborn infant. *Pediatrics* 84: 205–210.

Handler, A. et al. (1991). Cocaine use during pregnancy: Perinatal outcomes. *Am. J. Epidemiol.* 133: 818–825.

Hannig, V. L. and Phillips, J. A. (1991). Maternal cocaine abuse and fetal anomalies: Evidence for teratogenic effects of cocaine. *South. Med. J.* 84: 498–499.

Harsham, J., Keller, J. H., and Disbrow, D. (1994). Growth patterns of infants exposed to cocaine and other drugs *in utero*. *J. Am. Diet. Assoc.* 94: 999–1007.

Harvey, J. A. (2004). Cocaine effects on the developing brain: Current status. *Neurosci. Biobehav. Rev.* 27: 751–764.

Heffelfinger, A., Craft, S., and Shyken, J. (1997). Visual attention in children with prenatal cocaine exposure. *J. Int. Neuropsychol. Soc.* 3: 237–245.

Heier, L. A. et al. (1991). Maternal cocaine abuse: The spectrum of radiologic abnormalities in the neonate CNS. *AJNR Am. J. Neuroradiol.* 12: 951–956.

Holzman, C. and Paneth, N. (1994). Maternal cocaine use during pregnancy and perinatal outcomes. *Epidemiol. Rev.* 16: 315–334.

Howell, L.L. et al. (2001). Fetal development in rhesus monkeys exposed prenatally to cocaine. *Neurotoxicol. Teratol.* 23: 133–140.

Hoyme, H. E. et al. (1988). Maternal cocaine use and fetal vascular disruption. *Am. J. Hum. Genet.* 43 (Suppl.): A56.

Hoyme, H. E. et al. (1990). Prenatal cocaine exposure and fetal vascular disruption. *Pediatrics* 85: 743–747.

Hulse, G. K. et al. (1997a). Maternal cocaine use and low birth weight newborns: A meta-analysis. *Addiction* 92: 1561–1570.

Hulse, G. K. et al. (1997b). Assessing the relationship between maternal cocaine use and abruptio placentae. *Addiction* 92: 1547–1551.

Hume, R. F. et al. (1989). *In utero* cocaine exposure: Observations of fetal behavioral state may predict neonatal outcome. *Am. J. Obstet. Gynecol.* 161: 685–690.

Hume, R. F. et al. (1994). Ultrasound diagnosis of fetal anomalies associated with *in utero* cocaine exposure: Further support for cocaine-induced vascular disruption teratogenesis. *Fetal Diagn. Ther.* 9: 239–245.

Hurt, H. et al. (1995a). Cocaine-exposed children: Follow-up through 30 months. *J. Dev. Behav. Pediatr.* 16: 29–35.

Hurt, H. et al. (1995b). Natal status of infants of cocaine users and control subjects: A prospective comparison. *J. Perinatol.* 15: 297–304.

Hutchings, D. E. and Dow-Edwards, D. (1991). Animal models of opiate, cocaine, and *Cannabis* use. *Clin. Perinatol.* 18: 1–22.

Hutchings, D. E., Fico, T. A., and Dow-Edwards, D. L. (1989). Prenatal cocaine — maternal toxicity, fetal effects and locomotor activity in rat offspring. *Neurotoxicol. Teratol.* 11: 65–69.

Isenberg, S. J., Spierer, A., and Inkelis, S. H. (1987). Ocular signs of cocaine intoxication in neonates. *Am. J. Ophthalmol.* 103: 211–214.

Kandall, S. R. (1991). Perinatal effects of cocaine and amphetamine use during pregnancy. *Bull. NY Acad. Med.* 67: 240–255.

Kapur, R. P., Shaw, C. M., and Shepard, T. H. (1991). Brain hemorrhages in cocaine-exposed human fetuses. *Teratology* 44: 11–18.

Kashiwagi, M. et al. (2003). Fetal bilateral renal agenesis, phocomelia, and single umbilical artery associated with cocaine abuse in early pregnancy. *Birth Defects Res. (A)* 67: 951–952.

Keith, L. G. et al. (1989). Substance abuse in pregnant women — recent experience at the Perinatal Center for Chemical Dependence of Northwestern Memorial Hospital. *Obstet. Gynecol.* 73: 715–720.

Keller, R. W. and Snyder-Keller, A. (2000). Prenatal cocaine exposure. *Ann. NY Acad. Sci.* 909: 217–232.

King, T. A. et al. (1995). Neurologic manifestations of *in utero* cocaine exposure in near-term and term infants. *Pediatrics* 96: 259–264.

Kistin, N. et al. (1996). Cocaine and cigarettes: A comparison of risks. *Paediatr. Perinat. Epidemiol.* 10: 269–278.

Kobori, J. A., Ferriero, D. M., and Golabi, M. (1989). CNS and craniofacial anomalies in infants born to cocaine abusing mothers. *Clin. Res.* 37: 196A.

Koren, G. et al. (1989). Bias against the null hypothesis: The reproductive hazards of cocaine. *Lancet* 2: 1440–1442.

Koren, G. et al. (1992). The perception of teratogenic risk of cocaine. *Teratology* 46: 567–571.

Kramer, L. D. et al. (1990). Neonatal cocaine-related seizures. *J. Child Neurol.* 5: 60–64.

Kushnick, T., Robinson, M., and Tsao, C. (1972). 45,X Chromosome abnormality in the offspring of a narcotic addict. *Am. J. Dis. Child.* 124: 772–773.

Lacy, C. F. et al. (2004). *Drug Information Handbook (Pocket), 2004–2005,* Lexi-Comp, Inc., Hudson, OH.

Landy, H. J. and Hinson, J. (1988). Placental abruption associated with cocaine use: Case report. *Reprod. Toxicol.* 1: 203–205.

LeBlanc, P. E. et al. (1987). Effects of intrauterine exposure to alkaloidal cocaine ("crack"). *Am. J. Dis. Child.* 141: 937–938.

Lester, B. M. (2000). Prenatal cocaine exposure and child outcome: A model for the study of the infant at risk. *Isr. J. Psychiatry Relat. Sci.* 37: 223–235.

Lester, B. M. and Tronick, E. (1994). The effect of prenatal cocaine exposure and child outcome: Lessons from the past. *Infant Ment. Health J.* 15: 107–120.

Lester, B. M. et al. (1991). Neurobehavioral syndromes in cocaine exposed newborn infants. *Child. Dev.* 62: 694–705.

Lester, B. M., Freier, K., and LaGasse, L. (1995). Prenatal cocaine exposure and child outcome: What do we really know? In *Mothers, Babies, and Cocaine: The Role of Toxins in Development*, M. Lewis and M. Bendersky, Eds., Lawrence Erlbaum, Hillsdale, NJ, pp. 19–39.

Lester, B. M. et al. (1996). Studies of cocaine exposed human infants. *NIDA Res. Monogr. Ser.* 164: 175–210.

Lester, B. M., LaGasse, L. L., and Seifer, R. (1998). Cocaine exposure and children: The meaning of subtle effects. *Science* 282: 633–634.

Lewis, B. A. et al. (2004). Four-year language outcomes of children exposed to cocaine *in utero*. *Neurotoxicol. Teratol.* 26: 617–627.

Lewis, M. W. et al. (2004). Neurological and developmental outcomes of prenatally cocaine-exposed offspring from 12 to 36 months. *Am. J. Drug Alcohol Abuse* 30: 299–320.

Lezcano, L. et al. (1994). Crossed renal ectopia associated with maternal alkaloid cocaine abuse: A case report. *J. Perinatol.* 4: 230–233.

Lidow, M. S. (2003). Consequences of prenatal cocaine exposure in nonhuman primates. *Brain Res. Dev. Brain Res.* 147: 23–36.

Lidow, M. S. and Song, Z. M. (2001). Primates exposed to cocaine *in utero* display reduced density and number of cerebral cortical neurons. *J. Comp. Neurol.* 435: 263–275.

Lindenberg, C. A. et al. (1991). A review of the literature on cocaine abuse in pregnancy. *Nurs. Res.* 40: 69–75.

Little, B. B. et al. (1989). Cocaine abuse during pregnancy: Maternal and fetal implications. *Obstet. Gynecol.* 73: 157–160.

Lopez, S. L. et al. (1995). Time of onset of necrotizing enterocolitis in newborn infants with known prenatal cocaine exposure. *Clin. Pediatr.* 34: 424–429.

Lutiger, B. et al. (1991). Relationship between gestational cocaine use and pregnancy outcome: A meta-analysis. *Teratology* 44: 405–414.

MacGregor, S. N. et al. (1987). Cocaine use during pregnancy: Adverse perinatal outcome. *Am. J. Obstet. Gynecol.* 157: 686–690.

Madden, J. D., Payne, T. F., and Miller, S. (1986). Maternal cocaine abuse and effect on the newborn. *Pediatrics* 77: 209–211.

Mahalik, M. P., Gautieri, R. F., and Mann, D. E. (1980). Teratogenic potential of cocaine hydrochloride in CF-1 mice. *J. Pharm. Sci.* 69: 703–706.

Markov, D., Jacquemyn, Y., and Leroy, Y. (2003). Bilateral cleft lip and palate associated with increased nuchal translucency and maternal cocaine abuse at 14 weeks of gestation. *Clin. Exp. Obstet. Gynecol.* 30: 109–110.

Martin, M. L. et al. (1992). Trends in rates of multiple vascular disruption defects, Atlanta, 1968–1989: Is there evidence of a cocaine teratogenic epidemic? *Teratology* 45: 647–653.

Martinez, J. M. et al. (1994). Body stalk anomaly associated with maternal cocaine abuse. *Prenat. Diagn.* 14: 669–672.

Mayes, I. C. (1996). Exposure to cocaine: Behavioral outcomes in preschool and school-age children. *NIDA Res. Monogr. Ser.* 164: 211–229.

Mayes, I. C. et al. (1998). Regulation of arousal and attention in preschool children exposed to cocaine prenatally. *Ann. NY Acad. Sci.* 846: 126–143.

McLenan, D. A. et al. (1994). Evaluation of the relationship between cocaine and intraventricular hemorrhage. *J. Natl. Med. Assoc.* 86: 281–287.

Meeker, J. E. and Reynolds, P. C. (1990). Fetal and newborn death associated with maternal cocaine use. *J. Anal. Toxicol.* 14: 379–382.

Messinger, D. S. et al. (2004). The Maternal Lifestyle Study: Cognitive, motor, and behavioral outcomes of cocaine-exposed and opiate-exposed infants through three years of age. *Pediatrics* 113: 1677–1685.

Morild, I. and Stajic, M. (1990). Cocaine and fetal death. *Forensic Sci. Int.* 47: 181–189.

Morrow, C. E. et al. (2001). Influence of prenatal cocaine exposure on full-term infant neurobehavioral functioning. *Neurotoxicol. Teratol.* 23: 533–544.

Morrow, C. E. et al. (2003). Influence of prenatal cocaine exposure on early language development: Longitudinal findings from four months to three years of age. *J. Dev. Behav. Pediatr.* 24: 39–50.

Morrow, C. E. et al. (2004). Expressive and receptive language functioning in preschool children with prenatal cocaine exposure. *J. Pediatr. Psychol.* 29: 543–554,

Murphy, S. G. and Hoff, S. F. (1990). The teratogenicity of cocaine. *J. Clin. Exp. Neuropsychol.* 12: 69.

Myers, B. J. et al. (2003). Prenatal cocaine exposure and infant performance on the Brazelton Neonatal Behavioral Assessment Scale. *Subst. Use Misuse* 38: 2065–2096.

National Pregnancy and Health Survey. (1995). *A National Institute on Drug Abuse Report.* National Institutes of Health, Department of Health and Human Services, GPO, Washington, D.C.

Needlman, R. et al. (1995). Neurophysiological effects of prenatal cocaine exposure: Comparison of human and animal investigations. In *Mothers, Babies, and Cocaine: The Role of Toxins in Development*, M. Lewis and M. Bendersky, Eds., Lawrence Erlbaum, Hillsdale, NJ, pp. 229–250.

Neerhof, M. G. et al. (1989). Cocaine abuse during pregnancy — peripartum prevalence and perinatal outcome. *Am. J. Obstet. Gynecol.* 161: 633–638.

Neuspiel, D. R. and Hamel, S. C. (1991). Cocaine and infant behavior. *J. Dev. Behav. Pediatr.* 12: 55–64.

Noland, J. S. et al. (2003). Prenatal cocaine/polydrug exposure and infant performance on an executive functioning task. *Dev. Neuropsychol.* 24: 499–517.

Nulman, I. et al. (2001). Neurodevelopment of adopted children exposed *in utero* to cocaine: The Toronto Adoption Study. *Clin. Invest. Med.* 24: 129–137.

Oro, A. S. and Dixon, S. D. (1987). Perinatal cocaine and methamphetamine exposure: Maternal and neonatal correlates. *J. Pediatr.* 111: 571–578.

Ostrea, E. M., Ostrea, A. R., and Simpson, P. M. (1997). Mortality within the first 2 years in infants exposed to cocaine, opiate, or cannabinoid during gestation. *Pediatrics* 100: 79–83.

Petitti, D. B. and Coleman, C. (1990). Cocaine and the risk of low birth-weight. *Am. J. Public Health* 80: 25–28.

Plessinger, M. A. and Woods, J. R. (1991). The cardiovascular effects of cocaine use in pregnancy. *Reprod. Toxicol.* 5: 99–113.

Porat, R. and Brodsky, N. (1991). Cocaine: A risk factor for necrotizing enterocolitis. *J. Perinatol.* 11: 30–32.

Rajegowda, B. et al. (1991). Does cocaine (CO) increase congenital urogenital abnormalities (CUGA) in newborns. *Pediatr. Res.* 29: A71.

Ricci, B. and Molle, F. (1987). Ocular signs of cocaine intoxication in neonates. *Am. J. Ophthalmol.* 104: 550–551.

Richardson, G. A., Conroy, M. L., and Day, N. L. (1996). Prenatal cocaine exposure: Effects on the development of school-age children. *Neurotoxicol. Teratol.* 18: 627–634.

Robin, N. H. and Zackai, E. H. (1994). Unusual craniofacial dysmorphia due to prenatal alcohol and cocaine exposure. *Teratology* 50: 160–164.

Rosenak, D. et al. (1990). Cocaine: Maternal use during pregnancy and its effect on the mother, the fetus, and the infant. *Obstet. Gynecol. Surv.* 45: 348–359.

Rosenstein, B. J., Wheeler, J. S., and Heid, P. L. (1990). Congenital renal abnormalities in infants with *in utero* cocaine exposure. *J. Urol.* 144: 110–112.

Ryan, L., Ehrlich, S., and Finnegan, L. (1987). Cocaine abuse in pregnancy: Effects of the fetus and newborn. *Neurotoxicol. Teratol.* 9: 295–301.

Sarpong, S. and Headings, V. (1992). Sirenomelia accompanying exposure of the embryo to cocaine. *South. Med. J.* 85: 545–547.

Scafidi, F. A. et al. (1996). Cocaine-exposed preterm neonates show behavioral and hormonal differences. *Pediatrics* 97: 851–855.

Scanlon, J. W. (1991). The neuroteratology of cocaine: Background, theory, and clinical implications. *Reprod. Toxicol.* 5: 89–96.

Sehgal, S. et al. (1993). Morbidity of low-birthweight infants with intrauterine cocaine exposure. *J. Natl. Med. Assoc.* 85: 20–24.

Shaw, G. M. et al. (1991). Maternal use of cocaine during pregnancy and congenital cardiac anomalies. *J. Pediatr.* 118: 167–168.

Shaw, G. M., Velie, E. M., and Morland, K. B. (1996). Parental recreational drug use and risk for neural tube defects. *Am. J. Epidemiol.* 144: 1155–1160.

Sheinbaum, K. A. and Badell, A. (1992). Psychiatric management of two neonates with limb deficiencies and prenatal cocaine exposure. *Arch. Phys. Med. Rehabil.* 73: 385–388.

Silva-Araujo, A. and Tavares, M. A. (1996). Development of the eye after gestational exposure to cocaine. Vascular disruption in the retina of rats and humans. *Ann. NY Acad. Sci.* 801: 274–288.

Silva-Araujo, A. L. et al. (1996). Retinal hemorrhages associated with *in utero* exposure to cocaine. *Retina* 16: 411–418.

Sims, M. E. and Walther, F. J. (1989). Neonatal ultrasound casebook. *J. Perinatol.* 9: 349–350.

Singer, L. T., Garber, R., and Kliegman, R. (1991). Neurobehavioral sequelae of fetal cocaine exposure. *J. Pediatr.* 119: 667–671.

Singer, L., Arendt, R., and Minnes, S. (1993). Neurodevelopmental effects of cocaine. *Clin. Perinatol.* 20: 245–262.

Singer, L. T. et al. (1994). Increased incidence of intraventricular hemorrhage and developmental delay in cocaine-exposed, very low birth weight infants. *J. Pediatr.* 124: 765–771.

Singer, L. T. et al. (2000). Neurobehavioral outcomes of cocaine-exposed infants. *Neurotoxicol. Teratol.* 22: 653–666.

Singer, L. T. et al. (2001). Developmental outcomes and environmental correlates of very low birthweight, cocaine-exposed infants. *Early Hum. Dev.* 64: 91–103.

Singer, L. T. et al. (2002). Cognitive and motor outcomes of cocaine-exposed infants. *JAMA* 287: 1952–1960.

Slutsker, L. (1992). Risks associated with cocaine use during pregnancy. *Obstet. Gynecol.* 79: 778–789.

Smit, B. J. et al. (1994). Cocaine use in pregnancy in Amsterdam. *Acta Paediatr.* Suppl. 404: 32–35.

Smith, R. F. et al. (1989). Alterations in offspring behavior induced by chronic prenatal cocaine dosing. *Neurotoxicol. Teratol.* 11: 35–38.

Spinazzola, R. et al. (1992). Neonatal gastrointestinal complications of maternal cocaine abuse. *NY State J. Med.* 92: 22–23.

Sprauve, M. E. et al. (1997). Adverse perinatal outcome in parturients who use crack cocaine. *Obstet. Gynecol.* 89: 674–678.

Stafford, J. R. et al. (1994). Prenatal cocaine exposure and the development of the human eye. *Ophthalmology* 101: 301–308.

Streissguth, A. P. et al. (1991). Cocaine and the use of alcohol and other drugs during pregnancy. *Am. J. Obstet. Gynecol.* 164: 1239–1243.

Suchet, I. B. (1994). Schizencephaly: Antenatal and postnatal assessment with colour-flow Doppler imaging. *Can. Assoc. Radiol. J.* 45: 193–200.

Telsey, A. M., Merrit, T. A., and Dixon, S. D. (1988). Cocaine exposure in a term neonate: Necrotizing enterocolitis as a complication. *Clin. Pediatr.* 27: 547–550.

Tenorio, G. M. et al. (1988). Intrauterine stroke and maternal polydrug abuse. *Clin. Pediatr.* 27: 565–567.

Teske, M. P. and Trese, M. T. (1987). Retinopathy of prematurity-like fundus and persistent hyperplastic primary vitreous associated with maternal cocaine use. *Am. J. Ophthalmol.* 103: 719–720.

Torfs, C. P. et al. (1994). A population-based study of gastroschisis: Demographic, pregnancy, and life-style risk factors. *Teratology* 50: 44–53.

Townsend, R. R., Laing, F. C., and Jeffrey, R. B. (1988). Placental abruption associated with cocaine abuse. *AJRN Am. J. Roentgenol.* 150: 1339–1340.

Vandenanker, J. N. et al. (1991). Prenatal diagnosis of limb-reduction defects due to maternal cocaine use. *Lancet* 338: 1332.

Van Dyke, D. C. and Fox, A. A. (1990). Fetal drug exposure and its possible implications for learning in the preschool-age population. *J. Learn. Disabil.* 23: 160–163.

Vidaeff, A. C. and Mastrobattista, J. M. (2003). *In utero* cocaine exposure: A thorny mix of science and mythology. *Am. J. Perinatol.* 20: 165–172.

Viscarello, R. R. et al. (1992). Limb-body wall complex associated with cocaine abuse: Further evidence of cocaine's teratogenicity. *Obstet. Gynecol.* 80: 523–526.

Volpe, J. J. (1992). Effect of cocaine use on the fetus. *N. Engl. J. Med.* 327: 399–407.

Webster, W. S. et al. (1991). Fetal brain damage in the rat following prenatal exposure to cocaine. *Neurotoxicol. Teratol.* 13: 621-626.

Woods, N. S. et al. (1995). Cocaine use among pregnant women: Socioeconomic, obstetrical and psychological issues. In *Mothers, Babies, and Cocaine: The Role of Toxins in Development*, M. Lewis and M. Bendersky, Eds., Lawrence Erlbaum, Hillsdale, NJ, pp. 305–332.

Zuckerman, B. et al. (1989). Effects of maternal marijuana and cocaine use on fetal growth. *N. Engl. J. Med.* 320: 762–768.

32 Quinine

Chemical name: (8α,9R)-6′-Methoxycinchonan-9-ol

CAS #: 130-95-0

SMILES: c12c(C(C3N4CC(C(C3)CC4)C=C)O)ccnc1ccc(c2)OC

INTRODUCTION

Quinine is an alkaloid obtained from the plant genus *Cinchona*; dried bark of the tree contains ~0.8 to 4% of the chemical (*The Merck Index*, 2001). It has had many uses; it was marketed prior to the establishment of the U.S. Food and Drug Administration (FDA), in 1938, primarily in conjunction with other agents as an antimalarial drug, as a skeletal muscle relaxant, and as a flavoring agent in foods and beverages. Large doses are known to be abortifacient (Dannenberg et al., 1983; Smit and McFayden, 1998). The mechanisms of action of quinine are multiple: It intercalates into DNA, disrupting parasite replication and transcription as an antimalarial agent, and it affects calcium distribution within muscle fibers and decreases the excitability of the motor end-plate region as a neuromuscular agent (Lacy et al., 2004). The chemical is known by its generic name, and the salts of quinine are known by a variety of trade names, including but not limited to Quinamm®, Quine®, Quinsan®, Biquinate®, Dentojel®, Quiphile®, Quinaminoph®, and Quinbisan®. The drug has a pregnancy category of X. This classification is defended on the label where it is stated that the drug may cause fetal harm when administered to a pregnant woman. The label continues with the statement that congenital malformations in the human have been reported with its use, primarily with large doses (up to 30 g) for attempted abortion. In about half of these reports, the malformation was deafness related to auditory nerve hypoplasia. Among the other abnormalities reported were limb anomalies, visceral defects, and visual changes (see below). The label continues, "If this drug is used during pregnancy, or if the patient becomes pregnant while taking the drug, the patient should be apprised of the potential hazard to the fetus" (*PDR*, 2004).

DEVELOPMENTAL TOXICOLOGY

ANIMALS

Quinine has been tested in a variety of animal species for developmental toxicity. The results will be discussed here with reference to testing by the oral route, as that is the route of administration for humans. Quinine induced embryolethality and growth retardation but no malformations in mouse fetuses following doses of up to 500 mg/kg/day during various intervals in gestation (Tanimura, 1972). No developmental toxicity of any kind was evident in the rat when dams were administered up to 300 mg/kg quinine/day during gestation days 7 to 18 (Savini et al., 1971). In a 1938 experiment that was probably inadequately controlled, rabbit does given ~32 mg/day for 10 days in gestation elicited eighth nerve damage of the ear in the fetuses (West, 1938). This defect was observed in some human offspring exposed to the drug (see below). Cochlear damage leading to deafness was also observed in guinea pig fetuses whose dams were given ~1300 mg/day over varying periods of gestation in another 1938 experiment (Mosher, 1938). No significant developmental toxicity was recorded in two species of primates in which the mothers received doses over the range of 20 to 200 mg/kg/day for 3 days early in gestation (Tanimura and Lee, 1972).

HUMANS

A variety of birth defects associated with quinine administration were reported in the literature over the past 50 years, and 45 representatives are provided in Table 1. Review of the pertinent

---

TABLE 1
Reports of Malformations Attributed to Quinine Administration in Humans

Ref.
Taylor, 1933, 1934, 1935, 1937
Richardson, 1936
West, 1938
Forbes, 1940
Winckel, 1948
Ingalls and Prindle, 1949
Grebe, 1952
Kinney, 1953
Windorfer, 1953, 1961
Sylvester and Hughes, 1954
Reed et al., 1955
Uhlig, 1957
Fuhrmann, 1962
Kucera and Benasova, 1962
Robinson et al., 1963
Ferrier et al., 1964
Maier, 1964
McKinna, 1966
Zolcinski et al., 1966
Kup, 1966, 1967
Paufique and Magnard, 1969
Morgon et al., 1971
Nishimura and Tanimura, 1976
McGready et al., 1998

---

publications corroborates the information provided on the package label. A variety of malformations have been described, with no specific embryopathy identifiable. Central nervous system and limb defects appear to be the most common according to some observers (Nishimura and Tanimura, 1976), and in approximately one half of the malformed cases, hearing deficits and outright deafness were observed, apparently related to optic (eighth) nerve hypoplasia. Most of the cases of abnormalities were produced at abortifacient doses (of up to 3 g/day), higher than the therapeutic dose range of 200 to 1950 mg/day orally. Treatment, when provided in the reports, was usually confined to early pregnancy, although some cases occurred rather late in the gestation period. In contrast to these positive reports of malformation, several large studies or reviews found no clear association with teratogenicity by quinine (Mellin, 1964; Nishimura and Tanimura, 1976; Heinonen et al., 1977; Briggs et al., 2002).

Other classes of developmental toxicity were also recorded in association with quinine treatment, including death (Sadler et al., 1930; Kubata, 1939; Kinney, 1953; Mukherjee and Bhose, 1968; Dannenberg et al., 1983) and mental deficiency or retardation (Mautner, 1952; Reed et al., 1955), including the hearing and visual abnormalities discussed above as structural defects. Effects on growth are apparently not a common feature of the toxicity pattern.

One group of experts placed the magnitude of teratogenic risk due to quinine as moderate (for the high abortifacient-type dose levels) and unlikely for the low, therapeutic doses (Friedman and Polifka, 2000). The most complete review of quinine and its developmental toxicity potential, in 70 treated cases, was published by Dannenberg and associates (1983). Phillips-Howard and Wood (1996) published a recent review on the subject.

CHEMISTRY

Quinine is a larger compound of lower polarity. It is a hydrophobic molecule and can engage in hydrogen bonding interactions. The calculated physicochemical and topological propertes for quinine are listed below.

PHYSICOCHEMICAL PROPERTIES

Parameter	Value
Molecular weight	324.423 g/mol
Molecular volume	304.67 A^3
Density	1.071 g/cm^3
Surface area	359.06 A^2
LogP	2.312
HLB	5.069
Solubility parameter	24.501 J$^{(0.5)}$/cm$^{(1.5)}$
Dispersion	21.126 J$^{(0.5)}$/cm$^{(1.5)}$
Polarity	5.232 J$^{(0.5)}$/cm$^{(1.5)}$
Hydrogen bonding	11.253 J$^{(0.5)}$/cm$^{(1.5)}$
H bond acceptor	0.86
H bond donor	0.32
Percent hydrophilic surface	28.39
MR	94.006
Water solubility	−3.554 log (mol/M^3)
Hydrophilic surface area	101.93 A^2
Polar surface area	45.59 A^2
HOMO	−8.269 eV
LUMO	−0.666 eV
Dipole	1.865 debye

Topological Properties (Unitless)

Parameter	Value
x0	16.681
x1	11.707
x2	10.366
xp3	9.632
xp4	8.161
xp5	7.117
xp6	4.926
xp7	3.595
xp8	2.279
xp9	1.580
xp10	1.027
xv0	14.059
xv1	8.683
xv2	6.974
xvp3	5.820
xvp4	4.444
xvp5	3.521
xvp6	2.115
xvp7	1.350
xvp8	0.760
xvp9	0.461
xvp10	0.271
k0	33.125
k1	17.416
k2	7.319
k3	3.241
ka1	15.759
ka2	6.280
ka3	2.682

REFERENCES

Briggs, G. G., Freeman, R. K., and Yaffe, S. J. (2002). *Drugs in Pregnancy and Lactation. A Reference Guide to Fetal and Neonatal Risk*, Sixth ed., Lippincott Williams & Wilkins, Philadelphia.

Dannenberg, A. L., Dorfman, S. F., and Johnson, J. (1983). Use of quinine for self-induced abortion. *South. Med. J.* 76: 846–849.

Ferrier, P., Widgren, S., and Ferrier, S. (1964). Nonspecific pseudohermaphroditism: Report of two cases with cytogenetic investigations. *Helv. Paediatr. Acta* 19: 1–12.

Forbes, S. B. (1940). The etiology of nerve deafness with particular reference to quinine. *South. Med. J.* 33: 613–621.

Friedman, J. M. and Polifka, J. E. (2000). *Teratogenic Effects of Drugs. A Resource for Clinicians (TERIS)*, Second ed., Johns Hopkins University Press, Baltimore, MD.

Fuhrmann, W. (1962). Genetische und peristatische Ursachen ungeborener Angiokardiopathien. *Ergen. Inn. Med. Kinderheilkd.* 18: 47–115.

Grebe, H. (1952). Konnen abtreibungsversuche zu missbildungen fuhren? *Geburtschilfe Frauenheilkd.* 12: 333–339.

Heinonen, O. P., Slone, D., and Shapiro, S. (1977). *Birth Defects and Drugs in Pregnancy*, Publishing Sciences Group, Littleton, MA.

Ingalls, T. H. and Prindle, R. A. (1949). Esophageal atresia with tracheoesophageal fistula. *N. Engl. J. Med.* 240: 987–995.

Kinney, M. D. (1953). Hearing impairments in children. *Laryngoscope* 63: 220.

Kubata, T. (1939). [One case of the fetal death by a small dose of quinine]. *Nippon Fujinko Gakiwi Zasshi* 22: 128.

Kucera, J. and Benasova, D. (1962). Poruchy Nitrodelozniho Vyvoje Clovela Zpusobene Pokusem O. *Potrat. Cesk. Pediatr.* 17: 483–489.

Kup, J. (1966). Multiple missbildungen nach chinoneinnahme der schwangerschaft. *Munch. Med. Wochenschr.* 108: 2293–2294.

Kup, J. (1967). [Diaphragm defect following abortion attempt with quinine. Clinical case report]. *Munch. Med. Wochenschr.* 109: 2582–2583.

Lacy, C. F. et al. (2004). *Drug Information Handbook (Pocket), 2004–2005*, Lexi-Comp. Inc., Hudson, OH.

Maier, W. (1964). Unser derzeitiges Wissen uberaussere Schadigende Einflusse auf den Embryo und ange- borene Missbildungen. *Dtsch. Z. Gesamte Gerichtl. Med.* 55: 156–172.

Mautner, H. (1952). Pranatale vergiftungen. *Wien. Klin. Wochenschr.* 64: 646–647.

McGready, R. et al. (1998). Quinine and mefloquine in the treatment of multidrug-resistant *Plasmodium falciparum* malaria in pregnancy. *Ann. Trop. Med. Parasitol.* 92: 643–653.

McKinna, A. J. (1966). Quinine induced hypoplasia of the optic nerve. *Can. J. Ophthalmol.* 1: 261–266.

Mellin, G. W. (1964). Drugs in the first trimester of pregnancy and fetal life of *Homo sapiens. Am. J. Obstet. Gynecol.* 90: 1169–1180.

Morgon, A., Charachon, D., and Bringuier, N. (1971). Disorders of the auditory apparatus caused by embry- opathy or foetopathy. Prophylaxis and treatment. *Acta Otolaryngol. Suppl. (Stockh.)* 291: 1–27.

Mosher, H. P. (1938). Does animal experimentation show similar changes in the ear of mother and fetus after the ingestion of quinine by the mother? *Laryngoscope* 48: 361–395.

Mukherjee, S. and Bhose, L. N. (1968). Induction of labor and abortion with quinine infusion in intrauterine fetal death. *Am. J. Obstet. Gynecol.* 101: 853–854.

Nishimura, H. and Tanimura, T. (1976). *Clinical Aspects of the Teratogenicity of Drugs*, Excerpta Medica, American Elsevier, New York, pp. 140–143.

Paufique, L. and Magnard, P. (1969). [Retinal degeneration in 2 children following preventive antimalarial treatment of the mother during pregnancy]. *Bull. Soc. Ophthalmol. Fr.* 69: 466–467.

PDR® (*Physicians' Desk Reference®*). (2004). Medical Economics Co., Inc., Montvale, NJ.

Phillips-Howard, P. A. and Wood, D. (1996). The safety of antimalarial drugs in pregnancy. *Drug Saf.* 14: 131–145.

Reed, H., Briggs, J. N., and Martin, J. K. (1955). Congenital glaucoma, deafness, mental deficiency and cardiac anomaly following attempted abortion. *J. Pediatr.* 46: 182–185.

Richardson, S. (1936). The toxic effect of quinine on the eye. *South. Med. J.* 29: 1156–1164.

Robinson, G. C., Brummit, J. R., and Miller, J. R. (1963). Hearing loss in infants and preschool children. II. Etiological considerations. *Pediatrics* 32: 115–124.

Sadler, E. S., Dilling, W. J., and Gemmell, A. A. (1930). Further investigations into the death of the child following the induction of labour by means of quinine. *J. Obstet. Gynaecol. Br. Emp.* 37: 529–546.

Savini, E. C., Moulin, M. A., and Herrou, M. F. (1971). Experimental study of the effects of quinine on the fetus of rats, rabbits, and dogs. *Therapie* 26: 563–574.

Smit, J. A. and McFadyen, M. L. (1998). Quinine as an unofficial contraceptive — concerns about safety and efficacy. *S. Afr. Med. J.* 88: 865–866.

Sylvester, P. E. and Hughes, D. R. (1954). Congenital absence of both kidneys. A report of four cases. *Br. Med. J.* 1: 77–79.

Tanimura, T. (1972). Effects on macaque embryos of drugs reported or suspected to be teratogenic to humans. *Acta Endocrinol. Suppl. (Copenh.)* 166: 293–308.

Tanimura, T. and Lee, S. (1972). Discussion on the suspected teratogenicity of quinine to humans. *Teratology* 6: 122.

Taylor, H. M. (1933). Does quinine used in the induction of labour injure the ear of the fetus? *J. Fla. Med. Assoc.* 20: 20–22.

Taylor, H. M. (1934). Prenatal medication as a possible etiologic facor of deafness in the newborn. *South. Med. J.* 20: 790–803.

Taylor, H. M. (1935). Further observations on prenatal medication as a possible etiologic factor of deafness in the newborn. *South. Med. J.* 28: 125–130.

Taylor, H. M. (1937). Prenatal medication and its relation to the fetal ear. *Surg. Gynecol. Obstet.* 64: 542–546.

The Merck Index: An Encyclopedia of Chemicals, Drugs, and Biologicals, Thirteenth ed. (2001). Merck & Co., Inc., Whitehouse Station, NJ.

Uhlig, H. (1957). [Abnormalities in undesired children]. *Aerztl. Wochenschr.* 12: 61–64.

West, R. A. (1938). Effect of quinine upon auditory nerve. *Am. J. Obstet. Gynecol.* 36: 241–248.

Winckel, C. F. W. (1948). Quinine and congenital injuries of ear and eye of foetus. *J. Trop. Med. Hyg.* 51: 2–7.

Windorfer, A. (1953). Zum problem der missbildungen durch bewusste Keimund Fruchtschadigung. *Med. Klin.* 48: 293–297.

Windorfer, A. (1961). Uber die ursachen angeborener missbildungen. *Bundesgesundheitablatt* 6: 81–84.

Zolcinski, A., Heimroth, T., and Ujec, M. (1966). Quinine as a cause of dysplasia of the fetus. *Zentralbl. Gynaekol.* 88: 99–104.

33 Methylene Blue

Chemical name: 3,7-*Bis*(dimethylamino)phenothiazin-5-ium chloride

Alternate name: Methylthioninium chloride

CAS #: 61-73-4

SMILES: c12c(nc3c([s⁺]1)cc(N(C)C)cc3)ccc(c2)N(C)C

INTRODUCTION

Methylene blue is a vital dye that has therapeutic utility as an antidote for cyanide poisoning and drug-induced methemoglobinemia. As a dye, it is also used as a marker in various tissues and amniotic fluid. The chemical is used in the latter to identify by amniocentesis midtrimester anatomic and pathologic structures in twin pregnancies, especially to diagnose premature rupture of membranes. For nondye uses, the chemical acts by hastening the conversion of methemoglobin to hemoglobin in treating methemoglobinemia, and combines with cyanide to form cyanmethemoglobin, preventing its interference with the cytochrome system in cyanide poisoning (Lacy et al., 2004). It is known by the trade name Urolene Blue®, as well as by a host of other names, including C. I. Basic Blue 9, Swiss blue, and Solvent Blue 8, among others. It has a pregnancy category of C (inferring "risk cannot be ruled out") except when used intra-amniotically, where the category changes to D. The latter is presumably due to concern over whether the chemical can harm the fetus in pregnant women (see below).

DEVELOPMENTAL TOXICOLOGY

ANIMALS

Only one animal developmental toxicity study has been published following intra-amniotic (IA) injection of methylene blue (the concern in humans). In that study, increased fetal death and malformations occurred in rats at a dose of 5µl of 1 to 4% dye in water administered on gestation day 16, an amount equal to that used in humans (Piersma et al., 1991). The chemical was teratogenic and embryolethal in the mouse following subcutaneous injection of 35 mg/kg/day and higher on a single day of gestation (Tiboni and Lamonaca, 2000).

TABLE 1
Reports of Intestinal Malformation Associated with Methylene Blue Intra-amniotic Injections in Humans

Ref.

Moorman-Voestermons et al., 1990
Nicolini and Monni, 1990
Pruggmayer, 1991
Lopes et al., 1991
Fish and Chazen, 1992
Treffers, 1992
Lancaster et al., 1992
Van der Pol et al., 1992
Gluer, 1995

HUMANS

In the human, the use of methylene blue IA injections in pregnancy resulted in a number of case reports of malformation, as tabulated by >60 representative cases provided in Table 1. The malformations observed were multiple, occlusive intestinal defects in the form of atresia or stenosis of the ileum or jejunum. Dolk (1991), in a review of use of the dye in amniocentesis in 11 European countries, reported 119 cases of atresia/stenosis, but considered that the malformation was not much more prevalent than in cases in which the dye was not used.

The malformations occurred in these cases following IA instillation at approximately 16 weeks of pregnancy (second trimester) according to one reviewer (Cragan, 1999). They usually occurred in one twin, with the other twin unaffected, and the higher incidences of malformation (and death) were associated with higher concentrations of the dye (usual marking dose is 0.125 to 0.25% up to 1% solution of 1 to 10 ml in saline). One investigator suggested that the smaller doses would be adequate to confirm the status of fetal membranes without causing hemolysis (Plunkett, 1973). The mechanism for these adverse effects is most likely due to vascular disruption caused by arterial constriction induced by the dye (Cragan, 1999).

Other developmental toxicity associated with the malformations is also apparent. In several studies by a group of investigators, death of one or both twins occurred in greater incidence (32% versus 15%, odds ratio [OR] = 4.63, 95% confidence interval [CI] 0.93 to 23.13) at the lower concentrations of 0.125 to 0.25% and the higher concentration (OR = 14.98, 95% CI 3.40 to 66.08) of 1.0% than when methylene blue was not used (Kidd et al., 1996). Among stillbirths reported in another study, the death rate was nearly twice (5.3% versus 3.1%) as high as among nonexposed fetuses (Kidd et al., 1997). Functional alterations in the form of neonatal anemia, hyperbilirubinemia, and jaundice were recorded in a number of infants born of mothers following IA injection of the dye (Cowett et al., 1976; Serota et al., 1979; Spahr et al., 1980; Kirsch and Cohen, 1980; Crooks, 1982; McEnerney and McEnerney, 1983; Vincer et al., 1987; Poinsot et al., 1988; Fish and Chazen, 1992). Blue staining of the newborn occurs commonly. Growth alterations are not a component of the adverse findings (Kidd et al., 1997).

In summary, the developmental toxicity profile of methylene blue IA injection during pregnancy consists of malformation (intestinal atresia, stenosis) and death and functional deficit (neonatal hemolytic anemia and jaundice) in some cases. One group of experts placed the magnitude of teratogenic risk from IA injections of methylene blue during pregnancy as moderate to high (Friedman and Polifka, 2000). Several reviews on this subject were published (Dolk, 1991; Cragan, 1999; Bailey, 2003; Briggs et al., 2005). Suggestions were made to utilize markers other than methylene blue to alleviate the toxicity demonstrated by methylene blue during pregnancy (McFadyen, 1992). Ultrasound and the dyes Indigo carmine and Evans blue were proposed in this regard.

CHEMISTRY

Methylene blue is a positively charged human developmental toxicant. It is hydrophobic and of average size in comparison to the other compounds. It possesses a relatively low polar surface area. Methylene blue is a weak hydrogen bond acceptor. The calculated physicochemical and topological properties are given below.

PHYSICOCHEMICAL PROPERTIES

Parameter	Value
Molecular weight	284.405 g/mol
Molecular volume	259.50 A^3
Density	1.121 g/cm^3
Surface area	304.49 A^2
LogP	3.299
HLB	6.320
Solubility parameter	24.053 J$^{(0.5)}$/cm$^{(1.5)}$
Dispersion	21.639 J$^{(0.5)}$/cm$^{(1.5)}$
Polarity	6.750 J$^{(0.5)}$/cm$^{(1.5)}$
Hydrogen bonding	8.048 J$^{(0.5)}$/cm$^{(1.5)}$
H bond acceptor	0.34
H bond donor	0.04
Percent hydrophilic surface	33.83
MR	87.049
Water solubility	−2.136 log (mol/M^3)
Hydrophilic surface area	103.00 A^2
Polar surface area	19.37 A^2
HOMO	−11.618 eV
LUMO	−5.801 eV
Dipole	0.655 debye

TOPOLOGICAL PROPERTIES (UNITLESS)

Parameter	Value
x0	14.276
x1	9.542
x2	9.146
xp3	7.457
xp4	5.913
xp5	5.256
xp6	3.924
xp7	2.962
xp8	2.166
xp9	1.646
xp10	0.920
xv0	12.866
xv1	7.220
xv2	6.173
xvp3	4.254
xvp4	2.984
xvp5	2.358
xvp6	1.486

Continued.

Parameter	Value
xvp7	1.002
xvp8	0.613
xvp9	0.408
xvp10	0.196
k0	19.398
k1	14.917
k2	6.012
k3	3.122
ka1	13.491
ka2	5.127
ka3	2.576

REFERENCES

Bailey, B. (2003). Are there teratogenic risks associated with antidotes used in the acute management of poisoned pregnant women? *Birth Defects Res. (A)* 67: 133–140.

Briggs, G. G., Freeman, R. K., and Yaffe, S. J. (2005). *Drugs in Pregnancy and Lactation. A Reference Guide to Fetal and Neonatal Risk*, Seventh ed., Lippincott Williams & Wilkins, Philadelphia.

Cowett, R. M. et al. (1976). Untoward neonatal effect of intraamniotic administration of methylene blue. *Obstet. Gynecol.* 48 (Suppl.): 74S–75S.

Cragan, J. D. (1999). Teratogen update: Methylene blue. *Teratology* 60: 42–48.

Crooks, J. (1982). Hemolytic jaundice in a neonate after intraamniotic injection of methylene blue. *Arch. Dis. Child.* 57: 872–886.

Dolk, H. (1991). Methylene blue and atresia or stenosis of ileum and jejunum. EUROCAT Working Group. *Lancet* 338: 1021–1022.

Fish, W. H. and Chazen, E. M. (1992). Toxic effects of methylene blue on the fetus. *Am. J. Dis. Child.* 46: 1412–1413.

Friedman, J. M. and Polifka, J. E. (2000). *Teratogenic Effects of Drugs. A Resource for Clinicians (TERIS)*, Second ed., Johns Hopkins University Press, Baltimore, MD.

Gluer, S. (1995). Intestinal atresia following intraamniotic use of dyes. *Eur. J. Pediatr. Surg.* 5: 240–242.

Kidd, S. A. et al. (1996). Fetal death after exposure to methylene blue dye during mid-trimester amniocentesis in twin pregnancy. *Prenat. Diagn.* 16: 39–47.

Kidd, S. A. et al. (1997). A cohort study of pregnancy outcome after amniocentesis in twin pregnancy. *Paediatr. Perinat. Epidemiol.* 11: 200–213.

Kirsch, I. R. and Cohen, H. J. (1980). Heinz body hemolytic anemia from the use of methylene blue in neonates. *J. Pediatr.* 96: 276–278.

Lacy, C. F. et al. (2004). *Drug Information Handbook (Pocket), 2004–2005*, LexiComp. Inc., Hudson, OH.

Lancaster, P. A. L. et al. (1992). Intraamniotic methylene blue and intestinal atresia in twins. *J. Perinatol. Med.* 20 (Suppl. 1): 262.

Lopes, P. et al. (1991). [Ileal stenosis of the methylene blue staining of the amniotic fluid during early amniocentesis]. *Presse Med.* 20: 1568–1569.

McEnerney, J. K. and McEnerney, L. N. (1983). Unfavorable neonatal outcome after intraamniotic injection of methylene blue. *Obstet. Gynecol.* 61 (Suppl.): 35S–37S.

McFadyen, I. (1992). The dangers of intra-amniotic methylene blue. *Br. J. Obstet. Gynaecol.* 99: 89–90.

Moorman-Voestermons, C. G., Heij, H. A., and Vos, A. (1990). Jejunal atresia in twins. *J. Pediatr. Surg.* 25: 638–639.

Nicolini, U. and Monni, G. (1990). Intestinal obstruction in babies exposed *in utero* to methylene blue. *Lancet* 336: 1258–1259.

Piersma, A. H. et al. (1991). Embryotoxicity of methylene blue in the rat. *Teratology* 43: 458–459.

Plunkett, G. D. (1973). Neonatal complications. *Obstet. Gynecol.* 41: 476–477.

Poinsot, J. et al. (1988). Neonatal hemolytic anemia after intraamniotic injection of methylene blue. *Arch. Fr. Pediatr.* 45: 657–660.

Pruggmayer, M. (1991). [Intra-amnion administration of methylene blue within the scope of amniocentesis in multiple pregnancy — significant association with congenital small intestine obstruction]. *Geburtshilfe Frauenheilkd.* 51: 161.

Serota, F. T., Bernbaum, J. C., and Schwartz, E. (1979). The methylene-blue baby. *Lancet* 2: 1142–1143.

Spahr, R. C. et al. (1980). Intraamniotic injection of methylene blue leading to methemoglobinemia in one of twins. *Int. J. Gynaecol. Obstet.* 17: 477–478.

Tiboni, G. M. and Lamonaca, D. (2000). Transplacental exposure to methylene blue causes anterior neural tube closure defects in the mouse. *Teratology* 61: 477.

Treffers, P. E. (1992). [Methylene blue and pregnancy; revision of a calamity]. *Ned. Tijdschr.* 136: 1285–1286.

Van der Pol, J. G. et al. (1992). Jejunal atresia related to the use of methylene blue in genetic amniocentesis in twins. *Br. J. Obstet. Gynaecol.* 99: 1141–1143.

Vincer, M. J. et al. (1987). Methylene-blue-induced hemolytic anemia in a neonate. *Can. Med. Assoc. J.* 136: 503–504.

34 Warfarin

Chemical name: 4-Hydroxy-3-(3-oxo-1-phenylbutyl)-2*H*-1-benzopyran-2-one

CAS #: 81-81-2

SMILES: C1(C(c2ccccc2)CC(C)=O)=C(c3c(OC1=O)cccc3)O

INTRODUCTION

Warfarin is a coumarin derivative used as an anticoagulant in the prophylaxis and treatment of various thromboses and thromboembolic disorders. Mechanistically, it interferes with the hepatic synthesis of a number of vitamin-K-dependent coagulation factors (Lacy et al., 2004). In fact, an association between deficiency of these factors and the malformative phenotype induced by warfarin (see below) has been made (Pauli et al., 1987). The drug is available commercially by prescription under the trade name Coumadin® and is one of the top 100 most-often-prescribed drugs in the United States (www.rxlist.com). It has a pregnancy risk category of D. The package label has a pregnancy statement that the drug is contraindicated in women who are or may become pregnant, because the drug passes through the placental barrier and may cause fetal hemorrhage *in utero*. Furthermore, there have been reports of birth malformations in children born to mothers who were treated with warfarin during pregnancy. The label continues with the malformation descriptions and warnings about spontaneous abortion and stillbirth, low birth weight, and growth retardation attendant with its use (*PDR*, 2002; see below).

DEVELOPMENTAL TOXICOLOGY

ANIMALS

In laboratory animal studies, there are few published reports following oral administration, the pertinent and usual method of administration in humans, the exception being studies with the rat. In this species, oral doses of 50 or 100 mg/kg/day plus 10 mg/kg vitamin K1 during organogenesis caused hemorrhage in about one quarter of the offspring, resulting in central nervous system, facial, and limb defects (Howe and Webster, 1989). This regimen is a model for some of the concordant effects in the human (Howe and Webster, 1992, 1993; see below).

HUMANS

In humans, warfarin is associated with several different phenotypes induced by it. It is established that there are two distinct types of defects associated with the coumarin drugs, especially warfarin, dependent on the time administered during pregnancy (Hall, 1976).

Early Effects

A characteristic embryopathy, described as the warfarin embryopathy, the fetal warfarin syndrome, or coumarin embryopathy (because other coumarins in the class may be involved, see below), occurs after early, first trimester use (Schardein, 2000). Initially, the abnormalities were given the diagnosis "chondrodystrophy punctata," but later distinction was made, and this was determined to be a different entity altogether from the genetic disorder of that name. As summarized, the characteristic abnormalities of the embryopathy are of the skeleton. The most common consistent feature is a hypoplastic nose. The other common feature is bony abnormalities of the axial and appendicular skeleton, the most prominent being radiological stippling, particularly of the vertebral column, most dramatically in the lumbosacral area. Kyphoscoliosis, abnormal skull development, and brachydactyly have been irregularly observed as associated skeletal defects. Other, nonskeletal abnormalities reported in association with the syndrome include ophthalmological malformations, developmental delay, low birth weight (premature birth), mental retardation, nail hypoplasia, hypotonia, ear anomalies, hypertelorism, and death. A tabulation of over 70 cases identified as reporting the warfarin embryopathy among the malformations reported is given in Table 1.

There is a significant risk to first trimester treatment with warfarin or any other coumarin. Contemporaneous terminology is *coumarin embryopathy*, because almost identical defects are observed following treatment with other coumarins: acenocoumarol, phenindione, and phenprocoumon. It was estimated that about 10% of infants born alive to mothers who took warfarin during pregnancy have the warfarin embryopathy (Hall et al., 1980). However, another group of investigators stated that the magnitude of teratogenic risk is small to moderate (Friedman and Polifka, 2000). The critical period of exposure for embryopathy appears to be the sixth to ninth weeks of gestation (Hall et al., 1980). Many reports indicate that doses in the therapeutic range of 2 to 10 mg/day were effective in eliciting the embryopathy.

Late Effects

In contrast to the warfarin embryopathy of nasal hypoplasia and bony defects described above resulting from first trimester exposures, there are central nervous system (CNS) malformations including those of the eyes, visualized in offspring of mothers treated later in gestation, in the second or third trimesters. The 14 or so reported cases are tabulated in Table 2. The CNS malformations appear to be of two types: (1) dorsal midline malformations characterized by agenesis of corpus callosum, Dandy-Walker defects, and cerebellar atrophy, and (2) ventral midline defects characterized by optic atrophy (Hall et al., 1980). Many are deformations related to hemorrhage. The frequency of CNS structural anomalies among liveborn infants whose mothers took warfarin during late pregnancy was estimated at about 3% (Hall et al., 1980). These abnormalities may be observed in association with the features of the embryopathy or in otherwise unaffected infants whose mothers took warfarin during pregnancy (Friedman and Polifka, 2000).

Many cases of spontaneous abortion and death were also reported in the absence of embryopathy (Epstein, 1959; Kenmure, 1968; Palacios-Macedo et al., 1969; Radnich and Jacobs, 1970; Harrison and Roschke, 1975; Ibarra-Perez et al., 1976; Chen et al., 1982; Sheikhzadeh et al., 1983).

It was speculated (Shaul et al., 1975; Shaul and Hall, 1977) that microhemorrhage in the vascular embryonic cartilage might eventually result in scarring and calcification and thus be evidenced by stippling at birth in the case of the embryopathy. This is unlikely, because clotting factors affected by vitamin K antagonists are not yet demonstrable in the embryo at the 6- to 9-

TABLE 1
Developmental Toxicity Profile of Warfarin Following Exposure to Humans in Early Pregnancy

Case Number	Malformations	Growth Retardation	Death	Functional Deficit	Ref.
1	Embryopathy, eyes			✔	DiSaia, 1966
2	Embryopathy, limbs			✔	Kerber et al., 1968
3[a]	Embryopathy, brain and lungs (pathology)		✔		Ikonen et al., 1970
4	Embryopathy, limbs				Holmes et al., 1972 (Baker case)
5[a]	Embryopathy, brain, ears, palate, skull, vessels			✔	Tejani, 1973
6[a]	Embryopathy, eyes, ears, limbs		✔		Becker et al., 1975
7[a]	Embryopathy, limbs, digits				Fourie and Hay, 1975
8[a]	Embryopathy, limbs, eyes				Pettifor and Benson, 1975a
9[a]	Embryopathy				Pettifor and Benson, 1975a
10	Embryopathy				Pettifor and Benson, 1975b (Hirsh case)
11[a]	Embryopathy				Shaul et al., 1975
12, 13	Embryopathy				Shaul et al., 1975
14, 15[b]	Brain		✔		Warkany and Bofinger, 1975
16[a]	Embryopathy, muscle			✔	Pauli et al., 1976; Collins et al., 1977 (Kranzler case)
17	Embryopathy, eyes				Richman and Lahman, 1976
18	Embryopathy, brain, eyes	✔		✔	Holzgreve et al., 1976
19	Embryopathy, limbs		✔		Barr and Burdi, 1976
20	Embryopathy				Shaul and Hall, 1977 (O'Connor case)
21	Embryopathy		✔		Abbott et al., 1977
22[a]	Embryopathy, limbs		✔		Raivio et al., 1977
23[b]	Spleen, digits				Cox et al., 1977
24	Embryopathy				Robinson et al., 1978
25	Embryopathy		✔		Gooch et al., 1978 (Wilroy and Summit case)
26[a]	Embryopathy, skull		✔		Smith and Cameron, 1979
27[a]	Embryopathy				Hall et al., 1980 (Madden case)
28	Embryopathy				Hall et al., 1980 (Lutz case)
29[a]	Embryopathy, limbs				Hall et al., 1980 (Johnson case)
30[a]	Embryopathy				Hall et al., 1980 (MacLeod case)
31[a]	Embryopathy				Hall et al., 1980 (Pauli case)
32	Embryopathy				Hall et al., 1980 (Pauli case)
33[a]	Embryopathy, brain			✔	Hall et al., 1980 (Pauli case)
34	Embryopathy, eyes			✔	Stevenson et al., 1980
35	Embryopathy, limbs				Whitfield, 1980
36	Embryopathy				Curtin and Mulhern, 1980
37[a]	Embryopathy				Baillie et al., 1980
38[a]	Embryopathy, inner ear			✔	Harrod and Sherrod, 1981
39[a]	Embryopathy, eyes				Harrod and Sherrod, 1981
40	Embryopathy				Sugrue, 1981 (O'Neill et al. case)

Continued.

TABLE 1 *(Continued)*
Developmental Toxicity Profile of Warfarin Following Exposure to Humans in Early Pregnancy

Case Number	Malformations	Growth Retardation	Death	Functional Deficit	Ref.
41[a,b]	Heart, lungs				Dean et al., 1981
42[a]	Embryopathy, brain				Schivazappa, 1982
43	Embryopathy		✔		Sheikhzadeh et al., 1983
44	Embryopathy				Galil et al., 1984
45[a]	Embryopathy, eyes			✔	Hill and Tennyson, 1984
46–48	Embryopathy				Salazar et al., 1984
49	Embryopathy, digits				Lamontagne and Leclerc, 1984
50[b]	Body wall				O'Donnell et al., 1985
51	Embryopathy, eyes			✔	Zakzouk, 1986
52[b]	Jaw, tongue, digits	✔			Ruthnum and Tolmie, 1987
53[a]	Embryopathy				Tamburrini et al., 1987
54	Embryopathy				Holmes, 1988
55	Embryopathy, renal, genital, digits				Hall, 1989
56[b]	Body wall, lungs		✔		Normann and Stray-Pedersen, 1989
57[a]	Embryopathy				Patil, 1991
58	Embryopathy				Born et al., 1992
59	Embryopathy	✔			Born et al., 1992
60	Embryopathy		✔		Born et al., 1992
61[a]	Embryopathy, brain				Mason et al., 1992
62[b]	Brain				Ville et al., 1993
63[b]	Heart (pathology)				Ville et al., 1993
64[a,b]	Brain, heart (pathology), eyes, lip/palate				Wong et al., 1993
65	Embryopathy, heart, viscera				Barker et al., 1994
66	Embryopathy, limbs, brain (pathology)		✔		Pati and Helmbrecht, 1994
67	Embryopathy				Lee et al., 1994
68[b]	Heart				Lee et al., 1994
69[a]	Embryopathy				Howe et al., 1997
70	Embryopathy, genitals				Takano et al., 1998
71[a]	Embryopathy, brain, limbs, heart				Wellesley et al., 1998
72	Embryopathy, limbs, digits				Wellesley et al., 1998
73[a]	Embryopathy, brain, limbs	✔			Tongsong et al., 1999
74	Embryopathy		✔		Vitale et al., 1999
75	Embryopathy				Sonoda, 2000
76	Embryopathy				Nagai, 2001
77, 78	Embryopathy, heart		✔		Cotrufo et al., 2002
79, 80	Embryopathy				Cotrufo et al., 2002
81[a]	Embryopathy, brain, face, heart, body wall, ears		✔		Chan et al., 2003
82	Embryopathy				Bradley et al., 2003
83[a]	Embryopathy, heart, digits	✔			Hou, 2004

[a] Also treated late in pregnancy.
[b] No embryopathy reported.

TABLE 2
Developmental Toxicity Profile of Warfarin Following Human Exposures in Late Pregnancy

Case Number	Malformations	Growth Retardation	Death	Functional Deficit	Ref.
1	CNS and lungs (pathology), embryopathy		✔		Ikonen et al., 1970
2[a]	CNS, embryopathy, ears, skull, palate, vessels			✔	Tejani, 1973
3	CNS, renal				Warkany and Bofinger, 1975
4	CNS, eyes			✔	Carson and Reid, 1976
5	CNS, eyes			✔	Sherman and Hall, 1976
6[b]	Embryopathy, muscle			✔	Pauli et al., 1976; Collins et al., 1977
7	CNS, embryopathy			✔	Hall et al., 1980 (Pauli case)
8	CNS			✔	Hall et al., 1980
9[a]	CNS, embryopathy				Schivazappa, 1982
10[a]	CNS				Kaplan et al., 1982; Kaplan, 1985
11[b]	Embryopathy				Hill and Tennyson, 1984
12[a]	CNS, embryopathy				Mason et al., 1992
13	CNS				Sheikhzadeh et al., 1993
14[a]	CNS, heart (pathology), lip/palate				Wong et al., 1993
15[a]	CNS, embryopathy	✔			Tongsong et al., 1999
16[a]	CNS, heart, body wall, face, ears, embryopathy		✔		Chan et al., 2003

Note: CNS = central nervous system.

[a] Also treated early in pregnancy.
[b] No central nervous system malformations.

week stage of development (Hall et al., 1980). It is suggested by other evidence that the abnormalities are due to a basic disorder in chondrogenesis, not to focal hemorrhage, through disorganization of the islands of cartilage that calcify in advance of the surrounding cartilage (Barr and Burdi, 1976). This may be accomplished by coumarin derivatives inhibiting posttranslational carboxylation of coagulation proteins at the molecular level (Hall et al., 1980), thereby decreasing the ability of proteins to bind calcium (Stenflo and Suttie, 1977; Price et al., 1981). Rather than microscopic bleeding, it is this inhibition of calcium binding by proteins that explains the bony abnormalities. More recent research demonstrates that it is quite probable that at least three manifestations in the human fetus can result through vitamin K deficiency mechanisms. These are (1) *warfarin embryopathy* as reported here, resulting from coumarin-induced vitamin K deficiency and vitamin-K-dependent proteins by inhibition of vitamin recycling in the embryo matrix gla protein (Price et al., 1981; Suttie, 1991); (2) epoxide reductase deficiency (the *pseudo-warfarin embryopathy*), due to an inborn deficiency of the vitamin K epoxide reductase enzyme (Gericke et al., 1978; Pauli, 1988; Pauli and Haun, 1993); and (3) intestinal malabsorption, due secondarily to disease processes interfering with metabolism of the vitamin (Menger et al., 1997). The latter two conditions can result in *phenocopies* of warfarin embryopathy. Warfarin-induced vitamin K deficiency during early pregnancy is also an established etiology for Binder syndrome (Jaillet et al., 2005), the latter a maxillonasal dysostosis characterized by midface and nasal hypoplasia, short terminal phalanges, and transient radiological features of chondrodystrophy punctata.

In summary, warfarin given during pregnancy in humans induces all classes of developmental toxicity, whether given early or late in gestation. Several review articles on coumarin embryopathy are available (Warkany, 1976; Bates and Ginsberg, 1997; van Driel et al., 2002).

CHEMISTRY

Warfarin is an average-sized hydrophobic human developmental toxicant. It is of average polarity in comparison to the other compounds. Warfarin can engage in hydrogen bonding interactions, primarily as a hydrogen bond acceptor. The calculated physicochemical and topological properties for this compound are shown in the following.

PHYSICOCHEMICAL PROPERTIES

Parameter	Value
Molecular weight	308.334 g/mol
Molecular volume	273.93 A^3
Density	1.181 g/cm^3
Surface area	318.14 A^2
LogP	3.140
HLB	5.341
Solubility parameter	26.134 $J^{(0.5)}/cm^{(1.5)}$
Dispersion	22.869 $J^{(0.5)}/cm^{(1.5)}$
Polarity	4.893 $J^{(0.5)}/cm^{(1.5)}$
Hydrogen bonding	11.664 $J^{(0.5)}/cm^{(1.5)}$
H bond acceptor	0.78
H bond donor	0.32
Percent hydrophilic surface	29.57
MR	86.729
Water solubility	−1.544 log (mol/M^3)
Hydrophilic surface area	94.08 A^2
Polar surface area	73.83 A^2
HOMO	−9.177 eV
LUMO	−1.200 eV
Dipole	6.248 debye

TOPOLOGICAL PROPERTIES (UNITLESS)

Parameter	Value
x0	16.397
x1	11.075
x2	10.049
xp3	8.075
xp4	7.242
xp5	5.971
xp6	4.016
xp7	2.581
xp8	1.833
xp9	1.214
xp10	0.626
xv0	12.653

Continued.

Parameter	Value
xv1	7.367
xv2	5.517
xvp3	3.862
xvp4	2.815
xvp5	1.968
xvp6	1.058
xvp7	0.604
xvp8	0.351
xvp9	0.190
xvp10	0.079
k0	30.116
k1	17.811
k2	7.920
k3	3.984
ka1	15.340
ka2	6.281
ka3	2.994

REFERENCES

Abbott, A., Sibert, J. R., and Weaver, J. B. (1977). Chondrodysplasia punctata and maternal warfarin treatment. *Br. Med. J.* 1: 1639–1640.

Baillie, M., Allen, E. D., and Elkington, A. R. (1980). The congenital warfarin syndrome. A case report. *Br. J. Ophthalmol.* 64: 633–635.

Barker, D. P., Konje, J. C., and Richardson, J. A. (1994). Warfarin embryopathy with dextrocardia and situs inversus. *Acta Paediatr.* 83: 411.

Barr, M. and Burdi, A. R. (1976). Warfarin-associated embryopathy in a 17-week abortus. *Teratology* 14: 129–134.

Bates, S. M. and Ginsberg, J. S. (1997). Anticoagulants in pregnancy: Fetal effects. *Baillieres Clin. Obstet. Gynaecol.* 11: 479–488.

Becker, M. H. et al. (1975). Chondrodysplasia punctata: Is maternal warfarin therapy a factor? *Am. J. Dis. Child.* 129: 356–359.

Born, D. et al. (1992). Pregnancy in patients with prosthetic heart valves: The effects of anticoagulation on mother, fetus, and neonate. *Am. Heart J.* 124: 413–417.

Bradley, J. P., Kawamoto, H. K., and Taub, P. (2003). Correction of warfarin-induced nasal hypoplasia. *Plast. Reconstr. Surg.* 111: 1680–1687.

Carson, M. and Reid, M. (1976). Warfarin and fetal abnormality. *Lancet* 1: 1127.

Chan, K. Y., Gilbert-Barness, E., and Tiller, G. (2003). Warfarin embryopathy. *Pediatr. Pathol. Mol. Med.* 22: 277–283.

Chen, W. W. C. et al. (1982). Pregnancy in patients with prosthetic heart valves. An experience with 45 pregnancies. *Q. J. Med.* 51: 358–365.

Collins, P. et al. (1977). Relationship of maternal warfarin therapy in pregnancy to chondrodysplasia punctata: Report of a case. *Am. J. Obstet. Gynecol.* 127: 444–446.

Cotrufo, M. et al. (2002). Risk of warfarin during pregnancy with mechanical valve prostheses. *Obstet. Gynecol.* 99: 35–40.

Cox, D. R., Martin, L., and Hall, B. D. (1977). Asplenia syndrome after fetal exposure to warfarin. *Lancet* 2: 1134.

Curtin, T. and Mulhern, B. (1980). Foetal warfarin syndrome. *Ir. Med. J.* 73: 393–394.

Dean, H. et al. (1981). Warfarin treatment during pregnancy in patients with prosthetic mitral valves. *Acta Haematol. (Basel)* 66: 65–66.

DiSaia, P. (1966). Pregnancy and delivery of a patient with a Starr-Edwards mitral valve prosthesis. *Obstet. Gynecol.* 28: 469–472.

Epstein, W. A. (1959). Antepartum fetal death during anticoagulant therapy for thromboembolism. Report of three cases. *J. Mt. Sinai Hosp. NY* 26: 562–565.

Fourie, D. T. and Hay, I. T. (1975). Warfarin as a possible teratogen. *S. Afr. Med. J.* 49: 2081–2083.

Friedman, J. M. and Polifka, J. E. (2000). *Teratogenic Effects of Drugs. A Resource for Clinicians (TERIS)*, Second ed., Johns Hopkins University Press, Baltimore, MD.

Galil, A., Biale, Y., and Barziv, J. (1984). [Warfarin embryopathy]. *Harefuah* 107: 390–392.

Gericke, G. S., Van der Walt, A., and DeJong, G. (1978). Another phenocopy for chondrodysplasia punctata in addition to warfarin embryopathy? *S. Afr. Med. J.* 54: 6.

Gooch, W. M. et al. (1978). Warfarin embryopathy. Longitudinal and postmortem examination. *Clin. Res.* 26: 74a.

Hall, B. D. (1989). Warfarin embryopathy and urinary tract anomalies: Possible new association. *Am. J. Med. Genet.* 34: 292–293.

Hall, J. G. (1976). Warfarin and fetal abnormality. *Lancet* 1: 1127.

Hall, J. G., Pauli, R. M., and Wilson, K. M. (1980). Maternal and fetal sequelae of anticoagulation during pregnancy. *Am. J. Med.* 68: 122–140.

Harrison, E. C. and Roschke, E. J. (1975). Pregnancy in patients with cardiac valve prostheses. *Clin. Obstet. Gynecol.* 18: 107–123.

Harrod, M. J. E. and Sherrod, P. S. (1981). Warfarin embryopathy in siblings. *Obstet. Gynecol.* 57: 673–676.

Hill, R. M. and Tennyson, L. M. (1984). Drug-induced malformations in humans. In *Drug Use in Pregnancy*, L. Stern, Ed., Adis Health Science Press, Salgdwah, Australia, pp. 99–133.

Holmes, L. B. (1988). Human teratogens delineating the phenotypic effects, period of greatest sensitivity, the dose-response relationship and mechanisms of action. *Transplacental Effects Fetal Health* 81: 177–191.

Holmes, L. B. et al. (1972). *Mental Retardation: An Atlas of Diseases with Associated Physical Abnormalities*, Macmillan, New York, pp. 136–137.

Holzgreve, W., Carey, J. C., and Hall, B. D. (1976). Warfarin-induced fetal abnormalities. *Lancet* 2: 914–915.

Hou, J. W. (2004). Fetal warfarin syndrome. *Chang Gung Med. J.* 27: 691–695.

Howe, A. M. and Webster, W. S. (1989). An animal model for warfarin teratogenicity. *Teratology* 40: 259–260.

Howe, A. M. and Webster, W. S. (1992). The warfarin embryopathy: A rat model showing maxillonasal hypoplasia and other skeletal disturbances. *Teratology* 46: 379–390.

Howe, A. M. and Webster, W. S. (1993). Warfarin embryopathy: A rat model showing nasal hypoplasia and "stippling." *Teratology* 48: 185–186.

Howe, A. M. et al. (1997). Severe cervical dysplasia and nasal cartilage calcification following prenatal warfarin exposure. *Am. J. Med. Genet.* 71: 391–396.

Ibarra-Perez, C. et al. (1976). The course of pregnancy in patients with artificial heart valves. *Am. J. Med.* 61: 504–512.

Ikonen, E. et al. (1970). Mitral-valve prosthesis, warfarin anti-coagulation and pregnancy. *Lancet* 2: 1252.

Jaillet, J. et al. (2005). Biliary lithiasis in early pregnancy and abnormal development of facial and distal limb bones (Binder syndrome): A possible role for vitamin K deficiency. *Birth Defects Res. (A)* 73: 188–193.

Kaplan, L. C. (1985). Congenital Dandy-Walker malformation associated with frist trimester warfarin: A case report and literature review. *Teratology* 32: 333–337.

Kaplan, L. C., Anderson, G. G., and Ring, B. A. (1982). Congenital hydrocephalus and Dandy-Walker malformation associated with warfarin use during pregnancy. *Birth Defects Orig. Art. Ser.* 18: 79–83.

Kenmure, A. C. F. (1968). Pregnancy in a patient with a prosthetic mitral valve. *J. Obstet. Gynaecol. Br. Commonw.* 75: 581–582.

Kerber, I. J., Warr, O. S., and Richardson, C. (1968). Pregnancy in a patient with a prosthetic mitral valve. Associated with a fetal anomaly attributed to warfarin sodium. *JAMA* 203: 223–225.

Lacy, C. F. et al. (2004). *Drug Information Handbook (Pocket)*, 2004–2005, Lexi-Comp. Inc., Hudson, OH.

Lamontagne, J. M. and Leclerc, J. E. (1984). Warfarin embryopathy — a case report. *J. Otolaryngol.* 13: 127.

Lee, C.-N. et al. (1994). Pregnancy following cardiac prosthetic valve replacement. *Obstet. Gynecol.* 83: 353–360.

Mason, J. D., Jardine, A., and Gibbin, K. P. (1992). Foetal warfarin syndrome — a complex airway problem. Case report and review of the literature. *J. Laryngol. Otol.* 106: 1098–1099.

Menger, H. et al. (1997). Vitamin K deficiency embryopathy: A phenocopy of the warfarin embryopathy due to a disorder of embryonic vitamin K metabolism. *Am. J. Med. Genet.* 72: 129–134.

Nagai, T. (2001). [Fetal warfarin syndrome]. *Ryoikbetsu Shokogun Shirizu* 33: 706–707.

Normann, E. K. and Stray-Pedersen, B. (1989). Warfarin-induced fetal diaphragmatic hernia. Case report. *Br. J. Obstet. Gynaecol.* 96: 729–730.

O'Donnell, D., Meyers, A. M., and Sevitz, H. (1985). Pregnancy after renal transplantation. *Aust. N.Z. J. Med.* 15: 320–325.

Palacios-Macedo, X., Diaz-Devis, C., and Escudero, J. (1969). Fetal risk with the use of coumarin anticoagulant agents in pregnant patients with intracardiac ball valve prosthesis. *Am. J. Cardiol.* 24: 853–856.

Pati, S. and Helmbrecht, G. D. (1994). Congenital schizencephaly associated with *in utero* warfarin exposure. *Reprod. Toxicol.* 8: 115–120.

Patil, S. B. (1991). Warfarin embryopathy and heart disease. *Ann. Saudi Med.* 11: 359–360.

Pauli, R. M. (1988). Mechanism of bone and cartilage development in the warfarin embryopathy. *Pathol. Immunopathol. Res.* 7: 107–112.

Pauli, R. M. and Haun, J. M. (1993). Intrauterine effects of coumarin derivatives. *Dev. Brain Dysfunct.* 6: 229–247.

Pauli, R. M. et al. (1976). Warfarin therapy initiated during pregnancy and phenotypic chondrodysplasia punctata. *J. Pediatr.* 88: 506–508.

Pauli, R. M. et al. (1987). Association of congenital deficiency of multiple vitamin K-dependent coagulation factors and the phenotype of the warfarin embryopathy: Clues to the mechanism of teratogenicity of coumarin derivatives. *Am. J. Hum. Genet.* 41: 566–583.

PDR® (*Physicians' Desk Reference*®). (2002). Medical Economics Co., Inc., Montvale, NJ.

Pettifor, J. M. and Benson, R. (1975a). Congenital malformations associated with the administration of oral anticoagulants during pregnancy. *J. Pediatr.* 86: 459–462.

Pettifor, J. M. and Benson, R. (1975b). Teratogenicity of anticoagulants. *J. Pediatr.* 87: 838–839.

Price, P. A. et al. (1981). Developmental appearance of the vitamin K dependent protein of bone during calcification: Analysis of mineralizing tissues in the human, calf and rat. *J. Biol. Chem.* 256: 3781–3784.

Radnich, R. H. and Jacobs, W. M. (1970). Prosthetic heart valves. *Tex. Med.* 66: 58–61.

Raivio, K. O., Ikonen, E., and Saarikoski, S. (1977). Fetal risks due to warfarin therapy during pregnancy. *Acta Paediatr. Scand.* 66: 735–739.

Richman, E. M. and Lahman, J. E. (1976). Fetal anomalies associated with warfarin therapy initiated shortly prior to conception. *J. Pediatr.* 88: 509–510.

Robinson, M. J. et al. (1978). Fetal warfarin syndrome. *Med. J. Aust.* 1: 157.

Ruthnum, P. and Tolmie, J. L. (1987). Atypical malformations in an infant exposed to warfarin during the first trimester of pregnancy. *Teratology* 36: 299–301.

Salazar, E. et al. (1984). The problem of cardiac valve prostheses, anticoagulants, and pregnancy. *Circulation* 70 (Suppl. 1): 1169–1177.

Schardein, J. L. (2000). *Chemically Induced Birth Defects*, Third ed., Marcel Dekker, New York, pp. 124–128, 130.

Schivazappa, L. (1982). Fetal malformations caused by oral anticoagulants during pregnancy. Report of a case. *G. Ital. Cardiol.* 12: 897.

Shaul, W. L. and Hall, J. G. (1977). Multiple congenital anomalies associated with oral anticoagulants. *Am. J. Obstet. Gynecol.* 127: 191–198.

Shaul, W. L., Emery, H., and Hall, J. G. (1975). Chondrodysplasia punctata and maternal warfarin use during pregnancy. *Am. J. Dis. Child.* 129: 360–362.

Sheikhzadeh, A. et al. (1983). Congestive heart failure in valvular heart disease in pregnancies with and without valvular prostheses and anticoagulant therapy. *Clin. Cardiol.* 6: 465–470.

Sherman, S. and Hall, B. D. (1976). Warfarin and fetal abnormality. *Lancet* 1: 692.

Smith, M. F. and Cameron, M. D. (1979). Warfarin as teratogen. *Lancet* 1: 727.

Sonoda, T. (2000). [Fetal warfarin syndrome]. *Ryoikbetsu Shokogun Shirizu* 30: 105–107.

Stenflo, J. and Suttie, J. W. (1977). Vitamin-K-dependent formation of gamma-carboxyglutamic acid. *Annu. Rev. Biochem.* 46: 157–172.

Stevenson, R. E. et al. (1980). Hazards of oral anticoagulants during pregnancy. *JAMA* 243: 1549–1551.

Sugrue, D. (1981). Anticoagulation in pregnancy. Reply to Dr. Nageotte and associates. *Am. J. Obstet. Gynecol.* 141: 473.

Suttie, J. W. (1991). *Handbook of Vitamins. Nutritional, Biochemical and Clinical Aspects*, Second ed., Marcel Dekker, New York, pp. 145–194.

Takano, H. et al. (1998). Cervical spine abnormalities and instability with myelopathy in warfarin-related chondrodysplasia. *Pediatr. Radiol.* 28: 497–499.

Tamburrini, O., Bartolomeo-DeIuri, A., and DiGuglielmo, G. L. (1987). Chondrodysplasia punctata after warfarin. Case report with 18-month follow-up. *Pediatr. Radiol.* 17: 323–324.

Tejani, N. (1973). Anticoagulant therapy with cardiac valve prosthesis during pregnancy. *Obstet. Gynecol.* 42: 785–793.

Tongsong, T., Wanapirak, C., and Piyamongkol, W. (1999). Prenatal ultrasonographic findings consistent with fetal warfarin syndrome. *J. Ultrasound Med.* 18: 577–580.

Van Driel, D. et al. (2002). Teratogen update: Fetal effects after *in utero* exposure to coumarins overview of cases, followup findings, and pathogenesis. *Teratology* 66: 127–140.

Ville, Y. et al. (1993). Fetal intraventricular haemorrhage and maternal warfarin. *Lancet* 341: 1211.

Vitale, N. et al. (1999). Dose-dependent fetal complications of warfarin in pregnant woman with mechanical heart valves. *J. Am. Coll. Cardiol.* 33: 1637–1641.

Warkany, J. (1976). Warfarin embryopathy. *Teratology* 14: 205–209.

Warkany, J. and Bofinger, M. (1975). Le role de la coumadine dans les malformations congenitale. *Med. Hyg.* 33: 1454–1457.

Wellesley, D. et al. (1998). Two cases of warfarin embryopathy: A re-emergence of this condition? *Br. J. Obstet. Gynaecol.* 105: 805–806.

Whitfield, M. F. (1980). Chondrodysplasia punctata after warfarin in early pregnancy. Case report and summary of the literature. *Arch. Dis. Child.* 55: 139–142.

Wong, V., Cheng, C. H., and Chan, K. C. (1993). Fetal and neonatal outcome of exposure to anticoagulants during pregnancy. *Am. J. Med. Genet.* 45: 17–21.

Zakzouk, M. S. (1986). The congenital warfarin syndrome. *J. Laryngol. Otol.* 100: 215–219.

35 Phenobarbital

Chemical name: 5-Ethyl-5-phenyl-2,4,6(1*H*,3*H*,5*H*)-pyrimidinetrione

Alternate names: 5-Ethyl-5-phenylbarbituric acid, phenobarbitone, phenylethylmalonylurea

CAS #: 50-06-6

SMILES: C1(c2ccccc2)(C(NC(NC1=O)=O)=O)CC

INTRODUCTION

Phenobarbital is a barbiturate used therapeutically as a hypnotic, sedative, and anticonvulsant in the management of generalized tonic-clonic (grand mal) and partial seizures. It has been used for these purposes for almost 100 years. It shares the active chemical moiety with another anticonvulsant drug (and developmental toxicant), primidone. The drug acts by depressing the sensory cortex, decreasing motor activity, and altering cerebellar function (Lacy et al., 2004). Phenobarbital is obtained by prescription as Luminal® and Sulfoton® and by many other trade names. It has a pregnancy category of D, based on the package label that states "barbiturates can cause fetal damage when administered to a pregnant woman. Retrospective, case-controlled studies have suggested a connection between the maternal consumption of barbiturates and a higher than expected incidence of fetal abnormalities" (*PDR*, 2004). The label goes on to state that fetal blood levels approach maternal blood levels following parenteral administration.

DEVELOPMENTAL TOXICOLOGY

ANIMALS

In animal studies, phenobarbital has shown developmental toxicity in mice, rats, and rabbits by oral and parenteral routes of administration, those pertinent to the human condition. In mice, the drug induced cleft palate by the oral route — in the diet (Sullivan and McElhatton, 1975), by gavage (McElhatton and Sullivan, 1977), or via drinking water (Finnell et al., 1987) at doses approximating 50 mg/kg/day and greater over ten or more days during gestation. It also produced cleft palate but no other developmental toxicity in this species (mouse) subcutaneously at a higher dose of 175 mg/kg injected as single doses during gestation interval days 11 to 14 (Walker and Patterson, 1974). A second species, the rat, reacted somewhat differently. When given as 0.16% in the diet throughout pregnancy, only minor skeletal defects were apparent (McColl et al., 1963), while intramuscular doses of 50 mg/kg/day over 3 days late in gestation produced reduced postnatal learning capacity (Auroux, 1973). Finally, in a third species, the rabbit, gavage doses of 50 mg/kg/day for 8 days during organogenesis resulted in defects of the sternum and skull and increased fetal loss (McColl, 1966).

HUMANS

Published reports of the drug's use in humans showed variable results relating to developmental toxicity. It should be emphasized, however, that studies on anticonvulsants are difficult to interpret due to confounding factors, including multiple drug therapy, genetic constitution, or epilepsy effects themselves. In 1964, however, it first became apparent that phenobarbital might have teratogenic potential in humans, in a published report by Janz and Fuchs. Since then, a number of reports, excluding a number of single case reports, and especially those with monotherapy of the drug (providing additional credence to an association) have been published attesting to the probable malforming effect of the drug on the human fetus. A representative sampling of these reports is tabulated in Table 1. Cardiovascular malformations and facial clefts have been cited most commonly, although the pattern of the defect syndrome includes nail hypoplasia, typical facies characterized by depressed nasal bridge, epicanthal folds, and ocular hypertelorism, a syndrome not unlike that elicited by phenytoin in the fetal hydantoin syndrome (FHS; see Schardein, 2000). There is also evidence that some of the minor defects described in the FHS are also observed following phenobarbital exposures (Janz, 1982). Notably, the rat serves as a good model for both the structural and functional dysfunctions of phenobarbital (Vorhees, 1983). Based on a recent review of this literature, a mother using phenobarbital in combination with other antiepileptics has a two to three times greater risk for producing a child with malformations than does the general population (Briggs et al., 2005). It may also be the case that malformations are more severe and more frequent when phenobarbital is combined with other anticonvulsants, especially phenytoin, than if used as mono-

TABLE 1
Reports Associating Malformations to Phenobarbital
Treatment during Pregnancy in Humans

Ref.
Bethenod and Frederich, 1975
Seip, 1976[a]
Shapiro et al., 1976
Greenberg et al., 1977
Meinardi, 1977
Rothman et al., 1979
Nakane et al., 1980
Janz, 1982[a]
Robert et al., 1986[a]
Dansky and Finnell, 1991
Thakker et al., 1991[a]
Koch et al., 1992[a]
Jones et al., 1992[a]
Waters et al., 1994
Canger et al., 1999[a]
Arpino et al., 2000[a]
Holmes et al., 2001[a]
Briggs et al., 2005

[a] Monotherapy subjects included.

therapy. In contrast to these positive reports, a few researchers found no association between maternal use of phenobarbital and congenital abnormalities in the offspring (Mellin, 1964; Heinonen et al., 1977; Lakos and Czeizel, 1977; Czeizel et al., 1984, 1988; Bertollini et al., 1987), the latter a monotherapy study.

Another class of developmental toxicity, retarded growth, manifested in these reports as low birth weight, intrauterine growth retardation, and including a smaller than expected head circumference, was also reported in several publications (Seip, 1976; Hiilesmaa et al., 1981; Majewski and Steger, 1984; Dessens et al., 2000). In another of these reports, a statistically significant decrease in birth weight among 55 infants born of epileptic women treated with phenobarbital was reported (Mastroiacovo et al., 1988). Mortality of phenobarbital-exposed fetuses or infants was not reported to be a significant feature of the developmental toxicity profile of phenobarbital. A number of functional deficits were reported to be seen in phenobarbital-exposed infants. Included are impairments in intellectual development (Gaily et al., 1988), verbal intelligence (Reinisch et al., 1995), psychosexual development (Dessens et al., 1999), cognitive function (van der Pol et al., 1991; Dessens et al., 1998, 2000), general mental ability (Adams et al., 2004), and general development (Thorp et al., 1997, 1999). Several cases of mental retardation are also known (McIntyre, 1966; Berkowitz, 1979). Significantly higher mean apathy and optimality scores were also recorded in neurological assessments of infants prenatally exposed to phenobarbital (Koch et al., 1996). In contrast, no deficits in function were found in several other studies (Shapiro et al., 1976; Shankaran et al., 1996; Holmes et al., 2005). In most of the positive studies, results were based on monotherapy with phenobarbital. Hemorrhagic disease of the newborn following anticonvulsant pregnancy exposures including phenobarbital has been known for over 45 years (Schardein, 2000) but is not discussed in further detail here.

In summary, assessment of the developmental toxicity profile of phenobarbital, whether the drug was used alone or combined with other anticonvulsants, indicates a small but significant incidence of a syndrome of minor malformations, retarded growth, and functional impairment. Toxicity was produced within the recommended 300 mg to 1 to 2 g/day oral dosing when used as an anticonvulsant or within the 30 to 320 mg/day oral or parenteral dosing used as a sedative/hypnotic. First trimester treatment was the timetable. One group of experts places the magnitude of teratogenic risk at minimal to small (Friedman and Polifka, 2000). Several reviews on phenobarbital developmental toxicity are available (Lakos and Czeizel, 1977; Middaugh, 1986; Yaffe and Dorn, 1990; Yerby, 1994; Holmes et al., 2001).

CHEMISTRY

Phenobarbital is average in size. It is hydrophobic and can participate as a hydrogen bond donor and acceptor. Phenobarbital is of average polarity in comparison to the other human developmental toxicants. The calculated physicochemical and topological properties are listed below.

PHYSICOCHEMICAL PROPERTIES

Parameter	Value
Molecular weight	232.238 g/mol
Molecular volume	199.76 A^3
Density	1.074 g/cm^3
Surface area	248.30 A^2
LogP	1.375
HLB	11.672
Solubility parameter	24.751 J$^{(0.5)}$/cm$^{(1.5)}$
Dispersion	21.509 J$^{(0.5)}$/cm$^{(1.5)}$

Continued.

Parameter	Value
Polarity	8.573 $J^{(0.5)}/cm^{(1.5)}$
Hydrogen bonding	8.747 $J^{(0.5)}/cm^{(1.5)}$
H bond acceptor	1.13
H bond donor	0.56
Percent hydrophilic surface	57.09
MR	64.927
Water solubility	0.917 log (mol/M^3)
Hydrophilic surface area	141.76 A^2
Polar surface area	84.75 A^2
HOMO	−9.901 eV
LUMO	−0.217 eV
Dipole	2.055 debye

TOPOLOGICAL PROPERTIES (UNITLESS)

Parameter	Value
x0	12.466
x1	8.108
x2	7.180
xp3	6.324
xp4	5.658
xp5	3.539
xp6	2.311
xp7	1.175
xp8	0.708
xp9	0.259
xp10	0.149
xv0	9.319
xv1	5.334
xv2	3.862
xvp3	3.026
xvp4	2.064
xvp5	1.149
xvp6	0.562
xvp7	0.262
xvp8	0.114
xvp9	0.031
xvp10	0.013
k0	17.907
k1	13.432
k2	5.325
k3	2.291
ka1	11.630
ka2	4.195
ka3	1.691

REFERENCES

Adams, J., Holmes, L. B., and Janulewicz, P. (2004). The adverse effect profile of neurobehavioral teratogens: Phenobarbital. *Birth Defects Res. (A)* 70: 280.

Arpino, C. et al. (2000). Teratogenic effects of antiepileptic drugs: Use of an international database on malformations and drug exposures (MADRE). *Epilepsia* 41: 1436–1443.

Auroux, M. (1973). Effect of some drugs on the late development of the central nervous system in rats. Alteration of learning capacities in the offspring by administration of phenobarbital to the mother. *C. R. Soc. Biol. (Paris)* 167: 797–801.

Berkowitz, F. E. (1979). Fetal malformation due to phenobarbitone. A case report. *S. Afr. Med. J.* 55: 100–101.

Bertollini, R. et al. (1987). Anticonvulsant drugs in monotherapy. Effect on the fetus. *Eur. J. Epidemiol.* 3: 164–171.

Bethenod, M. and Frederich, A. (1975). [The children of drug-treated epileptics]. *Pediatrie* 30: 227–248.

Briggs, G. G., Freeman, R. K., and Yaffe, S. J. (2005) *Drugs in Pregnancy and Lactation. A Reference Guide to Fetal and Neonatal Risk*, Seventh ed., Lippincott Williams & Wilkins, Philadelphia.

Canger, R. et al. (1999). Malformations in offspring of women with epilepsy: A prospective study. *Epilepsia* 40: 1231–1236.

Czeizel, A. et al. (1984). Pregnancy outcome and health conditions of offspring of self-poisoned pregnant women. *Acta Paediatr. Hung.* 25: 209–236.

Czeizel, A. et al. (1988). A study of adverse effects on the progeny after intoxication during pregnancy. *Arch. Toxicol.* 62: 1–7.

Dansky, L. V. and Finnell, R. H. (1991). Parental epilepsy, anticonvulsant drugs, and reproductive outcome: Epidemiologic and experimental findings spanning three decades; 2: Human studies. *Reprod. Toxicol.* 5: 301–335.

Dessens, A. et al. (1998). Prenatal exposure to anticonvulsant drugs and spatial ability in adulthood. *Acta Neurobiol. Exp. (Wars.)* 58: 221–225.

Dessens, A. B. et al. (1999). Prenatal exposure to anticonvulsants and psychosexual development. *Arch. Sex Behav.* 28: 31–44.

Dessens, A. B. et al. (2000). Association of prenatal phenobarbital and phenytoin exposure with small head size at birth and with learning problems. *Acta Paediatr.* 89: 533–541.

Finnell, R. H. et al. (1987). Strain differences in phenobarbital-induced teratogenesis in mice. *Teratology* 35: 177–185.

Friedman, J. M. and Polifka, J. E. (2000). *Teratogenic Effects of Drugs. A Resource for Clinicians (TERIS)*, Second ed., Johns Hopkins University Press, Baltimore, MD.

Gaily, E., Kantola-Sorsa, E., and Granstrom, M. L. (1988). Intelligence of children of epileptic mothers. *J. Pediatr.* 113: 677–684.

Greenberg, G. et al. (1977). Maternal drug histories and congenital abnormalities. *Br. Med. J.* 2: 853–856.

Heinonen, O. P., Slone, D., and Shapiro, S. (1977). *Birth Defects and Drugs in Pregnancy*, Publishing Sciences Group, Littleton, MA.

Hiilesmaa, V. K. et al. (1981). Fetal head growth retardation associated with maternal antiepileptic drugs. *Lancet* 2: 165–166.

Holmes, L. B. et al. (2001). The teratogenicity of anticonvulsant drugs. *N. Engl. J. Med.* 334: 1132–1138.

Holmes, L. B. et al. (2005). The correlation of deficits in IQ with midface and digit hypoplasia in children exposed to anticonvulsant drugs. *J. Pediatr.* 146: 118–122.

Janz, D. (1982). Antiepileptic drugs and pregnancy: Altered utilization patterns and teratogenesis. *Epilepsia* 23 (Suppl. 1): S53–S63.

Janz, D. and Fuchs, U. (1964). Are anti-epileptic drugs harmful when given during pregnancy? *Ger. Med. Monatsschr.* 9: 20–22.

Jones, K. L., Johnson, K. A., and Chambers, C. C. (1992). Pregnancy outcome in women treated with phenobarbital monotherapy. *Teratology* 45: 452–453.

Koch, S. et al. (1992). Major and minor birth malformations and antiepileptic drugs. *Neurology* 42 (Suppl. 5): 83–88.

Koch, S. et al. (1996). Antiepileptic drug treatment in pregnancy: Drug side effects in the neonate and neurological outcome. *Acta Paediatr.* 84: 739–746.

Lacy, C. F. et al. (2004). *Drug Information Handbook (Pocket), 2004–2005*, Lexi-Comp., Inc., Hudson, OH.

Lakos, P. and Czeizel, E. (1977). A teratological evaluation of anticonvulsant drugs. *Acta Paediatr. Acad. Sci. Hung.* 18: 145–153.

Majewski, F. and Steger, M. (1984). Fetal head growth retardation associated with maternal phenobarbital/primidone and/or phenytoin therapy. *Eur. J. Pediatr.* 141: 188–189.

Mastroiacovo, P., Bertollini, R., and Licata, D. (1988). Fetal growth in the offspring of epileptic women: Results of an Italian multicentric cohort study. *Acta Neurol. Scand.* 78: 110–114.

McColl, J. D. (1966). Teratogenicity studies. *Appl. Ther.* 8: 48–52.

McColl, J. D., Globus, M., and Robinson, S. (1963). Drug induced skeletal malformations in the rat. *Experientia* 19: 183–184.

McElhatton, P. R. and Sullivan, F. M. (1977). Comparative teratogenicity of six antiepileptic drugs in the mouse. *Br. J. Pharmacol.* 59: 494P–495P.

McIntyre, M. S. (1966). Possible adverse drug reaction. *JAMA* 197: 62–63.

Meinardi, H. (1977). Teratogenicity of antiepileptic drugs. *Tijdschr. Kindergeneeskd.* 45: 87–91.

Mellin, G. W. (1964). Drugs in the first trimester of pregnancy and fetal life of *Homo sapiens. Am. J. Obstet. Gynecol.* 90: 1169–1180.

Middaugh, L. D. (1986). Phenobarbital during pregnancy in mouse and man. *Neurotoxicology* 7: 287–301.

Nakane, Y. et al. (1980). Multiinstitutional study on the teratogenicity and fetal toxicity of antiepileptic drugs: A report of the Collaborative Study Group in Japan. *Epilepsia* 21: 663–680.

PDR® (*Physicians' Desk Reference®*). (2004). Medical Economics Co., Inc., Montvale, NJ.

Reinisch, J. M. et al. (1995). *In utero* exposure to phenobarbital and intelligence deficits in adult men. *JAMA* 2274: 1518–1525.

Robert, E. et al. (1986). Evaluation of drug therapy and teratogenic risk in a Rhone-Alps district population of pregnant epileptic women. *Eur. Neurol.* 25: 436–443.

Rothman, K. J. et al. (1979). Exogenous hormones and other drug exposures of children with congenital heart disease. *Am. J. Epidemiol.* 109: 433–439.

Schardein, J. L. (2000). *Chemically Induced Birth Defects*, Third ed., Marcel Dekker, New York, pp. 186, 204.

Seip, M. (1976). Growth retardation, dysmorphic facies and minor malformations following massive exposure to phenobarbitone *in utero. Acta Paediatr. Scand.* 65: 617–621.

Shankaran, S. et al. (1996). Antenatal phenobarbital therapy and neonatal outcome. II. Neurodevelopmental outcome at 36 months. *Pediatrics* 97: 649–652.

Shapiro, S. et al. (1976). Anticonvulsants and parental epilepsy in the development of birth defects. *Lancet* 1: 272–275.

Sullivan, F. M. and McElhatton, P. R. (1975). Teratogenic activity of the antiepileptic drugs phenobarbital, phenytoin, and primidone in mice. *Toxicol. Appl. Pharmacol.* 34: 271–282.

Thakker, J. C. et al. (1991). Hypoplasia of nails and phalanges: A teratogenic manifestation of phenobarbitone. *Indian Pediatr.* 28: 73–75.

Thorp, J. A. et al. (1997). Does *in utero* phenobarbital lower IQ: Followup on the intracranial hemorrhage prevention trial. *Am. J. Obstet. Gynecol.* 176: S117.

Thorp, J. A. et al. (1999). Does perinatal phenobarbital exposure effect developmental outcome at age 2? *Am. J. Perinatol.* 16: 51–60.

van der Pol, M. C. et al. (1991). Antiepileptic medication in pregnancy: Late effects on the children's central nervous system development. *Am. J. Obstet. Gynecol.* 164: 121–128.

Vorhees, C. V. (1983). Fetal anticonvulsant syndrome in rats: Dose- and period-response relationships of prenatal diphenylhydantoin, trimethadione and phenobarbital exposure on the structural and functional development of the offspring. *J. Pharmacol. Exp. Ther.* 227: 274–287.

Walker, B. E. and Patterson, A. (1974). Induction of cleft palate in mice by tranquilizers and barbiturates. *Teratology* 10: 159–164.

Waters, C. H. et al. (1994). Outcomes of pregnancy associated with antiepileptic drugs. *Arch. Neurol.* 51: 250–253.

Yaffe, S. J. and Dorn, L. D. (1990). Effects of prenatal treatment with phenobarbital. *Dev. Pharmacol. Ther.* 15: 215–223.

Yerby, M. S. (1994). Pregnancy, teratogenesis, and epilepsy. *Neurol. Clin.* 12: 749–771.

36 Trimethoprim

Chemical name: 2,4-Diamino-5-(3,4,5-trimethoxybenzyl)pyrimidine

CAS #: 738-70-5

SMILES: c1(Cc2c(nc(nc2)N)N)cc(c(c(c1)OC)OC)OC

INTRODUCTION

Trimethoprim (TMP) is an antibiotic used in the treatment of urinary tract infections due to susceptible strains, acute exacerbations of chronic bronchitis in adults, and superficial ocular infections, and it is combined with other agents for the treatment of toxoplasmosis. The mechanism of action of TMP is through inhibition of folic acid reductase to tetrahydrofolate, thereby inhibiting microbial growth (Lacy et al., 2004). The drug is known by a variety of trade names including Proloprim®, Trimanyl®, Uretrim®, Primsol®, and many others. One of its popular combination products as an antibacterial agent is with the sulfonamide sulfamethoxazole, known as Septra® or Co-trimoxazole®. It is available by prescription, and it has a pregnancy category of C. The package label shows no warning but states that beause the drug may interfere with folic acid metabolism, it should be used during pregnancy only if the potential benefit justifies the potential risk to the fetus (*PDR*, 2005).

DEVELOPMENTAL TOXICOLOGY

ANIMALS

Laboratory studies in animals have been fairly limited and confined to oral studies, the route used for the drug in humans. Over a wide range of doses of 200 to 2000 mg/kg/day over 9 days during organogenesis in the rat, TMP induced malformations at maternally toxic levels (Udall, 1969). In the rabbit, the drug caused fetal resorption but no malformations at doses of 500 mg/kg/day for 9 days during gestation according to the package label, and in mice, no developmental toxicity of any kind was observed under the conditions utilized (Elmazar and Nau, 1993). TMP combined with sulfamethoxazole given orally at low doses (up to 32 mg/kg for 1 to 3 days in gestation) to hamsters caused embryotoxicity (Haliniarz and Sikorski, 1979) and malformations in rats and rabbits administered higher doses (600 mg/kg/day) on 7 or 8 days during organogenesis (Helm et al., 1976).

HUMANS

In the human, several reports suggested that TMP may have teratogenic potential. In a large, multicenter case-control study, an association was apparent in two groups of second and third month pregnancy exposures to folic acid antagonists, including TMP monotherapy: cardiovascular defects and oral clefts (relative risk [RR] = 3.4, 95% confidence interval [CI] 1.8 to 6.4) in one group and cardiovascular and urinary tract abnormalities (RR = 2.6, 95% CI 1.1 to 6.1) in the other group (Hernandez-Diaz et al., 2000). A breakdown of these results showed 12 cases of cardiovascular defects especially associated with TMP exposure (RR = 4.2, 95% CI 1.5 to 11.5; see Hernandez-Diaz and Mitchell, 2001). Malformations associated with the combination of TMP and sulfamethoxazole were more frequently reported. A citation is made of 2296 newborns exposed during the first trimester to the combined product and known to the U.S. Food and Drug Administration (FDA) in which 126 major birth defects were observed — the number greater than expected (Briggs et al., 2005). As with the study cited above with TMP alone, cardiovascular defects were particularly prominent (37 versus 23 in controls). Those conducting another study of 351 treated subjects of the combined drugs compared to 443 controls reported a higher rate of multiple congenital abnormalities of those exposed during the second and third months of pregnancy; urinary tract and cardiovascular malformations were again the primary defects observed (Czeizel et al., 2001; Czeizel, 2002). Reported in another publication was an association of the combined product to oral clefts, hypospadias, cardiovascular malformations, and neural tube defects (Hernandez-Diaz et al., 2001, 2004). Exposures covering the first trimester and doses in the usual therapeutic range of 200 mg/day or up to 15 to 20 mg/kg/day (orally) were used in these reports. In contrast to these positive reported studies, a number of investigators found no significant associations to either TMP alone or the combined TMP–sulfamethoxazole therapy (Williams et al., 1969; Gonzalez-Ochoa, 1971; Brumfitt and Pursell, 1973; Colley and Gibson, 1982; Bailey, 1984; Soper and Merrill-Nach, 1986; Cruikshank and Warenski, 1989; Czeizel, 1990; Seoud et al., 1991). In none of the reports identified above were any of the other classes of developmental toxicity mentioned as significant findings. One group of experts considered the magnitude of teratogenic potential of TMP to be unlikely (Friedman and Polifka, 2000), while another group considered that the drug might increase the risk for neural tube defects, congenital heart abnormalities, and oral clefts (Shepard et al., 2002). Whatever the future proves, whether or not TMP is a human developmental toxicant, at present, the evidence is tentative, but the risk cannot be ignored.

CHEMISTRY

Trimethoprim is a polar molecule of average size. It is slightly hydrophilic and can participate in hydrogen bonding as both an acceptor and donor. The calculated physicochemical and topological properties of trimethoprim are shown in the following sections.

PHYSICOCHEMICAL PROPERTIES

Parameter	Value
Molecular weight	290.322 g/mol
Molecular volume	258.77 A^3
Density	1.320 g/cm^3
Surface area	320.50 A^2
LogP	−0.472
HLB	7.111
Solubility parameter	26.875 J$^{(0.5)}$/cm$^{(1.5)}$
Dispersion	22.353 J$^{(0.5)}$/cm$^{(1.5)}$

Continued.

Parameter	Value
Polarity	7.830 $J^{(0.5)}/cm^{(1.5)}$
Hydrogen bonding	12.699 $J^{(0.5)}/cm^{(1.5)}$
H bond acceptor	1.73
H bond donor	1.07
Percent hydrophilic surface	37.27
MR	79.303
Water solubility	−0.921 log (mol/M^3)
Hydrophilic surface area	119.44 A^2
Polar surface area	105.51 A^2
HOMO	−8.026 eV
LUMO	0.506 eV
Dipole	0.587 debye

TOPOLOGICAL PROPERTIES (UNITLESS)

Parameter	Value
x0	15.405
x1	10.083
x2	8.638
xp3	7.249
xp4	5.991
xp5	4.639
xp6	2.789
xp7	1.864
xp8	1.244
xp9	0.817
xp10	0.487
xv0	12.213
xv1	6.244
xv2	4.349
xvp3	3.036
xvp4	2.028
xvp5	1.358
xvp6	0.692
xvp7	0.396
xvp8	0.211
xvp9	0.107
xvp10	0.048
k0	25.358
k1	17.355
k2	8.022
k3	4.260
ka1	15.488
ka2	6.703
ka3	3.409

REFERENCES

Bailey, R. R. (1984). Single-dose antibacterial treatment for bacteriuria in pregnancy. *Drugs* 27: 183–186.

Briggs, G. G., Freeman, R. K., and Yaffe, S. J. (2005). *Drugs in Pregnancy and Lactation. A Reference Guide to Fetal and Neonatal Risk*, Seventh ed., Lippincott Williams & Wilkins, Philadelphia.

Brumfitt, W. and Pursell, R. (1973). Trimethoprim — sulfamethoxazole in the treatment of bacteriuria in women. *J. Infect. Dis.* 128 (Suppl.): 657–665.

Colley, D. P. and Gibson, K. J. (1982). Study of the use in pregnancy of cotrimoxazole sullfamethizole. *Aust. J. Pharm.* 63: 570–575.

Cruikshank, D. P. and Warenski, J. C. (1989). First-trimester maternal *Listeria monocytogenes* sepsis and chorioamnionitis with normal neonatal outcome. *Obstet. Gynecol.* 73: 469–471.

Czeizel, A. (1990). A case-control analysis of the teratogenic effects of co-trimoxazole. *Reprod. Toxicol.* 4: 305–313.

Czeizel, A. E. (2002). Folic acid antagonists (trimethoprim-sulfonamides and sulfonamides) during pregnancy and the risk of orofacial clefts. *Reprod. Toxicol.* 16: 90.

Czeizel, A. E. et al. (2001). The teratogenic risk of trimethoprim-sulfonamides: A population based case-control study. *Reprod. Toxicol.* 15: 637–646.

Elmazar, M. M. A. and Nau, H. (1993). Trimethoprim potentiates valproic acid-induced neural tube defects (NTDs) in mice. *Reprod. Toxicol.* 7: 249–254.

Friedman, J. M. and Polifka, J. E. (2000). *Teratogenic Effects of Drugs. A Resource for Clinicians (TERIS)*, Second ed., Johns Hopkins University Press, Baltimore, MD.

Gonzalez-Ochoa, A. (1971). Trimethoprim and sulfamethoxazole in pregnancy. *JAMA* 217: 1244.

Haliniarz, W. and Sikorski, R. (1979). Study of the embryotoxicity of the preparation biseptol-polfa in pregnant hamsters. *Ginekol. Pol.* 50: 481–486.

Helm, F. et al. (1976). [Investigations on the effect of the combination sulfamoxazole-trimethoprim (CN 3123) on fertility and fetal development in rats and rabbits]. *Arzneimittelforschung* 26: 643–651.

Hernandez-Diaz, S. and Mitchell, A. A. (2001). Folic acid antagonists during pregnancy and risk of birth defects. *N. Engl. J. Med.* 344: 934–935.

Hernandez-Diaz, S. et al. (2000). Folic acid antagonists during pregnancy and the risk of birth defects. *N. Engl. J. Med.* 343: 1608–1614.

Hernandez-Diaz, S. et al. (2001). Neural tube defects in relation to use of folic acid antagonists during pregnancy. *Am. J. Epidemiol.* 1153: 961–968.

Hernandez-Diaz, S. et al. (2004). Teratogen update: Trimethoprim teratogenicity. *Birth Defects Res. (A)* 70: 276.

Lacy, C. F. et al. (2004). *Drug Information Handbook (Pocket), 2004–2005*, Lexi-Comp., Inc., Hudson, OH.

PDR® (*Physicians' Desk Reference®*). (2005). Medical Economics Co., Inc., Montvale, NJ.

Seoud, M. et al. (1991). Brucellosis in pregnancy. *J. Reprod. Med.* 36: 441–445.

Shepard, T. H. et al. (2002). Update on new developments in the study of human teratogens. *Teratology* 65: 153–161.

Soper, D. E. and Merrill-Nach, S. (1986). Successful therapy of penicillinase-producing *Neisseria gonorrhoeae* pharyngeal infection during pregnancy. *Obstet. Gynecol.* 68: 290–291.

Udall, V. (1969). Toxicology of sulphonamide-trimethoprim combinations. *Postgrad. Med. J.* 45 (Suppl.): 42–45.

Williams, J. D. et al. (1969). The treatment of bacteriuria in pregnant women with sulfamethoxazole and trimethoprim: A microbiological clinical and toxicological study. *Postgrad. Med. J.* 45 (Suppl.): 71–76.

37 Methyltestosterone

Chemical name: (17β)-17-Hydroxy-17-methylandrost-4-en-3-one

CAS # 58-18-4

SMILES: C12C(C3(C(CC1) =CC(CC3) =O)C)CCC4(C2CCC4(C)O)C

INTRODUCTION

Methyltestosterone (MT) is the 17-methyl-substituted synthetic derivative of testosterone, which as an androgen has primary therapeutic value in males, treating hypogonadism, delayed puberty, and impotence. Secondarily, it has had palliative value in treating metastatic breast cancer in females. The drug stimulates receptors in organs and tissues to promote growth and development of male sex organs and maintains secondary sex characteristics in androgen-deficient males (Lacy et al., 2004). It is available by prescription under various trade names, including Android®, Oreton Methyl®, Testred®, and Virilon®, among other names. MT has a pregnancy category risk factor of X. Indicated on the package label of the drug is that it is contraindicated in women who are or may become pregnant. It is stated that "when administered to pregnant women, androgens cause virilization of the external genitalia of the female fetus (see below). The virilization includes clitoromegaly, abnormal vaginal development, and fusion of genital folds to form a scrotal-like structure." The statement on the label is continued: "the degree of masculinization is related to the amount of drug given and the age of the fetus, and is most likely to occur in the female fetus when the drugs are given in the first trimester. If the patient becomes pregnant while taking these drugs, she should be apprised of the potential hazard to the fetus" (*PDR*, 2004).

DEVELOPMENTAL TOXICOLOGY

ANIMALS

In laboratory animal studies, MT elicits masculinization (virilization, pseudohermaphroditism) of female fetuses following prenatal administration in rats (Jost, 1960) and rabbits (Jost, 1947) of 10 µg/day by the subcutaneous route, and intersex puppies upon dietary administration of up to 150 µg/kg/day to bitches over several months (Shane et al., 1969). Oral administration, the route used in human therapy, also causes virilization of female offspring in rats given 0.5 or 1 mg/kg/day for 4 days late in gestation (Kawashima et al., 1977).

TABLE 1
Reports Attributing Virilization of Females to Methyltestosterone in Humans

Ref.	Number of Cases
Hayles and Nolan, 1957	1
Wilkins et al., 1958	1 (Foxworthy case)
Moncrieff, 1958	2
Gold and Michael, 1958	1
Nellhaus, 1958	1
Black and Bentley, 1959	1
Jones and Wilkins, 1960	1
Bisset et al., 1966	1
Serment and Ruf, 1968	2 (Dewhurts and deTomi cases)

HUMANS

In human females, masculinization of the external genitalia of offspring, as described on the package label and summarized by Schardein (2000) from the description largely from Wilkins et al. (1958), is provided in 11 case reports as shown in Table 1. The cases represented generally fell within the recommended therapeutic doses of 25 to 200 mg/day orally (female indication), with total doses of up to 2 g of drug. The degree of virilization appears to be related to the dosage of the drug administered: The greater the dose, the greater the degree of effect. Treatment intervals ranged from the third gestational week until term. The time of treatment in gestation is correlated with the type of anomaly observed. For instance, labioscrotal fusion is exhibited only in those instances in which the hormone is administered prior to the 13th week of gestation (Grumbach et al., 1959). More precisely, the degree of fusion is directly related to the quantity of drug given the mother between the 8th and 13th weeks of pregnancy (Grumbach and Ducharme, 1960). This is due to differentiation of the external genitalia, which occurs from 2 $\frac{1}{2}$ to 3 months in the developing fetus (Glenister and Hamilton, 1963). A similar effect does not occur in male offspring. No other classes of developmental toxicity were observed in association with the genital anatomical defects. One group of experts placed the magnitude of teratogenic risk for virilization of female fetuses as moderate and as undetermined for nongenital congenital anomalies (Friedman and Polifka, 2000). Of the latter, no significant reports have been published.

CHEMISTRY

Methyltestosterone is a large hydrophobic molecule. It is of low polarity in comparison to the other human developmental toxicants. Methyltestosterone can participate in donor/acceptor hydrogen bonding. The calculated physicochemical and topological properties are shown below.

PHYSICOCHEMICAL PROPERTIES

Parameter	Value
Molecular weight	302.457 g/mol
Molecular volume	303.37 A^3
Density	0.942 g/cm^3

Continued.

Parameter	Value
Surface area	376.56 A^2
LogP	4.268
HLB	1.397
Solubility parameter	21.268 J$^{(0.5)}$/cm$^{(1.5)}$
Dispersion	18.888 J$^{(0.5)}$/cm$^{(1.5)}$
Polarity	3.482 J$^{(0.5)}$/cm$^{(1.5)}$
Hydrogen bonding	9.134 J$^{(0.5)}$/cm$^{(1.5)}$
H bond acceptor	0.58
H bond donor	0.23
Percent hydrophilic surface	12.42
MR	88.562
Water solubility	−2.442 log (mol/M^3)
Hydrophilic surface area	46.78 A^2
Polar surface area	40.46 A^2
HOMO	−10.022 eV
LUMO	−0.131 eV
Dipole	3.578 debye

TOPOLOGICAL PROPERTIES (UNITLESS)

Parameter	Value
x0	15.751
x1	10.278
x2	10.915
xp3	9.944
xp4	7.967
xp5	6.754
xp6	5.104
xp7	3.879
xp8	2.848
xp9	2.061
xp10	1.344
xv0	14.322
xv1	9.242
xv2	9.238
xvp3	8.524
xvp4	6.803
xvp5	5.428
xvp6	4.074
xvp7	3.006
xvp8	2.108
xvp9	1.384
xvp10	0.841
k0	29.533
k1	15.523
k2	4.762
k3	1.977
ka1	14.930
ka2	4.466
ka3	1.830

REFERENCES

Bisset, W. H., Bain, A. D., and Gauld, I. K. (1966). Female pseudohermaphrodite presenting with bilateral cryptorchidism. *Br. Med. J.* 1: 279–280.

Black, J. A. and Bentley, J. F. R. (1959). Effect on the foetus of androgens given during pregnancy. *Lancet* 1: 21.

Friedman, J. M. and Polifka, J. E. (2000). *Teratogenic Effects of Drugs. A Resource for Clinicians (TERIS)* Second ed., Johns Hopkins University Press, Baltimore, MD.

Glenister, T. W. and Hamilton, W. J. (1963). The embryology of sexual differentiation in relation to the possible effects of administering steroid hormones during pregnancy. *J. Obstet. Gynaecol. Br. Commonw.* 70: 13–19.

Gold, A. P. and Michael, A. F. (1958). Testosterone-induced female pseudohermaphroditism. *J. Pediatr.* 52: 279–283.

Grumbach, M. M. and Ducharme, J. R. (1960). The effects of androgens on fetal sexual development. Androgen-induced female pseudohermaphroditism. *Fertil. Steril.* 11: 157–180.

Grumbach, M. M., Ducharme, J. R., and Moloshok, R. E. (1959). On the fetal masculinizing action of certain oral progestins. *J. Clin. Endocrinol. Metab.* 19: 1369–1380.

Hayles, A. B. and Nolan, R. B. (1957). Female pseudohermaphroditism: Report of a case of an infant born of a mother receiving methyltestosterone during pregnancy. *Proc. Staff Meet. Mayo Clin.* 32: 41–44.

Jones, H. W. and Wilkins, L. (1960). The genital anomaly associated with prenatal exposure to progestogens. *Fertil. Steril.* 11: 148–156.

Jost, A. (1947). Recherches sur la differenciation sexuelle de l'embryon de lapin. 2. Action des androgens de synthese sur l'histogenese genitale. *Arch. Anat. Microsc. Morphol. Exp.* 36: 242–270.

Jost, A. (1960). The action of various sex steroids and related compounds on the growth and sexual differentiation of the fetus. *Acta Endocrinol. Suppl. (Copenh.)* 50: 119–123.

Kawashima, K. et al. (1977). Virilizing activities of various steroids in female rat fetuses. *Endocrinol. Jpn.* 24: 77–81.

Lacy, C. F. et al. (2004). *Drug Information Handbook (Pocket), 2004–2005*, Lexi-Comp., Inc., Hudson, OH.

Moncrieff, A. (1958). Non-adrenal female pseudohermaphroditism associated with hormone administration in pregnancy. *Lancet* 2: 267–268.

Nellhaus, G. (1958). Artificially-induced female pseudohermaphroditism. *N. Engl. J. Med.* 258: 935–938.

PDR® (Physicians' Desk Reference®). (2004). Medical Economics Co., Inc., Montvale, NJ.

Schardein, J. L. (2000). *Chemically Induced Birth Defects*, Third ed., Marcel Dekker, New York, pp. 286–289.

Serment, H. and Ruf, H. (1968). Les dangers pour le produit de conception de medicaments administers a la femme enceinte. *Bull. Fed. Soc. Gynecol. Obstet. Lang. Fr.* 20: 69–76.

Shane, B. S. et al. (1969). Methyl testosterone-induced female pseudohermaphroditism in dogs. *Biol. Reprod.* 1: 41–48.

Wilkins, L. et al. (1958). Masculinization of female fetus associated with administration of oral and intramuscular progestins during gestation: Nonadrenal pseudohermaphroditism. *J. Clin. Endocrinol. Metab.* 18: 559–585.

38 Disulfiram

Chemical name: Tetraethylthioperoxydicarbonic diamide

Alternate name: Teturamin

CAS #: 97-77-8

SMILES: N(C(SSC(N(CC)CC)=S)=S)(CC)CC

INTRODUCTION

Disulfiram is a thiuram derivative that is used therapeutically as an antialcoholic agent in the management of chronic alcoholism. Mechanistically, it interferes with aldehyde dehydrogenase and when taken concomitantly with alcohol, the serum acetaldehyde levels are increased, causing uncomfortable symptoms and is the basis for postwithdrawal long-term care of alcoholism (Lacy et al., 2004). The agent also has industrial uses. For medicinal purposes, disulfiram is available by prescription under the trade name Antabuse®. It has a pregnancy category risk factor of C (risk cannot be ruled out).

DEVELOPMENTAL TOXICOLOGY

ANIMALS

Animal studies by the oral route have been conducted in mice, rats, hamsters, and guinea pigs. The agent is not teratogenic in any of these species, but increased resorption in mice at 10 mg/kg/day when given throughout gestation (Thompson and Folb, 1985), increased resorption in rats at 100 mg/day for 10 days in organogenesis (Salgo and Oster, 1974), reduced brain weight in guinea pigs at 125 mg/kg/day for 4 days late in gestation (Harding and Edwards, 1993), and no significant developmental toxicity in hamsters given up to 1000 mg/kg on a single day in mid-gestation (Robens, 1969) have been recorded. The mechanism of this embryotoxicity, where reported, is believed to be via copper chelation (Salgo and Oster, 1974).

HUMANS

Experience in humans with disulfiram has indicated potential teratogenic effects. The conclusion to be made with respect to this possibility is tenuous, however, because the drug is used in treating alcoholics, and the effects reported may be due to or influenced by the known teratogenic properties

TABLE 1
Developmental Toxicity Profile of Disulfiram in Humans

Case Number	Malformations	Growth Retardation	Death	Functional Deficit	Ref.
1	None		✔		Favre-Tissot and Delatour, 1965
2, 3	Limbs				Favre-Tissot and Delatour, 1965
4	Limbs				Nora et al., 1977
5	Vertebrae, limbs, t-e fistula				Nora et al., 1977
6	(Pierre Robin syndrome), heart				Dehaene et al., 1984
7	None		✔		Jones et al., 1991
8	Palate	✔			Reitnauer et al., 1997
9	Limbs	✔			Reitnauer et al., 1997

of alcohol itself. Nonetheless, several studies were published that reported malformations and other classes of developmental toxicity, as shown in Table 1. While no clearly identifiable constellation of defects is evident from these reports, limb malformations (reduction defects, clubfoot) were produced in common, with five cases reported. It should be mentioned that one report described malformations in an infant similar to those seen with fetal alcohol syndrome (FAS), associated with alcohol consumption, and even though the mother in the case denied alcohol use at the time, the case is too problematic to associate it with intake of disulfiram (Gardner and Clarkson, 1981). Another case was reported in which a child with FAS phenotype was from a pregnancy in which the mother also took an overdose of disulfiram in the second trimester (Czeizel et al., 1997). It, too, is not considered disulfiram induced. A total of 37 normal infants were reported from first trimester treatment with disulfiram in other reports (Favre-Tissot and Delatour, 1965; Nora et al., 1977; Hamon et al., 1991; Jones et al., 1991; Helmbrecht and Hoskins, 1993; Briggs et al., 2005). The two cases each of intrauterine growth retardation and spontaneous abortion/stillbirth do not appear to provide strong enough evidence to associate them with disulfiram exposure. No functional deficits of any kind were reported in any of the publications. The cases reported included administration of the drug at the therapeutic dose of 500 mg/day or less (orally), and treatments were in the first trimester. In summary, while the evidence is not compelling that disulfiram is a potent developmental toxicant in humans, the published case reports of malformations cannot be overlooked. One group of experts considers the teratogenic risk to be undeterminable based on the evidence at the time of writing (Friedman and Polifka, 2000).

CHEMISTRY

Disulfiram is an average-sized highly hydrophobic compound. It can act as a hydrogen bond acceptor. The calculated physicochemical and topological properties for this chemical are listed below.

PHYSICOCHEMICAL PROPERTIES

Parameter	Value
Molecular weight	296.546 g/mol
Molecular volume	261.84 A^3
Density	1.066 g/cm^3

Continued.

Parameter	Value
Surface area	351.27 A^2
LogP	6.128
HLB	6.592
Solubility parameter	22.587 $J^{(0.5)}/cm^{(1.5)}$
Dispersion	20.259 $J^{(0.5)}/cm^{(1.5)}$
Polarity	6.570 $J^{(0.5)}/cm^{(1.5)}$
Hydrogen bonding	7.521 $J^{(0.5)}/cm^{(1.5)}$
H bond acceptor	0.27
H bond donor	0.00
Percent hydrophilic surface	35.01
MR	87.366
Water solubility	−0.854 log (mol/M^3)
Hydrophilic surface area	122.98 A^2
Polar surface area	6.48 A^2
HOMO	−8.290 eV
LUMO	−2.566 eV
Dipole	2.546 debye

TOPOLOGICAL PROPERTIES (UNITLESS)

Parameter	Value
x0	12.552
x1	7.599
x2	5.685
xp3	4.784
xp4	3.184
xp5	1.158
xp6	0.702
xp7	0.540
xp8	0.222
xp9	0.111
xp10	0.000
xv0	13.294
xv1	7.951
xv2	5.791
xvp3	5.134
xvp4	3.545
xvp5	1.854
xvp6	0.951
xvp7	0.707
xvp8	0.225
xvp9	0.112
xvp10	0.000
k0	12.041
k1	16.000
k2	9.074
k3	5.778
ka1	16.800
ka2	9.792
ka3	6.360

REFERENCES

Briggs, G. G., Freeman, R. K., and Yaffe, S. J. (2005). *Drugs in Pregnancy and Lactation. A Reference Guide to Fetal and Neonatal Risk*, Seventh ed., Lippincott Williams & Wilkins, Philadelphia.

Czeizel, A. E., Tomczik, M., and Timor, L. (1997). Teratologic evaluation of 178 infants born to mothers who attempted suicide by drugs during pregnancy. *Obstet. Gynecol.* 90: 195–201.

Dehaene, P., Titran, M., and Dubois, D. (1984). Pierre Robin syndrome and cardiac malformations in a newborn. Was disulfiram taken during pregnancy responsible? *Presse Med.* 13: 1394.

Favre-Tissot, M. and Delatour, P. (1965). Psychopharmacologie et teratogenese a propos du disulfiram: Essai experimental. *Ann. Medicopsychol.* 123: 735–740.

Friedman, J. M. and Polifka, J. E. (2000). *Teratogenic Effects of Drugs. A Resource for Clinicians (TERIS)*, Second ed., Johns Hopkins University Press, Baltimore, MD.

Gardner, R. J. M. and Clarkson, J. E. (1981). A malformed child whose previously alcoholic mother had taken disulfiram. *N.Z. Med. J.* 93: 184–186.

Hamon, B. et al. (1991). Grossesse chez les maladies traitees par le disulfirame. *Presse Med.* 20: 1092.

Harding, A. J. and Edwards, M. J. (1993). Retardation of prenatal brain growth of guinea pigs by disulfiram. *Cong. Anom.* 33: 197–202.

Helmbrecht, G. D. and Hoskins, I. A. (1993). First trimester disulfiram exposure: Report of two cases. *Am. J. Perinatol.* 10: 5–7.

Jones, K. L., Chambers, C. C., and Johnson, K. A. (1991). The effect of disulfiram on the unborn baby. *Teratology* 43: 438.

Lacy, C. F. et al. (2004). *Drug Information Handbook (Pocket), 2004–2005*, Lexi-Comp., Inc., Hudson, OH.

Nora, A. H., Nora, J. J., and Blu, J. (1977). Limb-reduction anomalies in infants born to disulfiram treated alcoholic mothers. *Lancet* 2: 664.

Reitnauer, P. J. et al. (1997). Prenatal exposure to disulfiram implicated in the cause of malformations in discordant monozygotic twins. *Teratology* 56: 358–362.

Robens, J. F. (1969). Teratologic studies of carbaryl, diazinon, norea, disulfiram and thiram in small laboratory animals. *Toxicol. Appl. Pharmacol.* 15: 152–163.

Salgo, M. P. and Oster, G. (1974). Fetal resorption induced by disulfiram in rats. *J. Reprod. Fertil.* 39: 375–377.

Thompson, P. A. C. and Folb, P. I. (1985). The effects of disulfiram on the experimental C3H mouse embryo. *J. Appl. Toxicol.* 5: 1–10.

39 Valproic Acid

Chemical name: 2-Propylpentanoic acid

Alternate name: Dipropylacetic acid

CAS #: 99-66-1

SMILES: C(CCC)(CCC)C(O)=O

INTRODUCTION

Valproic acid (VPA) has had therapeutic utility as a popular anticonvulsant for over 25 years, particularly for petit mal and complex absence seizures. It is also used as an antimanic and antimigraine agent. Its mechanism of action is by increasing the availability of γ-aminobutyric acid (GABA), an inhibitory neurotransmitter to brain neurons that enhances the action of GABA or mimics its action at postsynaptic receptor sites (Lacy et al., 2004). VPA is available as a prescription drug by a variety of trade names, including Depakene®, Convulex®, and Mylproin®, among other names, and it is also available in the sodium valproate form as Depakine® and Epilim®, among other names. It has a pregnancy category risk factor of D. The package label contains a "black box" warning stating that "valproate can produce teratogenic effects such as neural tube defects (e.g., spina bifida). Accordingly, the use of valproate products in women of childbearing potential requires that the benefits of its use be weighed against the risk of injury to the fetus" (see below; *PDR*, 2005). The label further states that its usage in pregnancy, according to published and unpublished reports, may produce teratogenic effects. Multiple reports in the clinical literature indicate that the use of antiepileptic drugs during pregnancy (including VPA) results in an increased incidence of birth defects in offspring. Also stated on the label is that the Centers for Disease Control (CDC) estimated the risk of VPA-exposed women having children with spina bifida to be approximately 1 to 2%. Other congenital anomalies (e.g., craniofacial defects, cardiovascular malformations, and anomalies involving various body systems), compatible and incompatible with life, were reported. Animal studies (mice, rats, rabbits, monkeys) demonstrated valproate-induced teratogenicity, as well as intrauterine growth retardation and death following prenatal exposure to valproate (see following). Valproic acid readily crosses the placenta in humans, and the range of cord blood:maternal serum ratios of total VPA is on the order of 1.4:2.4 (cited, Briggs et al., 2005). Therapeutic levels of 50 to 100 μg/ml in the serum are thought to be adequate to control seizures from amounts administered ranging up to 2500 mg/day orally (Lacy et al., 2004).

TABLE 1
Developmental Toxicity Profile of Valproic Acid in Representative Laboratory Animal Species

Species	Effective Doses (mg/kg/day)	Developmental Toxicity Effects[a]	Ref.
Mouse	75–600	M	Miyagawa et al., 1971; Whittle, 1976
Rat	150–800	M, G, D, F	Vorhees, 1987; Binkerd et al., 1988
Gerbil	151	M, F	Chapman and Cutler, 1989
Rabbit	315	M	Whittle, 1976
Primate (rhesus)	20–600	M, G, D	Hendrickx et al., 1988

[a] M = malformation, G = growth retardation, D = death, F = functional/behavioral deficit.

DEVELOPMENTAL TOXICOLOGY

ANIMALS

The drug is developmentally toxic in at least five species of laboratory animals. The profile of the drug in the various species by the oral route (the route administered in humans) is shown in Table 1. All four classes of developmental toxicity have been reported. In addition, intraperitoneal administration is effective in eliciting developmental toxicity in mice (Brown et al., 1980), rats (Kao et al., 1981), and hamsters (Moffa et al., 1984). The parent drug, not metabolites, has been implicated as the teratogen, at least in mice (Nau, 1986). The calcium salt of valproate is equally effective as the sodium salt in producing developmental toxicity in rats (Ong et al., 1983) and rabbits (Petrere et al., 1986). Both rats (Briner and Lieske, 1995) and mice (Ehlers et al., 1992) serve as concordant models for the primary malformation (spina bifida) observed in the human (see below).

HUMANS

In the human, VPA has a history of adverse developmental effects, and the discussion following will center on these effects by class. The drug is unusual in that its teratogenicity in humans was predicted from animal studies, without any knowledge of mechanism (Brown et al., 1980; Kao et al., 1981).

Malformations

Valproic acid, first marketed in Europe in 1967, appeared to be without adversity to development over the initial 13 years following marketing. Then, in the early 1980s, Dalens and her associates made the initial association of VPA to birth defects (Dalens et al., 1980; Dalens, 1981). They reported an infant who died at 19 days of age, was growth retarded, and who had multiple malformations of the face and brain, heart, and skeleton, among other defects. The mother of the infant had taken 1000 mg/day of VPA throughout gestation. These observations were followed by a number of case reports and other publications attesting to the malformative effects of the drug when administered to a pregnant mother during gestation. It should be mentioned here that like other anticonvulsants used in treating epilepsy, multiple drug therapy, characteristics of epilepsy itself, and other factors make the interpretation of the toxicity profile of the monotherapy of a given drug tenuous. Nonetheless, the developmental toxicity of VPA has now been established firmly, both with monotherapy and when combined with other anticonvulsants. Case reports too numerous to categorize here as well as a large number of clinically descriptive studies and epidemiological

---

TABLE 2
**Representative Clinical and Epidemiological
Studies with Congenital Malformations Attributed
to Valproic Acid in Humans**

Nau et al., 1981	Kallen et al., 1989
Robert, 1982	Martinez-Frias, 1990
Jeavons, 1982	Battino et al., 1992
Granstrom, 1982	Dravet et al., 1992
Jager-Roman et al., 1982	Lindhout et al., 1992a, 1992b
Robert and Rosa, 1983	Kaneko et al., 1992
Koch et al., 1983	Omtzigt et al., 1992a, 1992b, 1992c
Robert et al., 1983	Raymond et al., 1993
Mastroiacovo et al., 1983	Kaneko et al., 1993
DiLiberti et al., 1984	Guibaud et al., 1993
Robert et al., 1984a, 1984b	Thisted and Ebbesen, 1993
Hanson et al., 1984	Christianson et al., 1994
Lindhout and Meinardi, 1984	Koch et al., 1996
Koch et al., 1985	Espinasse et al., 1996
Bertollini et al., 1985	Samren et al., 1997
Lindhout and Schmidt, 1986	Bradai and Robert, 1998
Jager-Roman et al., 1986	Canger et al., 1999
Weinbaum et al., 1986	Rodriguez-Pinella et al., 2000
Winter et al., 1987	Moore et al., 2000
Ardinger et al., 1988	Arpino et al., 2000
Robert, 1988	Alsdorf et al., 2004
Oakeshott and Hunt, 1989	Wide et al., 2004
Martinez-Frias et al., 1989	

---

studies have appeared, with well over 300 cases now recorded; a representative number of pertinent cases are provided in Table 2, linking VPA exposure during pregnancy with malformations, as will be described. Especially important was the finding of neural tube defects, particularly spina bifida, as provided in a history of VPA teratogenicity relived by the discoverer (Robert, 1988). Neural tube defects also include meningocele, meningomyelocele, lipomyelomeningocele, and microcephaly. The high prevalence of neural tube defects in reports of VPA exposures suggested to some investigators that they may represent a pharmacogenetic abnormality (Duncan et al., 2001). Results in twin pregnancies also suggest a genetic component to the malformations (Hockey et al., 1996; Malm et al., 2002). The constellation of defects observed in the various reports following the reports of neural tube defects and later termed the "fetal valproate syndrome" (DiLiberti et al., 1984) include a characteristic facial phenotype comprising hypertelorism, short nose, thin upper lip and thick lower lip, epicanthal folds, orofacial clefts, midface hypoplasia, deficient orbital ridge, micrognathia, prominent forehead ridge, and small, low-set, posterior-angulated ears. Also mentioned as features in the syndrome are congenital heart disease, hypospadias, postnatal growth retardation and developmental delay (see below), musculoskeletal and limb reduction abnormalities (including radial-ray reductions, as reported by Brons et al. [1990], Verloes et al. [1990]), Sharony et al. [1993], Ylagen and Budorick [1994], and Langer et al. [1994]). The phenotype was verified in 1988 (Ardinger et al.), and it is supposedly recognizable by mid-pregnancy (Serville et al., 1989). One study of 178 cases of malformation confirmed specific association between only the drug and spina bifida, preaxial limb defects, and hypospadias (cited from a personal communication [Robert], Schardein, 2000). Another report assigned the following incidences of malformation types from a review of 69 studies observed over a recent 12 yr interval: 62% musculoskeletal, 30% skin, 26%

cardiovascular, 22% genital, 16% pulmonary, and 3% neural tube (Kozma, 2000). Abnormalities less commonly seen that have been potentially considered as features of the syndrome include craniosynostoses (Lajeunie et al., 1998, 2001; Chabrolle et al., 2001; Assencio-Ferreira et al., 2001), eye defects (McMahon and Braddock, 2001; Boyle et al., 2001; Hornby and Welham, 2003), omphalocele (Boussemart et al., 1995), aplasia cutis congenita (Hubert et al., 1994), hypertrichosis and gum hypertrophy (Stoll et al., 2003), pancreatitis (Grauso-Eby et al., 2003), lung hypoplasia (Janas et al., 1998), vascular anomalies (Mo and Laduscans, 1999; Anoop and Sasidharan, 2003), and liver toxicity (Felding and Rane, 1984; Legius et al., 1987), but low frequencies of these findings have not yet been proven to be definitive features. More complete descriptions of the malformative aspects are provided in other publications (Friedman and Polifka, 2000; Schardein, 2000; Briggs et al., 2005).

The syndrome is apparently elicited by first trimester exposures, and the higher therapeutic range of doses (>1500 mg/day) is most often associated with the affected cases. The general consensus is that VPA induces malformations at an incidence in the range of 1 to 2%. Another estimate places the increased risk two- to threefold beyond that expected for other anticonvulsant drugs. One group of experts places the magnitude of teratogenic risk as moderate, with neural tube defects small to moderate (Friedman and Polifka, 2000). A more recent study calculated a relative risk of 7.3 (95% confidence interval [CI] 4.4 to 12.2) among offspring of 149 first trimester VPA-exposed women (Alsdorf et al., 2004), suggesting that the drug is more toxic developmentally than previously supposed. Polytherapy of VPA, with carbamazepine and primidone, is considered by some the most risky anticonvulsant treatment regimen for production of malformations (Murasaki et al., 1988).

The teratogenic mechanism of action of VPA has been theorized, and the following information is summarized largely from an NRC publication (NRC, 2000). VPA is one of the very few agents for which there are proposed mechanisms for its teratogenic activity, there are strict structural requirements for its teratogenicity, and a plausible structure–activity relationship has been suggested. In this regard, the teratogenic mechanism of VPA appears to be multifaceted, and suggested mechanisms include effects on the cytoskeleton and cell motility (Walmod et al., 1998, 1999), varying aspects of zinc (Bui et al., 1998), methionine (Alonso-Aperte et al., 1999), and homocysteine and glutathione metabolism (Hishida and Nau, 1998), peroxisome proliferation-activated receptor interaction (Lampen et al., 1999), and gene expression (Wlodarczyk et al., 1996; Finnell et al., 1997; Okada et al., 2004). Based on VPA's enantiomers, analogs, and metabolites, structural requirements according to Nau (1994) include the following:

1. A free carboxylic acid is required. Amides such as valpromide are inactive (Radatz et al., 1998; Spiegelstein et al., 1999), as are stable esters.
2. The C2 carbon must be bonded to one hydrogen and two alkyl chains, as well as the carboxyl group. Substituting the hydrogen with any group abolished activity, and a single chain or unsaturated derivatives (e.g., 2-en-VPA) are also inactive.
3. Activity is greatest when the two alkyl chains are unbranched and contain three carbons (Bojic et al., 1996, 1998).
4. Introducing a side-chain double or triple bond terminally between C3 and C4 enhances teratogenicity, but in any other position, it abolishes activity.
5. When one side chain has terminal unsaturation, C2 is asymmetric, and the enantiomers have markedly different potencies. In the cases of both 4-en-VPA and 4-yn-VPA, the S-enantiomer is more potent than the racemate, and the R-enantiomer is virtually inactive (Hauck and Nau, 1992; Andrews et al., 1995, 1997).
6. Finally, with respect to SAR relationships with VPA, it appears that these are not due to pharmacokinetic differences, as shown by direct measurements of tissue levels and by the activities of VPA and its analogs in embryo culture (Brown et al., 1987; Nau, 1994), and their consistency across species (Andrews et al., 1995).

According to Bojic et al. (1998), the teratogenic effect of valproids requires an interaction with a specific site, at which one alkyl chain becomes located in a hydrophobic pocket, thus enabling ionic bonding of the carboxyl group and interaction of the second chain with a region that favors the high electron density of terminal unsaturation. An elaboration of neural tube defects by VPA mechanistically through loss of heterozygosity of genes critical to development was recently advanced (Defoort et al., 2005).

Growth Retardation

Intrauterine or postnatal growth retardation appears to be an associated component of the developmental toxicity profile of VPA, being recorded in a significant number of case reports and clinical studies (Dalens et al., 1980; Nau et al., 1981; Jager-Roman et al., 1982, 1986; Granstrom, 1982; Koch et al., 1983; Bailey et al., 1983; Felding and Rane, 1984; DiLiberti et al., 1984; Hanson et al., 1984; Weinbaum et al., 1986; Leguis et al., 1987; Ardinger et al., 1988).

Death

Intrauterine death, abortion, or postnatal mortality do not appear to be associated with the pattern of developmental toxicity observed with VPA.

Functional Deficit

A number of dysfunctional parameters, especially affective disorders (Robert-Gnansia, 2004) were associated with the other features of the developmental pattern caused by VPA. One such effect is developmental delay (Hanson et al., 1984; Ardinger et al., 1988; Dean et al., 2002). The latter study found developmental delay or neurologic abnormalities in 71% of affected cases in their study. Behavioral disturbances or neurological dysfunction, especially hyperexcitability, was reported in studies by another group of investigators (Koch et al., 1985, 1996). Withdrawal manifestations, including irritability, jitteriness, hypotonia, and seizures, were described by others (Thisted and Ebbesen, 1993; Clayton-Smith and Donnai, 1995). Abnormal psychomotor development was observed by other researchers (Jager-Roman et al., 1982). One of the first suggestions of adverse effects on childrens' behavior was made by Moore et al. (2000). From a case series among 46 VPA-exposed children, 40% were hyperactive or exhibited poor concentration, 60% had two or more autistic features, and 60% had learning difficulties, speech delay, or gross motor delay. A recent report however, found no IQ deficits in children exposed to VPA (Holmes et al., 2005). Autism was reported in several other recent reports (Christianson et al., 1994; Williams and Hersh, 1997; Williams et al., 2001; Bescoby-Chambers et al., 2001; Dean et al., 2002). Cases of multiple blood disorders were reported (Majer and Green, 1987; Bruel et al., 2001).

A number of published articles reviewing the developmental toxicity and related aspects of valproic acid are available. These include those by Anonymous (1983), Rosa (1984), Kelly (1984), Kallen (1986), Lammer et al. (1987), Robert (1988), Martinez-Frias (1990), Cotariu and Zaidman (1991), Kaneko (1991), Dansky and Finnell (1991), Nau et al. (1991), Sharony et al. (1993), Yerby (1994), Clayton-Smith and Donnai (1995), Malone and D'Alton (1997), Schardein (2000), Friedman and Polifka (2000), Iqbal et al. (2001), Kallen (2004), Kultima et al. (2004), Merks et al. (2004), and Briggs et al. (2005).

CHEMISTRY

Valproic acid is a smaller human developmental toxicant. It is a hydrophobic compound. It is of low polarity in comparison to the other chemicals. Valproic acid can engage in hydrogen bonding. The calculated physicochemical and topological properties are shown below.

PHYSICOCHEMICAL PROPERTIES

Parameter	Value
Molecular weight	144.213 g/mol
Molecular volume	154.72 A^3
Density	0.856 g/cm^3
Surface area	214.96 A^2
LogP	2.810
HLB	4.722
Solubility parameter	19.099 J$^{(0.5)}$/cm$^{(1.5)}$
Dispersion	17.074 J$^{(0.5)}$/cm$^{(1.5)}$
Polarity	2.758 J$^{(0.5)}$/cm$^{(1.5)}$
Hydrogen bonding	8.104 J$^{(0.5)}$/cm$^{(1.5)}$
H bond acceptor	0.55
H bond donor	0.27
Percent hydrophilic surface	26.88
MR	42.413
Water solubility	0.799 log (mol/M^3)
Hydrophilic surface area	57.78 A^2
Polar surface area	40.46 A^2
HOMO	−11.108 eV
LUMO	1.268 eV
Dipole	1.604 debye

TOPOLOGICAL PROPERTIES (UNITLESS)

Parameter	Value
x0	7.983
x1	4.719
x2	3.581
xp3	2.262
xp4	1.553
xp5	0.955
xp6	0.144
xp7	0.000
xp8	0.000
xp9	0.000
xp10	0.000
xv0	6.761
xv1	3.947
xv2	2.612
xvp3	1.624
xvp4	1.088
xvp5	0.536
xvp6	0.144
xvp7	0.000
xvp8	0.000
xvp9	0.000
xvp10	0.000
k0	8.194
k1	10.000
k2	5.760
k3	4.480
ka1	9.630
ka2	5.418
ka3	4.162

REFERENCES

Alonso-Aperte, E. et al. (1999). Impaired methionine synthesis and hypermethylation in rats exposed to valproate during gestation. *Neurology* 52: 750–756.

Alsdorf, R. M. et al. (2004). Evidence of increased birth defects in the offspring of women exposed to valproate during pregnancy: Findings from the AED pregnancy registry. *Birth Defects Res. (A)* 70: 245.

Andrews, J. E. et al. (1995). Validation of an *in vitro* teratology system using chiral substances: Stereoselective teratogenicity of 4-yn-valproic acid in cultured mouse embryos. *Toxicol. Appl. Pharmacol.* 132: 310–316.

Andrews, J. E. et al. (1997). Stereoselective dysmorphogenicity of the enantiomers of the valproic acid analogue 2-N-propyl-4-pentynoic acid (4-yn-VPA): Cross-species evaluation in whole embryo culture. *Teratology* 55: 314–318.

Anonymous. (1983). Valproate: A new cause for birth defects — Report from Italy and follow-up from France. *MMWR* 32: 438–439.

Anoop, P. and Sasidharan, C. K. (2003). Patent ductus arteriosus in fetal valproate syndrome. *Indian J. Pediatr.* 70: 681–682.

Ardinger, H. H. et al. (1988). Verification of the fetal valproate syndrome phenotype. *Am. J. Med. Genet.* 29: 171–185.

Arpino, C. et al. (2000). Teratogenic effects of antiepileptic drugs: Use of an international database on malformations and drug exposure (MADRE). *Epilepsia* 41: 1436–1443.

Assencio-Ferreira, V. J. et al. (2001). [Metopic suture craniosynostosis: Sodium valproate teratogenic effect: Case report]. *Arq. Neuropsiquiatr.* 59: 417–420.

Bailey, C. J. et al. (1983). Valproic acid and fetal abnormality. *Br. Med. J.* 286: 190.

Battino, D. et al. (1992). Malformations in offspring of 305 epileptic women: A prospective study. *Acta Neurol. Scand.* 85: 204–207.

Bertollini, R., Mastroiacovo, P., and Segni, G. (1985). Maternal epilepsy and birth defects: A case-control study in the Italian Multicentric Registry of Birth Defects (IPIMC). *Eur. J. Epidemiol.* 1: 67–72.

Bescoby-Chambers, N., Forster, P., and Bates, G. (2001). Foetal valproate syndrome and autism: Additional evidence of an association. *Dev. Med. Child Neurol.* 43: 847.

Binkerd, P. E. et al. (1988). Evaluation of valproic acid (VPA) developmental toxicity and pharmacokinetics in Sprague-Dawley rats. *Fundam. Appl. Toxicol.* 11: 485–493.

Bojic, U. et al. (1996). Further branching of valproate-related carboxylic acids reduces the teratogenic activity, but not the anticonvulsant effect. *Chem. Res. Toxicol.* 9: 866–870.

Bojic, U. et al. (1998). Studies on the teratogen pharmacophore of valproic acid analogues: Evidence of interactions at a hydrophobic centre. *Eur. J. Pharmacol.* 354: 289–299.

Boussemart, T. et al. (1995). Omphalocele in a newborn baby exposed to sodium valproate *in utero*. *Eur. J. Pediatr.* 154: 220–221.

Boyle, N. J., Clarke, M. P., and Figueiredo, F. (2001). Reduced corneal sensation and severe dry eyes in a child with fetal valproate syndrome. *Eye* 15 (Part 5): 661–662.

Bradai, R. and Robert, E. (1998). [Prenatal ultrasonography diagnosis in the epileptic mother on valproic acid. Retrospective study of 161 cases in the central eastern France register of congenital malformations]. *J. Gynecol. Obstet. Biol. Reprod. (Paris)* 27: 413–419.

Briggs, G. G., Freeman, R. K., and Yaffe, S. J. (2005). *Drugs in Pregnancy and Lactation. A Reference Guide to Fetal and Neonatal Risk*, Seventh ed., Lippincott Williams & Wilkins, Philadelphia.

Briner, W. and Lieske, R. (1995). Arnold-Chiari-like malformation associated with a valproate model of spina bifida in the rat. *Teratology* 52: 306–311.

Brons, J. T. et al. (1990). Prenatal ultra-sonographic diagnosis of radial ray reduction malformations. *Prenat. Diagn.* 10: 279–288.

Brown, N. A., Kao, J., and Fabro, S. (1980). Teratogenic potential of valproic acid. *Lancet* 1: 660–661.

Brown, N. A., Coakley, M. E., and Clarke, D. O. (1987). Structure–teratogenicity relationships of valproic acid congeners in whole-embryo culture. In *Approaches to Elucidate Mechanisms in Teratogenesis*, F. Welsch, Ed., Hemisphere, Washington, D.C., pp. 17–31.

Bruel, H. et al. (2001). [Hyperammonia, hypoglycemia, and thrombocytopenia in a newborn after maternal treatment with valproate]. *Arch. Pediatr.* 8: 446–447.

Bui, L. M. et al. (1998). Altered zinc metabolism contributes to the developmental toxicity of 2-ethylhexanoic acid, 2-ethylhexanol and valproic acid. *Toxicology* 126: 9–21.

Canger, R. et al. (1999). Malformations in offspring of women with epilepsy: A prospective study. *Epilepsia* 40: 1231–1236.

Chabrolle, J. P. et al. (2001). [Metopic craniosynostosis probable effect of intrauterine exposure to maternal valproate treatment]. *Arch. Pediatr.* 8: 1333–1336.

Chapman, J. B. and Cutler, M. G. (1989). Effects of sodium valproate on development and social behavior in the Mongolian gerbil. *Neurotoxicol. Teratol.* 11: 193–198.

Christianson, A. L., Chesler, N., and Kromberg, J. G. R. (1994). Fetal valproate syndrome: Clinical and neurodevelopmental features in two sibling pairs. *Dev. Med. Child Neurol.* 36: 361–369.

Clayton-Smith, J. and Donnai, D. (1995). Fetal valproate syndrome. *J. Med. Genet.* 32: 724–727.

Cotariu, D. and Zaidman, J. L. (1991). Developmental toxicity of valproic acid. *Life Sci.* 48: 1341–1350.

Dalens, B. (1981). Possible teratogenicity of valproic acid — reply. *J. Pediatr.* 98: 509.

Dalens, B., Ranaud, E. J., and Gauline, J. (1980). Teratogenicity of valproic acid. *J. Pediatr.* 97: 332–333.

Dansky, L. V. and Finnell, R. H. (1991). Parental epilepsy, anticonvulsant drugs, and reproductive outcome: Epidemiologic and experimental findings spanning three decades. 2. Human studies. *Reprod. Toxicol.* 5: 301–335.

Dean, J. C. et al. (2002). Long term health and neurodevelopment in children exposed to antiepileptic drugs before birth. *J. Med. Genet.* 39: 251–259.

Defoort, E., Kim, P. M., and Winn, L. M. (2005). Valproic acid increases conservative recombination frequency: Implications for a mechanism of valproic acid-induced neural tube defects. *Toxicologist* 84: 463.

DiLiberti, J. H. et al. (1984). The fetal valproate syndrome. *Am. J. Med. Genet.* 19: 473–481.

Dravet, C. et al. (1992). Epilepsy, antiepileptic drugs, and malformations in children of women with epilepsy: A French prospective cohort study. *Neurology* 42 (Suppl. 5): 75–82.

Duncan, S. et al. (2001). Repeated neural tube defects and valproate monotherapy suggest a pharmacogenetic abnormality. *Epilepsia* 42: 750–753.

Ehlers, K. et al. (1992). Valproic acid-induced spina bifida: A mouse model. *Teratology* 45: 145–154.

Espinasse, M. et al. (1996). [Embryopathy due to valproate: A pathology only little known. Apropos of 4 cases]. *Arch. Pediatr.* 3: 896–899.

Felding, I. and Rane, A. (1984). Congenital liver damage after treatment of mother with valproic acid and phenytoin? *Acta Paediatr. Scand.* 73: 565–568.

Finnell, R. H. et al. (1997). Strain-dependent alterations in the expression of folate pathway genes following teratogenic exposure to valproic acid in a mouse model. *Am. J. Med. Genet.* 70: 303–311.

Friedman, J. M. and Polifka, J. E. (2000). *Teratogenic Effects of Drugs. A Resource for Clinicians (TERIS)*, Second ed., Johns Hopkins University Press, Baltimore, MD.

Granstrom, M.-L. (1982). Development of the children of epileptic mothers: Preliminary results from the prospective Helsinki study. In *Epilepsy, Pregnancy, and the Child*, D. Janz, L. Bossi, M. Dam, H. Helge, A. Richens, and D. Schmidt, Eds., Raven Press, New York, pp. 403–408.

Grauso-Eby, N. L. et al. (2003). Acute pancreatitis in children from valproic acid: Case series and review. *Pediatr. Neurol.* 28: 145–148.

Guibaud, S. et al. (1993). Prenatal diagnosis of spina bifida aperta after first-trimester valproate exposure. *Prenat. Diagn.* 13: 772–773.

Hanson, J. W. et al. (1984). Effects of valproic acid on the fetus. *Pediatr. Res.* 18: 306A.

Hauck, R. S. and Nau, H. (1992). The enantiomers of the valproic acid analogue 2-N-propyl-4-pentaynoic acid (4-yn-VPA): Asymmetric synthesis and highly stereoselective teratogenicity in mice. *Pharm. Res.* 9: 850–855.

Hendrickx, A. G. et al. (1988). Valproic acid developmental toxicity and pharmacokinetics in the rhesus monkey: An interspecies comparison. *Teratology* 28: 329–345.

Hishida, R. and Nau, H. (1998). VPA-induced neural tube defects in mice. I. Altered metabolism of sulfur amino acids and glutathione. *Teratog. Carcinog. Mutagen.* 18: 49–61.

Hockey, A. et al. (1996). Fetal valproate embryopathy in twins: Genetic modification of the response to a teratogen. *Birth Defects Orig. Art. Ser.* 30: 401–405.

Holmes, L. B. et al. (2005). The correlation of deficits in IQ with midface and digit hypoplasia in children exposed *in utero* to anticonvulsant drugs. *J. Pediatr.* 146: 118–122.

Hornby, S. J. and Welham, R. A. (2003). Congenital nasolacrimal duct obstruction requiring external dacryocystorhinostomies in a child with foetal valproate syndrome. *Eye* 17: 546–547.

Hubert, A. et al. (1994). Aplasia cutis congenita of the scalp in an infant exposed to valproic acid *in utero*. *Acta Paediatr.* 83: 789–790.

Iqbal, M. M., Sohhan, T., and Mahmud, S. Z. (2001). The effects of lithium, valproic acid, and carbamazepine during pregnancy and lactation. *J. Toxicol. Clin. Toxicol.* 39: 381–392.

Jager-Roman, E. et al. (1982). Somatic parameters, diseases, and psychomotor development in the offspring of epileptic parents. In *Epilepsy, Pregnancy, and the Child*, D. Janz, L. Bossi, M. Dam., H. Helge, A. Richens, and D. Schmidt, Eds., Raven Press, New York, pp. 425–432.

Jager-Roman, E. et al. (1986). Fetal growth, major malformations, and minor anomalies in infants born to women receiving valproic acid. *J. Pediatr.* 108: 997–1004.

Janas, M. S. et al. (1998). Lung hypoplasia — a possible teratogenic effect of valproate. *APMIS* 106: 300–304.

Jeavons, P. M. (1982). Sodium valproate and neural tube defects. *Lancet* 2: 1282–1283.

Kallen, B. (1986). Maternal epilepsy, antiepileptic drugs and birth defects. *Pathologica* 78: 757–768.

Kallen, B. (2004). Valproic acid is known to cause hypospadias in man but does not reduce anogenital distance or causes hypospadias in rats. *Basic Clin. Pharmacol. Toxicol.* 94: 51–54.

Kallen, B. et al. (1989). Anticonvulsant drugs and malformations: Is there a drug specificity? *Eur. J. Epidemiol.* 5: 31–36.

Kaneko, S. (1991). Antiepileptic drug therapy and reproductive consequences: Functional and morphologic effects. *Reprod. Toxicol.* 5: 179–198.

Kaneko, S. et al. (1992). Malformation in infants of mothers with epilepsy receiving antiepileptic drugs. *Neurology* 42 (Suppl. 5): 68–74.

Kaneko, S. et al. (1993). Teratogenicity of antiepileptic drugs and drug specific malformations. *Jpn. J. Psychol. Neurol.* 47: 306–308.

Kao, J. et al. (1981). Teratogenicity of valproic acid *in-vivo* and *in-vitro*. *Teratog. Carcinog. Mutagen.* 1: 367–382.

Kelly, T. E. (1984). Teratogenicity of anticonvulsant drugs. I. Review of the literature. *Am. J. Med. Genet.* 19: 413–434.

Koch, S. et al. (1983). Possible teratogenic effect of valproate during pregnancy. *J. Pediatr.* 103: 1007–1008.

Koch, S. et al. (1985). Neonatal behavior disturbances in infants of epileptic women treated during pregnancy. *Prog. Clin. Biol. Res.* 163B: 453–461.

Koch, S. et al. (1996). Antiepileptic drug treatment of pregnancy: Drug side effects in the neonate and neurological outcome. *Acta Paediatr.* 84: 739–746.

Kozma, C. (2000). Valproic acid embryopathy: Report of two siblings with further expansion of the phenotypic abnormalities and a review of the literature. *Am. J. Med. Genet.* 98: 168–175.

Kultima, L. et al. (2004). Valproic acid teratogenicity: A toxicogenomics approach. *Environ. Health Perspect.* 112: 1225–1235.

Lacy, C. F. et al. (2004). *Drug Information Handbook (Pocket), 2004–2005*, Lexi-Comp., Inc., Hudson, OH.

Lajeunie, E. et al. (1998). Syndromal and nonsyndromal primary trigonocephaly: Analysis of 237 patients. *Am. J. Med. Genet.* 75: 211–215.

Lajeunie, E. et al. (2001). Craniosynostosis and fetal exposure to sodium valproate. *J. Neurosurg.* 95: 778–782.

Lammer, E. J., Sever, L. E., and Oakley, G. P. Jr. (1987). Teratogen update: Valproic acid. *Teratology* 35: 465–473.

Lampen, A. et al. (1999). New molecular bioassays for the estimation of the teratogenic potency of valproic acid derivatives *in vitro*: Activation of the peroxisomal proliferators-activated receptor (PPARdelta). *Toxicol. Appl. Pharmacol.* 160: 238–249.

Langer, B. et al. (1994). Isolated fetal bilateral ray reduction associated with valproic acid usage. *Fetal Diagn. Ther.* 9: 155–156.

Legius, E., Jaeken, J., and Eggermont, E. (1987). Sodium valproate, pregnancy, and infantile fatal liver failure. *Lancet* 2: 1518–1519.

Lindhout, D. and Meinardi, H. (1984). Spina bifida and *in-utero* exposure to valproate. *Lancet* 2: 396.

Lindhout, D. and Schmidt, D. (1986). *In-utero* exposure to valproate and neural tube defects. *Lancet* 1: 1392–1393.

Lindhout, D. et al. (1992a). Antiepileptic drugs and teratogenesis in two consecutive cohorts: Changes in prescription policy paralleled by changes in pattern of malformations. *Neurology* 42 (Suppl. 5): 94–110.

Lindhout, D., Omtzigt, J. G. C., and Cornel, M. C. (1992b). Spectrum of neural-tube defects in 34 infants prenatally exposed to antiepileptic drugs. *Neurology* 42 (Suppl. 5): 111–118.

Majer, R. V. and Green, P. J. (1987). Neonatal afibrinogenaemia due to sodium valproate. *Lancet* 2: 740–741.

Malm, H. et al. (2002). Valproate embryopathy in three sets of siblings: Further proof of hereditary susceptibility. *Neurology* 59: 630–633.

Malone, F. D. and D'Alton, M. E. (1997). Drugs in pregnancy: Anticonvulsants. *Semin. Perinatol.* 21: 114–123.

Martinez-Frias, M. L. (1990). Clinical manifestation of prenatal exposure to valproic acid using case reports and epidemiologic information. *Am. J. Med. Genet.* 37: 277–282.

Martinez-Frias, M. L., Rodriguez-Pinilla, E., and Salvador, J. (1989). Valproate and spina bifida. *Lancet* 1: 611–612.

Mastroiacovo, P. et al. (1983). Maternal epilepsy, valproate exposure, and birth defects. *Lancet* 2: 1499.

McMahon, C. L., and Braddock, S. R. (2001). Septo-optic dysplasia as a manifestation of valproic acid embryopathy. *Teratology* 64: 83–86.

Merks, J. H. et al. (2004). Neuroblastoma, maternal valproic acid use, *in-vitro* fertilization and family history of mosaic chromosome 22: Coincidence or causal relationship? *Clin. Dysmorphol.* 13: 197–198.

Miyagawa, A. et al. (1971). Toxicity tests with sodium dipropyl acetate. Report Hofu Factory Kyowa Fermentation Industry, Ltd., Kyowa, Japan.

Mo, C. N. and Laduscans, E. J. (1999). Anomalous right pulmonary artery origins in association with the fetal valproate syndrome. *J. Med. Genet.* 36: 83–84.

Moffa, A. M. et al. (1984). Valproic acid, zinc and open neural tubes in 9-day old hamster embryos. *Teratology* 29: 47A.

Moore, S. J. et al. (2000). A clinical study of 57 children with fetal anticonvulsant syndromes. *J. Med. Genet.* 37: 489–497.

Murasaki, D. et al. (1988). Reexamination of the teratological effect of antiepileptic drugs. *Jpn. J. Psy. N.* 42: 592–593.

Nau, H. (1986). Valproic acid teratogenicity in mice after various administration and phenobarbital pretreatment regimens: The parent drug and not one of its metabolites assayed is implicated as teratogen. *Fundam. Appl. Toxicol.* 6: 662–668.

Nau, H. (1994). Valproic acid induced neural tube defects. In *Neural Tube Defects*, G. Bock and J. Marsh, Eds., John Wiley, Sussex, UK, pp. 144–160.

Nau, H. et al. (1981). Valproic acid and its metabolites: Placental transfer, neonatal pharmacokinetics, transfer via mothers milk and clinical status in neonates of epileptic mothers. *J. Pharmacol. Exp. Ther.* 219: 768–777.

Nau, H., Hauck, R.-S., and Ehlers, K. (1991). Valproic acid-induced neural tube defects in mouse and human: Aspects of chirality, alternative drug development, pharmacokinetics and possible mechanisms. *Pharmacol. Toxicol.* 69: 310–321.

NRC (National Research Council). (2000). *Scientific Frontiers in Developmental Toxicology and Risk Assessment*, National Academy Press, Washington, D.C., pp. 82–83.

Oakeshott, P. and Hunt, G. M. (1989). Valproate and spina bifida. *Br. Med. J.* 298: 1300–1301.

Okada, A. et al. (2004). Polycomb homologs are included in teratogenicity of valproic acid in mice. *Birth Defects Res. (A)* 70: 870–879.

Omtzigt, J. G. et al. (1992a). The risk of spina bifida aperta after first-trimester exposure to valproate in a prenatal cohort. *Neurol.* 42 (Suppl. 5): 119–125.

Omtzigt, J. G. C. et al. (1992b). The disposition of valproate and its metabolites in the late first trimester and early second trimester of pregnancy in maternal serum, urine, and amniotic fluid: Effect of dose, co-medication, and the presence of spina bifida. *Eur. J. Clin. Pharmacol.* 43: 381–388.

Omtzigt, J. G. C. et al. (1992c). Prenatal diagnosis of spina bifida aperta after first-trimester valproate exposure. *Prenat. Diagn.* 12: 893–897.

Ong, L. L. et al. (1983). Teratogenesis of calcium valproate in rats. *Fundam. Appl. Toxicol.* 3: 121–126.

PDR® (Physicians' Desk Reference®). (2005). Medical Economics Co., Inc., Montvale, NJ.

Petrere, J. A. et al. (1986). Teratogenesis of calcium valproate in rabbits. *Teratology* 34: 263–269.

Radatz, M. et al. (1998). Valnoctamide, valpromide and valnoic acid are much less teratogenic in mice than valproic acid. *Epilepsy Res.* 30: 41–48.

Raymond, G. V., Harvey, E. A., and Holmes, L. B. (1993). Valproate teratogenicity: Data from the maternal epilepsy study. *Teratology* 47: 393.

Robert, E. (1982). Valproic acid and spina bifida: A preliminary report — France. *MMWR* 31: 565–566.

Robert, E. (1988). Valproic acid as a human teratogen. *Congenital Anom.* 28 (Suppl.): S71–S80.

Robert, E. and Rosa, F. (1983). Valproate and birth defects. *Lancet* 2: 1142.

Robert, E., Robert, M., and Lapras, C. (1983). [Is valproic acid teratogenic?]. *Rev. Neurol.* 139: 445–447.

Robert, E., Tosa, F. W., and Robert, J. M. (1984a). Maternal anti-epileptic exposure rates for spina bifida, heart defects, facial cleft outcomes compared to rates for other defect outcomes in a birth defect registry in Lyon, France. *Teratology* 29: 31A.

Robert, E., Lofkvist, E., and Mauguiere, F. (1984b). Valproate and spina bifia. *Lancet* 2: 1392.

Robert-Gnansia, E. (2004). Risk for children exposed prenatally to valproic acid. *Reprod. Toxicol.* 19: 246–247.

Rodriguez-Pinilla, E. et al. (2000). Prenatal exposure to valproic acid during pregnancy and limb deficiencies: A case-control study. *Am. J. Med. Genet.* 90: 376–381.

Rosa, F. W. (1984). Teratogenesis in epilepsy: Birth defects with maternal valproic acid exposures. In *Advances in Epidemiology, XV,* Epilepsy Symposium, R. J. Porter, Ed., Raven Press, New York, pp. 309–314.

Samren, E. B. et al. (1997). Maternal use of antiepileptic drugs and the risk of major congenital malformations: A joint European prospective study of human teratogenesis associated with maternal epilepsy. *Epilepsia* 38: 981–990.

Schardein, J. L. (2000). *Chemically Induced Birth Defects,* Third ed., Marcel Dekker, New York, pp. 207–211.

Serville, F. et al. (1989). Fetal valproate phenotype is recognizable by mid pregnancy. *J. Med. Genet.* 26: 348–349.

Sharony, R. et al. (1993). Preaxial ray reduction defects as part of valproic acid embryofetopathy. *Prenat. Diagn.* 13: 909–918.

Spiegelstein, O. et al. (1999). Enantioselective synthesis and teratogenicity of propylisopropyl acetamide, a CNS-active chiral amide analogue of valproic acid. *Chirality* 11: 645–650.

Stoll, C. et al. (2003). Multiple congenital malformations including hypertrichosis with gum hypertrophy in a child exposed to valproic acid *in utero. Genet. Couns.* 14: 289–298.

Thisted, E. and Ebbesen, F. (1993). Malformations, withdrawal manifestations, and hypoglycaemia after exposure to valproate *in utero. Arch. Dis. Child.* 69: 288–291.

Verloes, A. et al. (1990). Proximal phocomelia and radial ray aplasia in fetal valproate syndrome. *Eur. J. Pediatr.* 149: 266–267.

Vorhees, C. V. (1987). Teratogenicity and developmental toxicity of valproic acid in rats. *Teratology* 35: 195–202.

Walmod, P. S. et al. (1998). Cell motility is inhibited by the antiepileptic compound, valproic acid and its teratogenic analogues. *Cell Motil. Cytoskeleton* 40: 220–237.

Walmod, P. S. et al. (1999). Antiepileptic teratogen valproic acid (VPA) modulates organization and dynamics of the actin cytoskeleton. *Cell Motil. Cytoskeleton* 42: 241–255.

Weinbaum, P. J. et al. (1986). Prenatal detection of a neural tube defect after fetal exposure to valproic acid. *Obstet. Gynecol.* 67: 315–335.

Whittle, B. A. (1976). Pre-clinical teratological studies on sodium valproate (Epilim) and other anticonvulsants. In *Clinical and Pharmacological Aspects of Sodium Valproate (Epilim) in the Treatment of Epilepsy. Proc. Symp. Nottingham, 1975,* N. J. Legg, Ed., MCS Consultants, Kent, UK, pp. 105–111.

Wide, K., Winbladh, B., and Kallen, B. (2004). Major malformations in infants exposed to antiepileptic drugs *in utero,* with emphasis on carbamazepine and valproic acid: A nation-wide, population-based register study. *Acta Paediatr.* 93: 174–176.

Williams, G. et al. (2001). Fetal valproate syndrome and autism: Additional evidence of an association. *Dev. Med. Child Neurol.* 43: 202–206.

Williams, P. G. and Hersh, J. H. (1997). A male with fetal valproate syndrome and autism. *Dev. Med. Child Neurol.* 39: 632–634.

Wlodarczyk, B. C. et al. (1996). Valproic acid-induced changes in gene expression during neurulation in a mouse model. *Teratology* 54: 284–297.

Winter, R. M. et al. (1987). Fetal valproate syndrome: Is there a recognizable phenotype? *J. Med. Genet.* 24: 692–695.

Yerby, M. S. (1994). Pregnancy, teratogenesis, and epilepsy. *Neurol. Clin.* 12: 749–771.

Ylagan, L. R. and Budorick, N. E. (1994). Radial ray aplasia *in utero*: A prenatal finding associated with valproic acid exposure. *J. Ultrasound Med.* 13: 408–411.

40 Carbon Disulfide

Alternate names: Carbon bisulfide, carbon disulphide

CAS # 75-15-0

SMILES: C(=S)=S

S ══════ S

INTRODUCTION

Carbon disulfide is a colorless liquid used as a solvent for a wide variety of chemicals and in the manufacture of rayon viscose fibers and cellophane. The chemical is toxic upon exposure to humans via inhalational or dermal routes. The threshold limit value-time-weighted average for carbon disulfide is 10 ppm (31 mg/m^3 skin absorption; see Hathaway and Proctor, 2004; ACGIH, 2005). The chemical is known by its generic name in the United States.

DEVELOPMENTAL TOXICOLOGY

ANIMALS

In laboratory animals, carbon disulfide is developmentally toxic and teratogenic in both rats and rabbits (the only two species tested) by the inhalational route of exposure (which is pertinent to human exposures). In the rat, exposures over the range of 50 to 2000 mg/m^3 throughout gestation induced gross and skeletal malformations and postnatal functional effects (Tabacova, 1976; Tabacova et al., 1978). In the rabbit, concentrations of 600 or 1200 ppm for 6 hours daily over 13 days in gestation caused malformations and fetal death and reduced fetal body weight; these doses were maternally toxic as well (Gerhart et al., 1991). Oral doses of 150 mg/kg/day administered for 14 days during gestation in this species (rabbit) also elicited similar developmental toxicity, while higher oral doses (600 mg/kg/day) administered over 10 days in gestation in the rat were maternally toxic but produced only fetotoxicity and no malformations (Price et al., 1984).

HUMANS

In the human, carbon disulfide has long been considered a *reproductive toxicant*, affecting spermatogenesis in man and menstrual disorders in women at high concentrations (Hathaway and Proctor, 2004). The chemical may also be a developmental toxicant, although the data reported in published studies are tenuous at best. Nonetheless, there are suggestive reports associating occupational exposures to carbon disulfide during pregnancy with increased malformation, spontaneous abortion, and functional alterations.

In a prospective epidemiological study conducted in China, of 682 female workers comprising 1112 pregnancies who were exposed for at least 6 months prior to and during pregnancy in rayon factories, the incidence of birth defects was significantly higher (relative risk [RR] = 2.02, 95%

confidence interval [CI], 1.13 to 3.60) compared to a similar control group of 745 nonexposed women, even after confounding factors were considered (Bao et al., 1991). The highest incidence of abnormalities noted were congenital heart defects (0.9%), inguinal hernias (0.7%), and central nervous system defects (0.5%), but there was no distinctive pattern or syndrome of defects. There also was no other class of developmental toxicity apparent in the exposed group, including incidences of spontaneous abortion, prematurity, stillbirth, low birth weight, or neonatal/perinatal death. There was also no specific association to exposure levels around the 10 mg/m^3 baseline. In an earlier, retrospective cohort study, 265 women exposed to generally lower concentrations of carbon disulfide in the range of 1.7 to 14.8 mg/m^3 prior to pregnancy for as long as 15 years showed no differences in the rate of congenital malformation compared to those of 291 nonexposed women (Zhou et al., 1988). Increased rates for spontaneous abortion, stillbirth, reduced birth weight, or premature or overdue deliveries were not observed.

Contrary to the larger study cited above with respect to spontaneous abortion, there are four rather imperfectly documented foreign reports that indicate increased spontaneous abortion among women exposed during pregnancy to carbon disulfide in viscose manufacturing plants in widely separated venues (Ehrhardt, 1967; Petrov, 1969; Bezvershenko, 1979; Hemminki et al., 1980). Specific details in the reports are lacking, including accurate exposure levels and subject information; thus, they cannot be considered definitive in any respect. Of interest, too, is the fact that confirming reports have not surfaced in almost 25 years. Nonetheless, the suggestion that spontaneous abortion may occur among occupationally exposed women in industry cannot be discounted.

One study reported neurobehavioral abnormalities among children prenatally exposed to carbon disulfide at concentration levels encountered in the workplace, said to be up to 0.33 mg/m^3 (Tabacova and Khinkova, 1981). These abnormalities were described as sensory, neurofunctional, and behavioral deviations as indicators of prenatal stress.

While the reported published data on carbon disulfide are scant, the established toxicity pattern of the chemical to other organ systems (e.g., coronary heart disease, central and peripheral nervous systems) is sufficient evidence that this chemical exhibits significant toxicity (Hathaway and Proctor, 2004) and may also include developmental toxicity. In fact, one official body in Europe (the German Commission for the Investigation of Health Hazards of Chemical Compounds in the Work Area) placed carbon disulfide as a Group B Developmental Toxicant, a classification indicating that a risk of damage to the developing embryo or fetus must be considered when pregnant women are exposed, especially if the exposure level is >10 ppm (Hofman, 1995). This judgment was apparently based on the study of Chinese women cited above.

Several pertinent reviews on the subject of carbon disulfide toxicity were published (Beauchamp et al., 1983; Stetkiewicz and Wronska-Nofer, 1998).

CHEMISTRY

Carbon disulfide is one of the smallest nonpolar human developmental toxicants. The calculated physicochemical and topological properties for this chemical are listed below.

PHYSICOCHEMICAL PROPERTIES

Parameter	Value
Molecular weight	76.143 g/mol
Molecular volume	53.36 A^3
Density	1.346 g/cm^3

Continued.

Parameter	Value
Surface area	76.24 A^2
LogP	0.030
HLB	21.540
Solubility parameter	26.923 $J^{(0.5)}/cm^{(1.5)}$
Dispersion	26.923 $J^{(0.5)}/cm^{(1.5)}$
Polarity	0.000 $J^{(0.5)}/cm^{(1.5)}$
Hydrogen bonding	0.000 $J^{(0.5)}/cm^{(1.5)}$
H bond acceptor	0.12
H bond donor	0.00
Percent hydrophilic surface	100.00
MR	21.704
Water solubility	2.525 log (mol/M^3)
Hydrophilic surface area	76.24 A^2
Polar surface area	0.00 A^2
HOMO	−9.512 eV
LUMO	−1.372 eV
Dipole	0.000 debye

TOPOLOGICAL PROPERTIES (UNITLESS)

Parameter	Value
x0	2.707
x1	1.414
x2	0.707
xp3	0.000
xp4	0.000
xp5	0.000
xp6	0.000
xp7	0.000
xp8	0.000
xp9	0.000
xp10	0.000
xv0	2.950
xv1	1.225
xv2	0.750
xvp3	0.000
xvp4	0.000
xvp5	0.000
xvp6	0.000
xvp7	0.000
xvp8	0.000
xvp9	0.000
xvp10	0.000
k0	0.829
k1	3.000
k2	2.000
k3	0.000
ka1	3.220
ka2	2.220
ka3	0.000

REFERENCES

ACGIH (American Conference of Government Industrial Hygienists). (2005). *TLVs® and BEIs®. Threshold Limit Values for Chemical Substances and Physical Agents & Biological Exposure Indices*, ACGIH, Cincinnati, p. 17.

Bao, Y.-S. et al. (1991). Birth defects in the offspring of female workers occupationally exposed to carbon disulfide in China. *Teratology* 43: 451–452.

Beauchamp, R. O. Jr. et al. (1983). A critical review of the literature on carbon disulfide toxicity. *Crit. Rev. Toxicol.* 11: 169–278.

Bezvershenko, A. S. (1979). *Environmental Health Criteria 10. Carbon Disulphide.* (cited by WHO).

Ehrhardt, W. (1967). Experience with the employment of women exposed to carbon disulphide. In *International Symposium on Toxicology of Carbon Disulphide*, Prague, 1966, Excerpta Medica Foundation, Amsterdam, p. 240.

Gerhart, J. M. et al. (1991). Developmental inhalation toxicity of carbon disulfide in rabbits. *Toxicologist* 11: 344.

Hathaway, G. J. and Proctor, N. H. (2004). *Proctor and Hughes' Chemical Hazards of the Workplace*, 5th ed., John Wiley & Sons, Hoboken, NJ, pp. 121–123.

Hemminki, K., Fransilia, E., and Nainio, H. (1980). Spontaneous abortions among female chemical workers in Finland. *Int. Arch. Occup. Environ. Health* 45: 123–126.

Hofman, A. (1995). Fundamentals and possibilities of classification of occupational substances as developmental toxicants. *Int. Arch. Occup. Environ. Health* 67: 139–145.

Petrov, M. (1969). [Some data on the course and termination of pregnancy in female workers in the viscose industry]. *Akush. Ginekol.* 3: 50–52.

Price, C. J. et al. (1984). Developmental toxicity of carbon disulfide in rabbits and rats. *Toxicologist* 4: 86.

Stetkiewicz, J. and Wronska-Nofer, T. (1998). Updating of hygiene standards for carbon disulfide, based on health risk assessment. *Int. J. Occup. Med. Environ. Health* 11: 129–143.

Tabacova, S. (1976). Further observations on the effect of carbon disulfide inhalation on rat embryo development. *Teratology* 14: 374–375.

Tabacova, S. and Khinkova, L. (1981). [Early behavioral and neurofunctional deviations following prenatal carbon disulfide exposure]. *Probl. Khig.* 6: 21–26.

Tabacova, S., Hinkova, L., and Balabaeva, L. (1978). Carbon disulphide teratogenicity and postnatal effects in rat. *Toxicol. Lett.* 2: 129–133.

Zhou, S. Y. et al. (1988). Effects of occupational exposure to low-level carbon disulfide (CS_2) on menstruation and pregnancy. *Ind. Health* 26: 203–214.

41 Norethindrone

Chemical name: (17α)-17-Hydroxy-19-norpregn-4-en-20-yn-3-one

Alternate names: 19-Norethisterone, norpregneninolone

CAS #: 68-22-4

SMILES: C12C(C(CC1)(C#C)O)(CCC3C2CCC4C3CCC(C=4)=O)C

INTRODUCTION

Norethindrone is a synthetic progestin derived from 19-nortestosterone that is used medicinally in the treatment of menstrual disorders and endometriosis. More commonly, it is used as an oral contraceptive when combined with an estrogen. It acts by inhibiting the release of pituitary gonadotropin, transforming proliferative to secretory endometrium, and by thickening cervical mucus (Weiner and Buhimschi, 2004). It is available commercially by prescription by a host of trade names, including Aygestin®, Micronor®, Norlutate®, Norlutin®, and Nor-QD®, among other names, for reproductive disorders, and as Ortho-Novum®, Norlestrin®, Norinyl®, and Brevicon®, among other names, as oral contraceptives also containing an estrogenic substance, either mestranol or ethinyl estradiol. Norethindrone has a pregnancy category of X. The risk is based on the contraindication on the package label that states that "estrogen or progestin may cause fetal harm when administered to a pregnant woman." Therefore, they should not be used during pregnancy (*PDR*, 2005; see below).

DEVELOPMENTAL TOXICOLOGY

ANIMALS

Laboratory animal studies have demonstrated masculinization (virilization) of female offspring in six species. Of those species tested, only the rabbit was resistant, in which only fetal resorption was observed at doses in the range of 0.25 to 2 mg/day at intervals ranging from 3 to 14 days in gestation (Allen and Wu, 1959). In mice, doses of norethindrone of 0.5 to 1 mg/kg/day either orally or parenterally variously in a 11-day interval in gestation produced up to 57% fetuses with virilization (Andrew et al., 1972). In the rat, female offspring were masculinized from oral doses of 5 or 10 mg/kg/day given for only 4 days late in gestation (Kawashima et al., 1977). As little as 0.05 mg/kg/day by subcutaneous injection for up to 7 days in gestation was sufficient to induce the defect in this species (Miyake et al., 1966). In beagle dogs, 2.5 to 5 mg orally from middle to late gestation caused masculinization of female puppies (Curtis and Grant, 1964). Subcutaneous injection of guinea pig dams with 1 mg norethindrone for 43 days in gestation resulted in virilization of some of their progeny (Foote et al., 1968). In primates, 25 mg drug given intramuscularly for 27 or 35 days in gestation produced virilization in both female and male offspring and resulted in 8/10 stillborn (Wharton and Scott, 1964).

HUMANS

Norethindrone administration during pregnancy in humans has resulted in approximately 80 cases of masculinized female offspring and a small number of cases of hypospadias (virilization) in male offspring, as tabulated in Table 1. Investigators reported incidences over a wide range from 0.3 to 18.5% among infants of women taking the drug during pregnancy (Bongiovanni and McPadden,

TABLE 1
Reports of Virilization Associated with Norethindrone in Humans

Ref.	Male	Female
Greenblatt and Jungck, 1958		✔
Grumbach et al., 1959		✔
Valentine, 1959		✔
Wilkins, 1960		✔
Mortimer, 1960		✔
Jones and Wilkins, 1960		✔
Magnus, 1960		✔
Thomsen and Napp, 1960		✔
Leibow and Gardner, 1960		✔
Jacobson, 1962		✔
Thierstein et al., 1962		✔
Greenstein, 1962		✔
Fine et al., 1963		✔
Overzier, 1963		✔
Hagler et al., 1963		✔
Anonymous, 1963		✔
Ehrhardt and Money, 1967		✔
Voorhess, 1967[a]		✔
Serment and Ruf, 1968		✔
Lewin and Isador, 1968[a]	✔	
Aarskog, 1970[a]	✔	
Aarskog, 1970	✔	
Dillon, 1970		✔
Shepard, 1975		✔
Apold et al., 1976	✔	
Stevenson, 1977[a]	✔	
Aarskog, 1979[a]	✔	
Aarskog, 1979	✔	
Beicher et al., 1992		✔
Briggs et al., 2002	✔	
Carmichael et al., 2004	✔	

[a] Also combined with estrogen (mestranol or ethinyl estradiol).

1960; Jacobson, 1962). As indicated above from the package label, nongenital malformations were not associated with the drug in significant numbers to be considered drug related. The restriction for use of progestins during pregnancy that existed earlier for nongenital malformations was lifted by the U.S. Food and Drug Administration (FDA) in 1999 (Brent, 2000). The genital anomalies were variously described as virilization, masculinization, and pseudohermaphroditism in females and hypospadias in males. The anomalies are virtually identical to those produced by androgenic agents. They were first discovered almost half a century ago (Jones, 1957; Wilkins et al., 1958) and were described in detail by others more recently (Keith and Berger, 1977; Schardein, 1980, 2000; Wilson and Brent, 1981). Basically, in females there is phallic enlargement (clitoral hypertrophy), with or without labioscrotal fusion, and the labia are usually enlarged. In some cases, masculinization may have progressed to the degree that labioscrotal fusion resulted in the formation of a urogenital sinus. There is usually a normal vulva, endoscopic evidence of a cervix, and a palpable, though sometimes infantile, uterus. In males, hypospadias (feminization, incomplete masculinization, or ambiguous genitalia) occurs anywhere from a subcoronal location to a site at the base of the penile shaft. It was proposed that the progestin interferes with the fusion of the urethral fold, leading to the hypospadias. In both females and males, the anomalies correlated with the time of drug exposure and the dose of the progestin (see following).

Most all cases cited occurred following the larger doses used for treating endometriosis, on the order of >15 mg/day (orally), rather than the lower doses of 0.4 to 2.5 mg/day more commonly used for contraception. Actual dose ranges used in the studies cited ranged from 10 to 40 mg/day and are lower than those used in animals to induce similar anomalies. They were produced in the cited cases from the fifth gestational week at the earliest and continuing throughout pregnancy in females, and in the interval from the third to the twentieth gestational week in males.

Interestingly, the genital malformations induced by norethindrone (or other progestins) were not described in the published scientific literature over the past 30 years with rare exceptions: Increased hypospadias was alluded to, although not by specific drug name, recently in progestin-treated subjects (Carmichael et al., 2004). No other class of developmental toxicity was associated with the genital defects. One group of experts places the magnitude of teratogenic risk for virilization of female fetuses at high doses to be small and at low doses to be none (Friedman and Polifka, 2000). No such estimate of risk was made for male subjects.

CHEMISTRY

Norethindrone is a larger than average human developmental toxicant. It is hydrophobic and of low polarity. Norethindrone can engage within hydrogen bonding interactions. The calculated physicochemical and topological properties are shown in the following.

PHYSICOCHEMICAL PROPERTIES

Parameter	Value
Molecular weight	298.425 g/mol
Molecular volume	295.80 A^3
Density	0.978 g/cm^3
Surface area	360.98 A^2
LogP	2.870
HLB	1.518
Solubility parameter	21.941 J$^{(0.5)}$/cm$^{(1.5)}$
Dispersion	19.477 J$^{(0.5)}$/cm$^{(1.5)}$
Polarity	3.691 J$^{(0.5)}$/cm$^{(1.5)}$
Hydrogen bonding	9.404 J$^{(0.5)}$/cm$^{(1.5)}$

Continued.

Parameter	Value
H bond acceptor	0.70
H bond donor	0.46
Percent hydrophilic surface	12.95
MR	86.560
Water solubility	-2.695 log (mol/M^3)
Hydrophilic surface area	46.74 A^2
Polar surface area	40.46 A^2
HOMO	-10.046 eV
LUMO	-0.152 eV
Dipole	4.038 debye

TOPOLOGICAL PROPERTIES (UNITLESS)

Parameter	Value
x0	15.535
x1	10.483
x2	10.263
xp3	9.772
xp4	8.030
xp5	6.554
xp6	5.055
xp7	3.952
xp8	2.978
xp9	2.222
xp10	1.436
xv0	13.476
xv1	8.918
xv2	8.303
xvp3	7.690
xvp4	6.434
xvp5	5.115
xvp6	3.743
xvp7	2.834
xvp8	2.030
xvp9	1.371
xvp10	0.805
k0	29.533
k1	15.523
k2	5.250
k3	2.111
ka1	14.518
ka2	4.712
ka3	1.850

REFERENCES

Aarskog, D. (1970). Clinical and cytogenetic studies in hypospadias. *Acta Paediatr. Scand.* Suppl. 203: 7–62.
Aarskog, D. (1979). Maternal progestins as a possible cause of hypospadias. *N. Engl. J. Med.* 300: 75–78.
Allen, W. M. and Wu, D. H. (1959). Effects of 17alpha-ethinyl-19-nortestosterone on pregnancy in rabbits. *Fertil. Steril.* 10: 424–438.

Andrew, F. D. et al. (1972). Teratogenicity of contraceptive steroids in mice. *Teratology* 5: 249.

Anonymous. (1963). General practitioner clinical trials. Drugs in pregnancy survey. *Practitioner* 191: 775–780.

Apold, J., Dahl, E., and Aarskog, D. (1976). The VATER association: Malformations of the male external genitalia. *Acta Paediatr. Scand.* 65: 150–152.

Beicher, N. A. et al. (1992). Norethisterone and gestational diabetes. *Aust. N.Z. J. Obstet. Gynecol.* 32: 233–238.

Bongiovanni, A. M. and McPadden, A. J. (1960). Steroids during pregnancy and possible fetal consequences. *Fertil. Steril.* 11: 181–186.

Brent, R. L. (2000). Nongenital malformations and exposure to progestational drugs during pregnancy; the final chapter of an erroneous allegation. *Teratology* 61: 449.

Briggs, G. G., Freeman, R. K., and Yaffe, S. J. (2002). *Drugs in Pregnancy and Lactation. A Reference Guide to Fetal and Neonatal Risk*, Sixth ed., Lippincott Williams & Wilkins, Philadelphia.

Carmichael, S. L. et al. (2004). Hypospadias and maternal intake of progestins and oral contraceptives. *Birth Defects Res. (A)* 70: 255.

Curtis, E. M. and Grant, R. P. (1964). Masculinization of female pups by progestogens. *J. Am. Vet. Med. Assoc.* 144: 395–398.

Dillon, S. (1970). Progestogen therapy in early pregnancy and associated congenital defects. *Practitioner* 205: 80–84.

Ehrhardt, A. A. and Money, J. (1967). Progestin-induced hermaphroditism: IQ and psychosexual identity in a study of 10 girls. *J. Sex Res.* 3: 83–100.

Fine, E., Levin, H. M., and McConnell, E. L. (1963). Masculinization of female infants associated with norethindrone acetate. *Obstet. Gynecol.* 22: 210–213.

Foote, W. D., Foote, W. C., and Foote, L. H. (1968). Influence of certain natural and synthetic steroids on genital development in guinea pigs. *Fertil. Steril.* 19: 606–615.

Friedman, J. M. and Polifka, J. E. (2000). *Teratogenic Effects of Drugs. A Resource for Clinicians (TERIS)*, Second ed., Johns Hopkins University Press, Baltimore, MD.

Greenblatt, R. B. and Jungck, E. C. (1958). Delay of menstruation with norethindrone, an orally given progestational compound. *JAMA* 166: 1461–1463.

Greenstein, N. M. (1962). Iatrogenic female pseudohermaphroditism. *Jewish Mem. Hosp. Bull. (NY)* 7: 191–195.

Grumbach, M. M., Ducharme, J. R., and Moloshok, R. E. (1959). On the fetal masculinizing action of certain oral progestins. *J. Clin. Endocrinol. Metab.* 19: 1369–1380.

Hagler, S. et al. (1963). Fetal effects of steroid therapy during pregnancy. *Am. J. Dis. Child.* 106: 586–590.

Jacobson, B. D. (1962). Hazards of norethindrone therapy during pregnancy. *Am. J. Obstet. Gynecol.* 84: 962–968.

Jones, H. W. (1957). Female hermaphroditism without virilization. *Obstet. Gynecol. Surv.* 12: 433–460.

Jones, H. W. and Wilkins, L. (1960). The genital anomaly associated with prenatal exposure to progestogens. *Fertil. Steril.* 11: 148–156.

Kawashima, K. et al. (1977). Virilizing activities of various steroids in female rat fetuses. *Endocrinol. Jpn.* 24: 77–81.

Keith, L. and Berger, G. S. (1977). The relationship between congenital defects and the use of exogenous progestational contraceptive hormones during pregnancy: A 20-year review. *Int. J. Gynaecol. Obstet.* 15: 115–124.

Leibow, S. G. and Gardner, L. E. (1960). Clinical conference — genital abnormalities associated with administration of progesteroids to their mothers. *Pediatrics* 26: 151–160.

Lewin, D. and Isador, P. (1968). [Hyperplasia of the interstitial tissue of the embryonal testis after ingestion of hormonal products by the mother]. *Bull. Fed. Soc. Gynecol. Obstet. Lang. Fr.* 20: 414–415.

Magnus, E. M. (1960). Female pseudohermaphroditism associated with administration of oral progestin during pregnancy. Report on a case. *Tidsskr. Nor. Leageforen.* 80: 92–93.

Miyake, Y. et al. (1966). [Biological activities of chlormadinone acetate. 2. Its effects on the pregnancy, fetal growth and parturition in rats]. *Folia Endocrinol. Jpn.* 41: 1154–1165.

Mortimer, P. E. (1960). Female pseudohermaphroditism due to progestogens. *Lancet* 2: 438–439.

Overzier, C. (1963). Induced pseudo-hermaphroditism. In *Intersexuality*, C. Overzier, Ed., Academic Press, New York, pp. 387–401.

PDR® (Physicians' Desk Reference®). (2005). Medical Economics Co., Inc., Montvale, NJ.

Schardein, J. L. (1980). Congenital abnormalities and hormones during pregnancy: A clinical review. *Teratology* 22: 251–270.

Schardein, J. L. (2000). *Chemically Induced Birth Defects*, Third ed., Marcel Dekker, New York, pp. 286–289, 298, 299.

Serment, H. and Ruf, H. (1968). Les dangers pour le produit de conception de medicaments administers a la femme enceinte. *Bull. Fed. Soc. Gynecol. Obstet. Lang. Fr.* 20: 69–76.

Shepard, T. H. (1975). Teratogenic drugs and therapeutic agents. In *Pediatric Therapy*, H. C. Shirkey, Ed., C.V. Mosby, St. Louis, p. 161.

Stevenson, R. E. (1977). *The Fetus and Newly Born Infant. Influence of the Prenatal Environment*, C.V. Mosby, St. Louis, p. 156.

Thierstein, S. T. et al. (1962). Habitual abortion. Progesterone-like hormones for prevention of fetal loss. *J. Kans. Med. Soc.* 63: 288–291.

Thomsen, K. and Napp, J. H. (1960). Nebenwirkungen bei hochdosierter Nortestosteronmedikation in der Graviditat. *Geburtschilfe Frauenheilkd.* 20: 508–513.

Valentine, G. H. (1959). Masculinization of a female foetus with oestrogenic effect. *Arch. Dis. Child.* 34: 495–497.

Voorhess, M. L. (1967). Masculinization of the female fetus associated with norethindrone-mestranol therapy during pregnancy. *J. Pediatr.* 71: 128–131.

Weiner, C. P. and Buhimschi, C. (2004). *Drugs for Pregnant and Lactating Women*, Elsevier Science, New York, pp. 701–702.

Wharton, L. R. and Scott, R. B. (1964). Experimental production of genital lesions with norethindrone. *Am. J. Obstet. Gynecol.* 89: 701–715.

Wilkins, L. (1960). Masculinization of female fetus due to use of orally given progestins. *JAMA* 172: 1028–1032.

Wilkins, L. et al. (1958). Masculinization of female fetus associated with administration of oral and intramuscular progestins during gestation: Nonadrenal pseudohermaphoditism. *J. Clin. Endocrinol. Metab.* 18: 559–585.

Wilson, J. G. and Brent, R. L. (1981). Are female sex hormones teratogenic? *Am. J. Obstet. Gynecol.* 141: 567–580.

42 Phenytoin

Chemical name: 5,5-Diphenyl-2,4-imidazolidinedione

Alternate name: Diphenylhydantoin

CAS #: 57-41-0

SMILES: C1(c2ccccc2)(c3ccccc3)NC(NC1=O)=O

INTRODUCTION

Phenytoin is a hydantoin anticonvulsant drug, long used (since 1938) as the sodium salt in the therapy for epilepsy in the management of generalized tonic-clonic (grand mal) and complex partial seizures. It also has utility in the prevention of seizures following head trauma. Mechanistically, it acts by stabilizing neuronal membranes and decreasing seizure activity by increasing efflux or decreasing influx of sodium ions across cell membranes in the motor cortex during generation of nerve impulses (Lacy et al., 2004). The drug is available by prescription under the trade names Dilantin®, Phenytek®, and Epanutin®, among others. It is a popular medication, being among the top 200 drugs most often prescribed in 2004 (www.rxlist.com). Phenytoin has a pregnancy category of D. The warning on the package label states that a number of reports suggest an association between the use of antiepileptic drugs by women and a higher incidence of birth defects in children born to these women (*PDR*, 2005). The label goes on to state, however, that these reports cannot be regarded as adequate to prove a definite cause and effect relationship. In addition to the reports of increased incidence of congenital malformations, such as cleft lip/palate and heart malformations, there have been more recent reports of a *fetal hydantoin syndrome*, according to the label, consisting of prenatal growth deficiency, microcephaly, and mental deficiency in children born to mothers who received phenytoin (and other agents; see below).

DEVELOPMENTAL TOXICOLOGY

ANIMALS

Phenytoin has been studied extensively in the laboratory with six species of animals assessed, representative studies of which are tabulated in Table 1. The pertinent route of administration is

TABLE 1
Representative Oral Developmental Studies in Animals with Phenytoin

Species	Developmental Toxicity Produced[a]	Details (mg/kg, treatment interval in gestation)	Ref.
Mouse	M, G, D	125, 3 days	Miller and Becker, 1975
Rat	G, D, F	100–200, 12 or 13 days	Elmazar and Sullivan, 1981; Vorhees and Minck, 1989
Rabbit	M, D	75, 12 days — maternal toxicity at higher doses	McClain and Langhoff, 1980
Hamster	—	Not known	Becker, 1972 (personal communication)
Cat	D	2, 13 days	Khera, 1979
Dog	—	Not known	Esaki, 1978
Primate (rhesus sp.)	M, D, F	10, 27 days or 4–12 μg/ml blood level, throughout gestation	Wilson, 1973; Phillips and Lockard, 1996

[a] M = malformation, G = growth retardation, D = death, F = functional deficit.

oral, as given to humans. The initial study published by Massey (1966) describes the results of testing in the mouse. In general, the drug is teratogenic by the oral route in only the mouse, rabbit, and primate species. The response in the rat was also teratogenesis, but only by several parenteral routes of administration. When malformations were elicited in rodents, they most often were of the skeleton (micromelia) and oral cavity (cleft lip or cleft palate). In rhesus monkeys, only a minor urinary tract anomaly was reported. None of the three metabolites of the drug were teratogenic, at least in the mouse (Harbison, 1969). There are marked strain differences in the teratogenic response (Johnston et al., 1979); the ability to metabolize phenytoin may be genetically determined (Millicovsky and Johnston, 1981). Both the mouse (Finnell et al., 1989) and the rat (Lorente et al., 1981) can be said to serve as animal models for the developmentally toxic effects in the human. While all classes of developmental toxicity have been observed following phenytoin treatment in the laboratory, postnatal functional changes were apparent only in the rat and the rhesus monkey, manifested by delays in motor development and persistent impairment of locomotor function (rats) and hyperexcitability (monkey).

HUMANS

In the human, a very large number of reports were published associating the administration of phenytoin to epileptic women during pregnancy to developmental toxicity of most all classes, with the exception of viability. This is of concern, because estimates made some time ago place the number of infants exposed to this drug at approximately 6000 annually in the United States alone (Hanson et al., 1976). The following discussion will be presented by class affected as provided in published scientific reports. The timetable for producing the effects ranges over a wide period, including the first trimester. Dosage associated with the effects, where provided in the reports, was generally within the therapeutic range of 6 to 20 mg/kg/day (oral or intravenous).

Malformation

In 1964, Janz and Fuchs reported the first (five) cases of congenital malformation related to phenytoin administration during pregnancy. Following that report, a very large number of publications attesting to induced malformations have appeared. It is probable that thousands of cases exist. In general, defects thought to be increased in incidence include cardiac defects, orofacial clefts (lip, palate), and skeletal anomalies, especially of the joints. These are believed to occur in two-to threefold greater incidence than in untreated women in the general population. All other associations with phenytoin treatment that have been made occur in incidence <1% and appear not to be drug related. Midface hypoplasia was shown to be the most common feature of the anticonvulsant embryopathy, occurring in 13% of infants exposed to phenytoin (Holmes et al., 2001). Hypoplastic distal phalanges and nails appear to serve as markers for the more severe associated abnormalities.

Accompanying the above defects are other major and minor anomalies appearing as a clustering of physical findings, described as an identifiable population of neonates, and otherwise characterized as "fetal hydantoin syndrome" (FHS). Comprising this syndrome are craniofacial features, appendicular defects, and deficiency of both growth and mentality. The craniofacial features present as distinct entities in the reported cases include short nose with low nasal bridge, inner epicanthic folds, ptosis, strabismus, hypertelorism, low-set or abnormal ears, wide mouth, wide fontanelles, and prominent lips. The appendicular defects include hypoplasia of nails and distal phalanges, finger-like thumb, abnormal palmar creases, and five or more digital arches. Some patients have a short or webbed neck with or without a low hairline, coarse hair, and skeletal abnormalities of the ribs, sternum, and spine, and widely spaced, hypoplastic nipples. The growth and functional abnormalities will be discussed below. The syndrome has been misdiagnosed in individual cases as Coffin-Siris or Noonan's syndromes, and the altered pattern of morphogenesis is distinct from other recognized disorders (Hanson and Smith, 1975). It should be noted that not all features of the syndrome are present in every case.

The incidence of these and other clinical findings in FHS are tabulated in Table 2. Representative pertinent publications referencing congenital malformations associated with phenytoin treatment are given in Table 3. About 10% of infants born to epileptic women who took phenytoin during pregnancy had FHS, according to several reviewing investigators (Kelly, 1984; Kelly et al., 1984; Hanson, 1986). It should be stated that objectively determining the adverse developmental effects caused by phenytoin, as well as any other anticonvulsant drug, is subject to difficulties, in that there are several underlying factors, including combination therapy of anticonvulsants, the role of epilepsy and its association with congenital anomalies, and familial factors. In humans, reaction to phenytoin, in particular, has been noted to have a genetic predisposition (Phelan et al., 1982; Strickler et al., 1985; Gaily et al., 1990a; Buehler et al., 1994). Genetic differences in susceptibility to phenytoin also exist in animals within different inbred strains (Finnell et al., 1989). Nonetheless, the current consensus is that FHS is a true and recognizable entity related directly to the use of either monotherapy or combined therapy of phenytoin in addition to the factors noted above (Kelly, 1984; Delgado-Escueta and Janz, 1992; Dravet et al., 1992; Lindhout and Omtzigt, 1992, 1994; Nulman et al., 1997; Olafsson et al., 1998; Sabers et al., 1998; Arpino et al., 2000; Holmes et al., 2001). It may be that the risk of malformation is even greater if other anticonvulsants are given along with phenytoin, as suggested by a number of investigators (Lindhout et al., 1984; Kaneko et al., 1988; Lindhout and Omtzigt, 1992; Nakane and Kaneko, 1992; Tanganelli and Regesta, 1992; Janz, 1994).

The mechanism of phenytoin teratogenicity has been the subject of numerous experimental studies. It appears to be necessary for phenytoin to be metabolized by cytochrome P450 (CYP) enzymes to reactive intermediates that form adducts with DNA or protein within the embryo (NRC, 2000). The most likely intermediate is an arene oxide. An alternative hypothesis suggests that phenytoin is metabolized by prostaglandin synthetase to a teratogenic intermediate that is a more stable oxepin that can be transported to the target tissue easier (Harbison et al., 1977; Martz et al., 1977; Spielberg et al., 1981; Hansen, 1991). This hypothesis is supported by the observation that phenytoin teratogenicity in mice can be mitigated by cotreatment with aspirin, an inhibitor of prostaglandin synthetase (Wells et al., 1989). Vitamin K supplementation was also shown to counter the teratogenic effect (Howe et al., 1995). Another hypothesis is that the mechanism is through folate deficiency (DeVore and Woodbury, 1977; Labadarios, 1979) or due to low oxygen delivery (Watkinson and Millicovsky, 1983). The teratogenicity may be initiated by pharmacologically induced embryonic hypoxic ischemia (Danielsson et al., 1997; Lyon et al., 2003). It was observed that phenytoin treatment in rodents decreases the expression of the mRNAs for a number of

TABLE 2
Clinical Findings Reported in 213 FHS Cases[a]

Clinical Findings	Frequency (%)
Craniofacial anomalies[b]	
Cleft lip, palate	3
High-arched palate	4
Low-set or abnormal ears	4
Ptosis (eyelid)	5
Short-webbed neck ± low hairline	6
Metopic sutural ridging	8
Wide fontanelles	10
Epicanthus	13
Strabismus	14
Hypertelorism	21
Short nose with low, broad nasal bridge	21
Microcephaly	29
Skeletal anomalies[c]	
Rib, sternal, or spinal abnormalities	1
Finger-like thumb	6
Abnormal palmar creases	7
Hypoplasia of nails and distal phalanges	11
Positional deformities	11
Growth alterations	
Prenatal growth deficiency	36
Postnatal growth deficiency	52
Functional deficits	
Motor or mental deficiency	25
Other anomalies[d]	
Hypospadias	1
Congenital heart disease	3
Hirsutism	3
Widely spaced hypoplastic nipples	4
Hernias	9

[a] Taken from Schardein, J. L., *Chemically Induced Birth Defects*, Third ed., Marcel Dekker, New York, 2000, after cases described by Hill, R. M. et al., *Am. J. Dis. Child.*, 127, 645–653, 1974; Hanson, J. W. and Smith, D. W., *J. Pediatr.*, 87, 285–290, 1975; Bethenod, M. and Frederich, A., *Pediatrie*, 30, 227–248, 1975; and Hanson, J. W. et al., *J. Pediatr.*, 89, 662–668, 1976.

[b] Wide mouth, prominent lips, broad alveolar ridge, cleft gum, and cranial asymmetry listed in single reports.

[c] Five or more digital arches listed in a single report.

[d] Coarse hair, undescended testes, osteoporosis, epidermal cyst, bifid sternum, and pyloric stenosis listed in a single report.

TABLE 3
Representative Published Reports Attributing Phenytoin Treatment during Pregnancy to Congenital Malformations in Humans

References Relating To:	
Fetal Hydantoin Syndrome (FHS)	**Other Malformations**
Meadow, 1968[a]	Melchior et al., 1967[b]
Hanson and Smith, 1975[c]	Mirkin, 1971
Hanson et al., 1976	Meyer, 1973
Hanson, 1976	Fedrick, 1973
Tunnessen and Lowenstein, 1976	Monson et al., 1973
Goodman et al., 1976	Loughnan et al., 1973
Leiber, 1976	Annegers et al., 1974
Zutel et al., 1977	Hill et al., 1974
Hanson and Smith, 1977	Barr et al., 1974
Pinto et al., 1977	Biale et al., 1975
Apt and Gaffney, 1977	Dabee et al., 1975
Smith, 1977	Anderson, 1976
Bustamente and Stumpff, 1978	Corcoran and Rizk, 1976
Yang et al., 1978	Lakos and Czeizel, 1977
Elefant, 1978	Prakash et al., 1978
Waller et al., 1978	Hassell et al., 1979
Wilson et al., 1978	Allen et al., 1980[d]
Wood and Young, 1979	Johnson and Goldsmith, 1981
Dieterich, 1979	Pai, 1982
Stankler and Campbell, 1980	Bartoshesky et al., 1982
Truog et al., 1980	Albengres and Tillement, 1983
Majewski et al., 1980	Kelly, 1984
Silver, 1981	Gaily et al., 1988
Michalodimitrakis et al., 1981	Gaily and Granstrom, 1989
Hampton and Krepostman, 1981	Friis, 1989
Nagy, 1981	Adams et al., 1990
Hanson and Buehler, 1982	Gaily, 1990
Kousseff and Root, 1982	Kaneko, 1991
Kelly et al., 1984	Koch et al., 1992
Hanson, 1986	Gaily and Granstrom, 1992
Kotzot et al., 1993	Marks et al., 1994
Sabry and Farag, 1996	Wester et al., 2002
Yalcinkaya et al., 1997	Orup et al., 2003
Ozkinay et al., 1998	
Trousdale, 1998	
Godbole et al., 1999	
Murki et al., 2003	
Holmes et al., 2005	

[a] First to describe features of FHS.
[b] First association of drug to malformations in English literature.
[c] Coined term "FHS."
[d] First reported FHS, hemorrhagic disease, and malignancy in one case.

important growth factors (Musselman et al., 1994). Whether the decrease is due to an effect on gene expression or a degradation of RNA by reactive intermediates of phenytoin is not known. One group of experts places the teratogenic risk of phenytoin administered during pregnancy as small to moderate (Friedman and Polifka, 2000).

Growth Retardation

Growth deficiency, both pre- and postnatally, is a common characteristic of FHS as noted in 36 to 52% of cases in several published reports in Table 2 and in many of the reports listed in Table 3.

Death

Fetal mortality has not been a characteristic of phenytoin-induced developmental toxicity in the human.

Functional Deficit

As indicated in Table 2, motor or mental deficiency in an incidence of 25% of cases, is an associated finding in FHS and other phenytoin-induced developmental toxicities. A number of different types of functional deficiencies were reported, to name a few, including developmental delay (Dabee et al., 1975; Gladstone et al., 1992); deficits in intelligence or IQ (Gaily et al., 1988, 1990a, 1990b; Vanoverloop et al., 1992; Scolnik et al., 1994; Holmes et al., 2005); neurological effects, including jitteriness (D'Souza et al., 1990); higher mean apathy scores (Koch et al., 1996); lower visual motor integration test scores (Vanoverloop et al., 1992); and other adverse effects on neurodevelopment, including learning problems (Dessens et al., 2000) and retarded psychomotor development (Wide et al., 2002), some of which are also included in the reports tabulated in Table 3.

Perinatal and neonatal hemorrhage were observed and recorded in a number of publications associated with phenytoin treatment, which is an example of another drug-related functional disorder (Kohler, 1966; Douglas, 1966; Solomon et al., 1972; Allen et al., 1980; McNinch and Tripp, 1991). Approximately eight reports of neural tumors were published in association with phenytoin administration; none were reported since the mid-1980s, and no causation has been considered for this association at the present time (see Schardein, 2000).

A number of published reviews on phenytoin and its effects during pregnancy have appeared (Leiber, 1976; Zutel et al., 1977; Lakos and Czeizel, 1977; Elefant, 1978; Hassell et al., 1979; Hanson and Buehler, 1982; Albengres and Tillement, 1983; Kelly, 1984; Hanson, 1986; Wells et al., 1997; Friedman and Polifka, 2000; Schardein, 2000; Briggs et al., 2005).

CHEMISTRY

Phenytoin is an average-sized hydrophobic compound. It is of average polarity in comparison to the other human developmental toxicants. It can participate in hydrogen bonding both as an acceptor and donor. The calculated physicochemical and topological properties for phenytoin are listed below.

PHYSICOCHEMICAL PROPERTIES

Parameter	Value
Molecular weight	252.272 g/mol
Molecular volume	219.52 A^3
Density	1.095 g/cm^3
Surface area	255.00 A^2
LogP	1.755
HLB	11.345

Continued.

Parameter	Value
Solubility parameter	24.974 $J^{(0.5)}/cm^{(1.5)}$
Dispersion	22.796 $J^{(0.5)}/cm^{(1.5)}$
Polarity	6.642 $J^{(0.5)}/cm^{(1.5)}$
Hydrogen bonding	7.741 $J^{(0.5)}/cm^{(1.5)}$
H bond acceptor	0.94
H bond donor	0.57
Percent hydrophilic surface	55.68
MR	73.436
Water solubility	-0.469 log (mol/M^3)
Hydrophilic surface area	141.97 A^2
Polar surface area	64.52 A^2
HOMO	-9.711 eV
LUMO	-0.080 eV
Dipole	2.944 debye

TOPOLOGICAL PROPERTIES (UNITLESS)

Parameter	Value
x0	13.295
x1	9.232
x2	8.228
xp3	7.106
xp4	6.462
xp5	5.091
xp6	2.910
xp7	2.099
xp8	1.299
xp9	0.658
xp10	0.289
xv0	10.090
xv1	5.980
xv2	4.390
xvp3	3.282
xvp4	2.405
xvp5	1.565
xvp6	0.765
xvp7	0.444
xvp8	0.219
xvp9	0.090
xvp10	0.033
k0	18.276
k1	13.959
k2	5.780
k3	2.492
ka1	11.772
ka2	4.422
ka3	1.779

REFERENCES

Adams, J., Vorhees, C. V., and Middaugh, L. D. (1990). Developmental neurotoxicity of anticonvulsants: Human and animal evidence on phenytoin. *Neurotoxicol. Teratol.* 12: 203–214.

Albengres, E. and Tillement, J. P. (1983). Phenytoin in pregnancy: A review of the reported risks. *Biol. Res. Preg. Perinatol.* 4: 71–74.

Allen, R. W. et al. (1980). Fetal hydantoin syndrome, neuroblastoma, and hemorrhagic disease. *Clin. Res.* 28: 115A.

Anderson, R. C. (1976). Cardiac defects in children of mothers receiving anticonvulsant therapy during pregnancy. *J. Pediatr.* 89: 318–319.

Annegers, J. F. et al. (1974). Do anticonvulsants have a teratogenic effect? *Arch. Neurol.* 31: 364–373.

Apt, L. and Gaffney, W. L. (1977). Is there a "fetal hydantoin syndrome"? *Am. J. Ophthalmol.* 84: 439–440.

Arpino, C. et al. (2000). Teratogenic effects of antiepileptic drugs: Use of an international database on malformations and drug exposure (MADRE). *Epilepsia* 41: 1436–1443.

Barr, M., Poznanski, A. K., and Schmickel, R. D. (1974). Digital hypoplasia and anticonvulsants during gestation: A teratogenic syndrome. *J. Pediatr.* 84: 254–256.

Bartoshesky, L. E. et al. (1982). Severe cardiac and ophthalmologic malformations in an infant exposed to diphenylhydantoin *in utero. Pediatrics* 69: 202–203.

Bethenod, M. and Frederich, A. (1975). [The children of drug-treated epileptics]. *Pediatrie* 30: 227–248.

Biale, Y., Lewenthal, H., and Aderet, N. B. (1975). Congenital malformations due to anticonvulsant drugs. *Obstet. Gynecol.* 45: 439–442.

Briggs, G. G., Freeman, R. K., and Yaffe, S. J. (2005). *Drugs in Pregnancy and Lactation. A Reference Guide to Fetal and Neonatal Risk,* Seventh ed., Lippincott Williams & Wilkins, Philadelphia.

Buehler, B. A., Rao, V., and Finnell, R. H. (1994). Biochemical and molecular teratology of fetal hydantoin syndrome. *Neurol. Clin.* 12: 741–748.

Bustamente, S. A. and Stumpff, L. C. (1978). Fetal hydantoin syndrome in triplets. *Am. J. Dis. Child.* 132: 978–979.

Corcoran, R. and Rizk, M. W. (1976). VACTERL congenital malformation and phenytoin therapy. *Lancet* 2: 960.

Dabee, V., Hart, A. G., and Hurley, R. M. (1975). Teratogenic effects of diphenylhydantoin. *Can. Med. Assoc. J.* 112: 75–77.

Danielsson, B. R. et al. (1997). Initiation of phenytoin teratogenesis: Pharmacologically induced embryonic bradycardia and arrhythmia resulting in hypoxia and possible free radical damage at reoxygenation. *Teratology* 56: 271–281.

Delgado-Escueta, A. V. and Janz, D. (1992). Consensus guidelines: Preconception counseling, management, and care of the pregnant woman with epilepsy. *Neurology* 42 (Suppl. 5): 149–160.

Dessens, A. B. et al. (2000). Association of prenatal phenobarbital and phenytoin exposure with small head size at birth and with learning problems. *Acta Paediatr.* 89: 533–541.

DeVore, G. R. and Woodbury, D. M. (1977). Phenytoin: An evaluation of several potential teratogenic mechanisms. *Epilepsia* 18: 387–396.

Dieterich, E. (1979). [Antiepileptic embryopathies]. *Ergeb. Inn. Med. Kinderheilkd.* 43: 93–107.

Douglas, H. (1966). Hemorrhage in the newborn. *Lancet* 1: 816–817.

Dravet, C. et al. (1992). Epilepsy, antiepileptic drugs, and malformations in children of women with epilepsy: A French prospective cohort study. *Neurology* 42 (Suppl. 5): 75–82.

D'Souza, S. W. et al. (1990). Fetal phenytoin exposure, hypoplastic nails, and jitteriness. *Arch. Dis. Child.* 65: 320–324.

Elefant, E. (1978). [Fetal hydantoin syndrome]. *Klin. Paediatr.* 190: 307–312.

Elmazar, M. M. A. and Sullivan, F. M. (1981). Effect of prenatal phenytoin administration on postnatal development of the rat: A behavioral teratology study. *Teratology* 24: 115–124.

Esaki, K. (1978). The beagle dog in embryotoxicity tests. *Teratology* 18: 129–130.

Fedrick, J. (1973). Epilepsy and pregnancy: A report from the Oxford Record Linkage Study. *Br. Med. J.* 2: 442–448.

Finnell, R. H., Abbott, L. C., and Taylor, S. M. (1989). The fetal hydantoin syndrome: Answers from a mouse model. *Reprod. Toxicol.* 3: 127–133.

Friedman, J. M. and Polifka, J. E. (2000). *Teratogenic Effects of Drugs. A Resource for Clinicians (TERIS),* Second ed., Johns Hopkins University Press, Baltimore, MD.

Friis, M. L. (1989). Facial clefts and congenital heart defects in children of parents with epilepsy: Genetic and environmental etiologic factors. *Acta Neurol. Scand.* 79: 433–459.

Gaily, E. (1990). Distal phalangeal hypoplasia in children with prenatal phenytoin exposure: Results of a controlled anthropometric study. *Am. J. Med. Genet.* 35: 574–578.

Gaily, E. and Granstrom, M.-L. (1989). A transient retardation of early postnatal growth in drug-exposed children of epileptic mothers. *Epilepsy Res.* 4: 147–155.

Gaily, E. and Granstrom, M.-L. (1992). Minor anomalies in children of mothers with epilepsy. *Neurology* 42 (Suppl. 5): 128–131.

Gaily, E. K., Kantola-Sorsa, E., and Granstrom, M.-L. (1988). Intelligence of children of epileptic mothers. *J. Pediatr.* 113: 677–684.

Gaily, E. K. et al. (1990a). Head circumference in children of epileptic mothers: Contributions of drug exposure and genetic background. *Epilepsy Res.* 5: 217–222.

Gaily, E. K., Kantola-Sorsa, E., and Granstrom, M.-L. (1990b). Specific cognitive dysfunction in children with epileptic mothers. *Dev. Med. Child Neurol.* 32: 403–414.

Gladstone, D. J. et al. (1992). Course of pregnancy and fetal outcome following maternal exposure to carbamazepine and phenytoin: A prospective study. *Reprod. Toxicol.* 6: 257–261.

Godbole, K. G. et al. (1999). Fetal hydantoin syndrome with rheumatic valvular heart disease. *Indian J. Pediatr.* 66: 290–293.

Goodman, R. M. et al. (1976). Congenital malformations in four siblings of a mother taking anticonvulsant drugs. *Am. J. Dis. Child.* 130: 884–887.

Hampton, G. R. and Krepostman, J. I. (1981). Ocular manifestations of the fetal hydantoin syndrome. *Clin. Pediatr.* 20: 475–478.

Hansen, D. K. (1991). The embryotoxicity of phenytoin: An update on possible mechanisms. *Proc. Soc. Exp. Med.* 197: 361–368.

Hanson, J. W. (1976). Fetal hydantoin syndrome. *Teratology* 13: 185–188.

Hanson, J. W. (1986). Teratogen update: Fetal hydantoin effects. *Teratology* 33: 349–353.

Hanson, J. W. and Buehler, B. A. (1982). Fetal hydantoin syndrome: Current status. *J. Pediatr.* 101: 816–818.

Hanson, J. W. and Smith, D. W. (1975). The fetal hydantoin syndrome. *J. Pediatr.* 87: 285–290.

Hanson, J. W. and Smith, D. W. (1977). Are hydantoins (phenytoins) human teratogens? *J. Pediatr.* 90: 674–675.

Hanson, J. W. et al. (1976). Risks to the offspring of women treated with hydantoin anticonvulsants, with emphasis on the fetal hydantoin syndrome. *J. Pediatr.* 89: 662–668.

Harbison, R. D. (1969). Studies on the mechanism of teratogenic action and neonatal pharmacology of diphenylhydantoin. Dissertation, State University of Iowa.

Harbison, R. D. et al. (1977). Proposed mechanism for diphenylhydantoin-induced teratogenesis. *Pharmacologist* 19: 179.

Hassell, T. M. et al. (1979). Summary of an international symposium on phenytoin-induced teratology and gingival pathology. *J. Am. Dent. Assoc.* 99: 652–655.

Hill, R. M. et al. (1974). Infants exposed *in utero* to antiepileptic drugs. A prospective study. *Am. J. Dis. Child.* 127: 645–653.

Holmes, L. B. et al. (2001). The teratogenicity of anticonvulsant drugs. *N. Engl. J. Med.* 344: 1132–1138.

Holmes, L. B. et al. (2005). The correlation of deficits in IQ with midface and digit hypoplasia in children exposed *in utero* to anticonvulsant drugs. *J. Pediatr.* 146: 118–122.

Howe, A. M. et al. (1995). Prenatal exposure to phenytoin, facial development, and a possible role for vitamin K. *Am. J. Med. Genet.* 58: 238–244.

Janz, D. (1994). Are antiepileptic drugs harmful when taken during pregnancy? *J. Perinat. Med.* 22: 367–377.

Janz, D. and Fuchs, U. (1964). Are anti-epileptic drugs harmful when given during pregnancy? *Ger. Med. Monatsschr.* 9: 20–22.

Johnson, R. B. and Goldsmith, L. A. (1981). Dilantin digit defects. *J. Am. Acad. Dermatol.* 5: 191–196.

Johnston, M. C., Sulik, K. K., and Dudley, K. H. (1979). Genetic and metabolic studies of the differential sensitivity of AJ and C57Bl/6J mice to phenytoin ("Dilantin")-induced cleft lip. *Teratology* 19: 33A.

Kaneko, S. (1991). Antiepileptic drug therapy and reproductive consequences: Functional and morphological effects. *Reprod. Toxicol.* 5: 179–198.

Kaneko, S. et al. (1988). Teratogenicity of antiepileptic drugs: Analysis of possible risk factors. *Epilepsia* 29: 459–467.

Kelly, T. E. (1984). Teratogenicity of anticonvulsant drugs. I. Review of the literature. *Am. J. Med. Genet.* 19: 413–434.

Kelly, T. E. et al. (1984). Teratogenicity of anticonvulsant drugs. II. A prospective study. *Am. J. Med. Genet.* 19: 435–443.

Khera, K. S. (1979). A teratogenicity study on hydroxyurea and diphenylhydantoin in cats. *Teratology* 20: 447–452.

Koch, S. et al. (1992). Major and minor birth malformations and antiepileptic drugs. *Neurology* 42 (Suppl. 5): 83–88.

Koch, S. et al. (1996). Antiepileptic drug treatment in pregnancy: Drug side effects in the neonate and neurological outcome. *Acta Paediatr.* 84: 739–746.

Kohler, H. G. (1966). Hemorrhage in the newborn of epileptic mothers. *Lancet* 1: 267.

Kotzot, D. et al. (1993). Hydantoin syndrome with holoprosencephaly: A possible rare teratogenic effect. *Teratology* 48: 15–19.

Kousseff, B. G. and Root, E. R. (1982). Expanding phenotype of fetal hydantoin syndrome. *Pediatrics* 70: 328–329.

Labadarios, D. (1979). Diphenylhydantoin during pregnancy. *S. Afr. Med. J.* 55: 154.

Lacy, C. F. et al. (2004). *Drug Information Handbook (Pocket), 2004–2005*, Lexi-Comp., Inc., Hudson, OH.

Lakos, P. and Czeizel, E. (1977). A teratological evaluation of anticonvulsant drugs. *Acta Paediatr. Acad. Sci. Hung.* 18: 145–153.

Leiber, B. (1976). [Embryopathic hydantoin syndrome]. *Monatsschr. Kinderheilkd.* 124: 634–637.

Lindhout, D. and Omtzigt, J. G. C. (1992). Pregnancy and the risk of teratogenicity. *Epilepsia* 33 (Suppl. 4): 541–548.

Lindhout, D. and Omtzigt, J. G. C. (1994). Teratogenic effects of antiepileptic drugs: Implications for the management of epilepsy in women of childbearing age. *Epilepsia* 35 (Suppl. 4): S19–S28.

Lindhout, D., Hoppener, R. J. E. A., and Meinardi, H. (1984). Teratogenicity of antiepileptic drug combinations with special emphasis on epoxidation (of carbamazepine). *Epilepsia* 25: 77–83.

Lorente, C. A., Tassinari, M. S., and Keith, D. A. (1981). The effects of phenytoin on rat development: An animal model system for fetal hydantoin syndrome. *Teratology* 24: 169–180.

Loughnan, P. M., Gold, H., and Vance, H. C. (1973). Phenytoin teratogenicity in man. *Lancet* 1: 70–72.

Lyon, H. M., Holmes, L. B., and Huang, T. (2003). Multiple congenital anomalies associated with *in utero* exposure of phenytoin: Possible hypoxic ischemic mechanism? *Birth Defects Res. (A)* 67: 993–996.

Majewski, F. et al. (1980). [Teratogenicity of anticonvulsant drugs]. *Dtsch. Med. Wochenschr.* 105: 719–723.

Marks, P., Mills, B., and West, L. (1994). Basilar invagination and mid-line skeletal abnormalities due to *in utero* exposure to phenytoin. *Br. J. Neurosurg.* 8: 365–368.

Martz, F., Failinger, C., and Blake, D. A. (1977). Phenytoin teratogenesis: Correlation between embryopathic effect and covalent binding of putative arene oxide metabolite in gestational tissue. *J. Pharmacol. Exp. Ther.* 203: 231–239.

Massey, K. M. (1966). Teratogenic effects of diphenylhydantoin sodium. *J. Oral Ther.* 2: 380–385.

McClain, R. M. and Langhoff, L. (1980). Teratogenicity of diphenylhydantoin in the New Zealand white rabbit. *Teratology* 21: 371–379.

McNinch, A. W. and Tripp, J. H. (1991). Hemorrhagic disease of the newborn in the British Isles: Two year prospective study. *Br. Med. J.* 303: 1105–1109.

Meadow, S. R. (1968). Anticonvulsant drugs and congenital abnormalities. *Lancet* 2: 1296.

Melchior, J. C., Svensmark, P., and Trolle, D. (1967). Placental transfer of phenobarbitone in epileptic women and elimination in newborns. *Lancet* 2: 860–861.

Meyer, J. G. (1973). The teratological effects of anticonvulsants and the effects on pregnancy and birth. *Eur. Neurol.* 10: 179–190.

Michalodimitrakis, M., Parchas, S., and Coutselins, A. (1981). Fetal hydantoin syndrome: Congenital malformation of the urinary tract — a case report. *Clin. Toxicol.* 18: 1095–1097.

Miller, R. P. and Becker, B. A. (1975). Teratogenicity of oral diazepam and diphenylhydantoin in mice. *Toxicol. Appl. Pharmacol.* 32: 53–61.

Millicovsky, G. and Johnston, M. C. (1981). Maternal hyperoxia greatly reduces the incidence of phenytoin-induced cleft lip and palate in A/J mice. *Science* 212: 671–672.

Mirkin, B. L. (1971). Diphenylhydantoin: Placental transport, fetal localization, neonatal metabolism, and possible teratogenic effects. *J. Pediatr.* 78: 329–337.

Monson, R. R. et al. (1973). Diphenylhydantoin and selected congenital malformations. *N. Engl. J. Med.* 289: 1049–1052.

Murki, S., Dutta, S., and Mahesh, V. (2003). Truncus arteriosus in fetal hydantoin syndrome — a new association? *Indian Pediatr.* 40: 69–70.

Musselman, A. C. et al. (1994). Preliminary evidence of phenytoin-induced alterations in embryonic gene expression in a mouse model. *Reprod. Toxicol.* 8: 383–395.

Nagy, R. (1981). Fetal hydantoin syndrome. *Arch. Dermatol.* 117: 593–595.

Nakane, Y. and Kaneko, S. (1992). Congenital anomalies in the offspring of epileptic mothers. *Senten Ijo* 32: 309–321.

NRC (National Research Council). (2000). *Scientific Frontiers in Developmental Toxicology and Risk Assessment*, National Academy Press, Washington, D.C., pp. 73–74.

Nulman, I. et al. (1997). Findings in children exposed *in utero* to phenytoin and carbamazepine monotherapy: Independent effects of epilepsy and medications. *Am. J. Med. Genet.* 68: 18–24.

Olafsson, E. et al. (1998). Pregnancies of women with epilepsy: A population-based study in Iceland. *Epilepsia* 39: 887–892.

Orup, H. I. et al. (2003). Craniofacial skeletal deviations following *in utero* exposure to the anticonvulsant phenytoin: Monotherapy and polytherapy. *Orthod. Craniofac. Res.* 6: 2–19.

Ozkinay, F. et al. (1998). Two siblings with fetal hydantoin syndrome. *Turk. J. Pediatr.* 40: 273–278.

Pai, G. S. (1982). Cardiac and ophthalmic malformations and *in utero* exposure to Dilantin. *Pediatrics* 70: 327–328.

PDR® (*Physicians' Desk Reference®*). (2005). Medical Economics Co., Inc., Montvale, NJ.

Phelan, M. C., Pellock, J. M., and Nance, W. E. (1982). Discordant expression of fetal hydantoin syndrome in heteropaternal dizygotic twins. *N. Engl. J. Med.* 307: 99–101.

Phillips, N. K. and Lockard, J. S. (1996). Infant monkey hyperexcitability after prenatal exposure to antiepileptic compounds. *Epilepsia* 37: 991–999.

Pinto, W., Gardner, L. I., and Rosenbaum, P. (1977). Abnormal genitalia as a presenting sign in two male infants with hydantoin embryopathy syndrome. *Am. J. Dis. Child.* 131: 452–455.

Prakash, P., Saxsena, S., and Raturi, B. M. (1978). Hypoplasia of nails and phalanges: A teratogenic manifestation of diphenylhydantoin sodium. *Indian Pediatr.* 15: 866–867.

Sabers, A. et al. (1998). Pregnancy and epilepsy: A retrospective study of 151 pregnancies. *Acta Neurol. Scand.* 97: 164–170.

Sabry, M. A. and Farag, T. I. (1996). Hand anomalies in fetal-hydantoin syndrome: From nail/phalangeal hypoplasia to unilateral acheiria. *Am. J. Med. Genet.* 62: 410–412.

Schardein, J. L. (2000). *Chemically Induced Birth Defects*, Third ed., Marcel Dekker, New York, pp. 194–195, 202.

Scolnik, D. et al. (1994). Neurodevelopment of children exposed *in utero* to phenytoin and carbamazepine monotherapy. *JAMA* 271: 767–770.

Silver, L. (1981). Hand abnormalities in the fetal hydantoin syndrome. *J. Hand Surg.* 6: 262–265.

Smith, D. W. (1977). Distal limb hypoplasia in the fetal hydantoin syndrome. *Birth Defects* 13: 355–359.

Solomon, G. E., Hilgartner, M. W., and Kutt, H. (1972). Coagulation defects caused by diphenylhydantoin. *Neurology* 22: 1165–1171.

Spielberg, S. P. et al. (1981). Anticonvulsant toxicity *in vitro*: Possible role of arene oxides. *J. Pharmacol. Exp. Ther.* 217: 386–389.

Stankler, L. and Campbell, A. G. (1980). Neonatal acne vulgaris: A possible feature of the fetal hydantoin syndrome. *Br. J. Dermatol.* 103: 453–455.

Strickler, S. M. et al. (1985). Genetic predisposition to phenytoin-induced birth defects. *Lancet* 2: 746–749.

Tanganelli, P. and Regesta, G. (1992). Epilepsy, pregnancy, and major birth anomalies: An Italian prospective, controlled study. *Neurology* 42 (Suppl. 5): 89–93.

Trousdale, R. T. (1998). Fetal hydantoin syndrome: An unusual course of hip dysplasia. *Orthopedics* 21: 210–212.

Truog, W. E., Feusner, J. H., and Baker, D. L. (1980). Association of hemorrhagic disease with the syndrome of persistent fetal circulation with the fetal hydantoin syndrome. *J. Pediatr.* 96: 112–114.

Tunnessen, W. W. and Lowenstein, E. H. (1976). Glaucoma associated with the fetal hydantoin syndrome. *J. Pediatr.* 89: 154–155.

Vanoverloop, D. et al. (1992). The effects of prenatal exposure to phenytoin and other anticonvulsants on intellectual function at 4 to 8 years of age. *Neurotoxicol. Teratol.* 14: 329–335.

Vorhees, C. V. and Minck, D. R. (1989). Long-term effects of prenatal phenytoin exposure on offspring behavior in rats. *Neurotoxicol. Teratol.* 11: 295–305.

Waller, P. H., Genstler, D. E., and George, C. C. (1978). Multiple systemic and periocular malformations associated with the fetal hydantoin syndrome. *Ann. Ophthalmol.* 10: 1568–1572.

Watkinson, W. P. and Millicovsky, G. (1983). Effect of phenytoin on maternal heart rate in A/J mice: Possible role in teratogenesis. *Teratology* 28: 1–8.

Wells, P. G. et al. (1989). Modulation of phenytoin teratogenicity and embryonic covalent binding by acetylsalicylic acid, caffeic acid, and α-phenyl-*N-t*-butylnitrone: Implications for bioactivation by prostaglandin synthetase. *Toxicol. Appl. Pharmacol.* 97: 192–202.

Wells, P. G. et al. (1997). Reactive intermediates. In *Drug Toxicity in Embryonic Development, Vol. 1. Handbook of Experimental Pharmacology*, R. J. Kavlock and G. P. Daston, Eds., Springer, Heidelberg, pp. 451–516.

Wester, U. et al. (2002). Chondrodysplasia punctata (CDP) with features of the tibia-metacarpal type and maternal phenytoin treatment during pregnancy. *Prenat. Diagn.* 22: 663–668.

Wide, K. et al. (2002). Psychomotor development in preschool children exposed to antiepileptic drugs *in utero*. *Acta Paediatr.* 91: 409–414.

Wilson, J. G. (1973). Present status of drugs as teratogens in man. *Teratology* 7: 3–16.

Wilson, R. S., Smead, W., and Char, F. (1978). Diphenylhydantoin teratogenicity: Ocular manifestations and related deformities. *J. Pediatr. Ophthalmol. Strabismus* 15: 137–140.

Wood, B. P. and Young, L. W. (1979). Pseudohyperphalangism in fetal Dilantin syndrome. *Radiology* 131: 371–372.

Yalcinkaya, C. et al. (1997). Polydactyly and fetal hydantoin syndrome: An additional component of the syndrome? *Clin. Genet.* 51: 343–345.

Yang, T.-S. et al. (1978). Diphenylhydantoin teratogenicity in man. *Obstet. Gynecol.* 52: 682–684.

Zutel, A. J. et al. (1977). [Drug-related prenatal syndromes]. *Rev. Hosp. Ninos B. Aires* 19: 281–289.

43 Etretinate

Chemical name: (*all-E*)-9-(4-Methoxy-2,3,6-trimethylphenyl)-3,7-dimethyl-2,4,6,8-nonatetraenoic acid ethyl ester

Alternate name: Ro-10-9359

CAS #: 54350-48-0

SMILES: c1(c(c(c(c(cc1C)OC)C)C)C=CC(=CC=CC(=CC(OCC)=O)C)C

INTRODUCTION

Etretinate is a synthetic analog of retinoic acid closely related to vitamin A that is used therapeutically in the treatment of severe recalcitrant psoriasis. It is a second-generation orally active retinoid that exerts its effects by binding to specific nuclear receptors and modulating gene expression (Hardman et al., 2001). It is available by prescription under the trade names Tegison® and Tigason®, but is gradually being replaced in the armamentarium by its active metabolite, acitretin, another developmental toxicant. Like the latter, etretinate has a pregnancy category of X. The package label contains a "CAUSES BIRTH DEFECTS. DO NOT GET PREGNANT" icon and a "black box" warning that the drug must not be used by females who are pregnant or who intend to become pregnant during therapy or at any time for at least 3 years following discontinuation of therapy (*PDR*, 2005; see below). Major human fetal abnormalities have been reported with the administration of etretinate. Potentially, any fetus exposed can be affected. The label continues with the statement that major fetal abnormalities associated with etretinate administration have been reported, including meningomyelocele, meningoencephalocele, multiple synostoses, facial dysmorphia, syndactyly, absence of terminal phalanges, malformations of hip, ankle, and forearm, low-set ears, high palate, decreased cranial volume, cardiovascular malformation, and alterations of the skull and cervical vertebrae.

DEVELOPMENTAL TOXICOLOGY

ANIMALS

In the laboratory, etretinate is a potent developmental toxicant and teratogen in all four animal species tested. Multiple malformations are induced by the oral route (the pertinent human route)

in hamsters (Williams et al., 1984), mice (Reiners et al., 1988), rats (Aikawa et al., 1982), and rabbits (Hummler and Schuepbach, 1981). Developmental toxicity including reduced fetal body weight is also recorded in some species, but maternally toxic doses have not been identified in any species. Teratogenic doses have ranged from 2 to 100 mg/kg/day over a treatment interval ranging from 1 to 11 days during organogenesis in the various species.

HUMANS

In the human, known to the U.S. Food and Drug Administration (FDA) from 1969 to 1990 were 21 cases of malformations associated with spontaneous abortion in some resulting offspring of women treated with etretinate either during pregnancy or following cessation of treatment with the drug (Rosa, 1991). The 24 separately published cases to date are tabulated in Table 1. Six cases described no malformations, only death. Some of the case reports are typical of the "retinoid embryopathy," while others varied in type and were inconsistent from those of other retinoids (e.g., isotretinoin). The types of malformations reported by the manufacturer are given on the package label. Some of the malformations encountered are concordant with the pattern observed in laboratory animal species, especially the mouse and rat. A number of cases of malformation were reported to occur long after treatment was ceased, including cases 1, 2, 5, 7, 8, and 23. This interval ranged from 4 months to almost 4 years in these cases. There are examples of normal infants born under these conditions, however (Vahlquist and Rollman, 1990), and the malformations attributable to etretinate have been challenged by others (Blake and Wyse, 1988). This is related to the fact that the drug is very slowly released over a prolonged period (of up to 2.9 yr) after treatment has been stopped (Anonymous, 1986). Because of this, one investigator recommended that women treated with etretinate should avoid conception indefinitely (Lammer, 1988). Others disagree with this suggestion (Greaves, 1988; Rinck et al., 1989). The mouse has been considered a model for human embryopathy (Lofberg et al., 1990).

Critical factors in the teratogenicity profile of etretinate are a dose of 0.75 to 1.5 mg/kg/day (the recommended human dose range) to include the first 10 weeks of pregnancy. As with the other developmentally toxic retinoids, infant death (stillbirth) or spontaneous abortion is an associated feature of the developmental toxicity pattern. Growth retardation and adverse functional effects are apparently not associated with the drug. The mechanism of teratogenicity by retinoids was investigated quite thoroughly, and the reader is referred to the review article on retinoic acid metabolism by the National Research Council (NRC, 2000). It appears that the receptors for retinoids are of

TABLE 1
Developmental Toxicity Profile of Etretinate in Humans

Case Number	Malformations	Growth Retardation	Death	Functional Deficit	Ref.
1–3	Skeleton				Happle et al., 1984
4–6	Brain		✔		Happle et al., 1984
7	Limbs				Grote et al., 1985; Kietzmann et al., 1986
8	Embryopathy	✔			Lammer, 1988
9	Embryopathy				Lambert et al., 1988
10–15	None		✔		Hopf and Mathias, 1988
16–21	Embryopathy				Hopf and Mathias. 1988
22	Embryopathy				Martinez-Tallo, 1989
23	Heart, kidney, ears				Verloes et al., 1990; Bonnivert et al., 1990
24	Embryopathy				Geiger et al., 1994

two types (RAR and RXR) of the nuclear hormone ligand-dependent, transcription-factor super-family, and the receptor specificity correlates, generally, with their teratogenic actions. When activated by exogenously added retinoic acid, the receptor affects gene expression at abnormal times and sites. Details of the process are available in that publication (NRC, 2000). One group of experts placed the magnitude of teratogenic risk as high (Friedman and Polifka, 2000). Several useful reviews of retinoid and etretinate toxicity are available (Reiners et al., 1988; Mitchell, 1992; Chan et al., 1996; Guillonneau and Jacqz-Aigrain, 1997; Monga, 1997).

CHEMISTRY

Etretinate is a large molecule with extended conjugation of double bonds. It is highly hydrophobic. Etretinate is of low polarity and has a low propensity for hydrogen bonding. The calculated physicochemical and topological properties are as follows.

PHYSICOCHEMICAL PROPERTIES

Parameter	Value
Molecular weight	354.489 g/mol
Molecular volume	358.57 A^3
Density	0.904 g/cm^3
Surface area	461.12 A^2
LogP	6.855
HLB	0.275
Solubility parameter	18.713 $J^{(0.5)}/cm^{(1.5)}$
Dispersion	17.807 $J^{(0.5)}/cm^{(1.5)}$
Polarity	1.876 $J^{(0.5)}/cm^{(1.5)}$
Hydrogen bonding	5.439 $J^{(0.5)}/cm^{(1.5)}$
H bond acceptor	0.39
H bond donor	0.04
Percent hydrophilic surface	7.54
MR	107.975
Water solubility	–5.080 log (mol/M^3)
Hydrophilic surface area	34.79 A^2
Polar surface area	38.69 A^2
HOMO	–7.670 eV
LUMO	–1.463 eV
Dipole	6.817 debye

TOPOLOGICAL PROPERTIES (UNITLESS)

Parameter	Value
x0	19.690
x1	12.294
x2	10.612
xp3	8.151
xp4	6.142
xp5	3.845
xp6	2.700
xp7	1.417
xp8	0.843
xp9	0.540

Continued.

Parameter	Value
xp10	0.357
xv0	16.973
xv1	8.826
xv2	6.303
xvp3	4.192
xvp4	2.574
xvp5	1.453
xvp6	0.871
xvp7	0.368
xvp8	0.168
xvp9	0.096
xvp10	0.056
k0	36.789
k1	24.038
k2	12.457
k3	8.280
ka1	21.812
ka2	10.692
ka3	6.900

REFERENCES

Aikawa, M. et al. (1982). Toxicity study of etretinate. III. Reproductive segment 2 study in rats. *Yokuri Chiryo* 9: 5095 passim 5143.

Anonymous. (1986). Etretinate approved. *FDA Drug Bull.* 16: 16–17.

Blake, K. D. and Wyse, R. K. H. (1988). Embryopathy in infant conceived one year after termination of maternal etretinate: A reappraisal. *Lancet* 2: 1254.

Bonnivert, J., Lamgotte, R., and Verloes, A. (1990). Etretinate embryotoxicity 7 months after discontinuation of treatment. *Am. J. Med. Genet.* 37: 437–438.

Chan, A. et al. (1996). Oral retinoids and pregnancy. *Med. J. Aust.* 165: 164–167.

Friedman, J. M. and Polifka, J. E. (2000). *Teratogenic Effects of Drugs. A Resource for Clinicians (TERIS)*, Second ed., Johns Hopkins University Press, Baltimore, MD.

Geiger, J. M., Boudin, M., and Sourot, J.-H. (1994). Teratogenic risk with etretinate and acitretin treatment. *Dermatology* 189: 109–116.

Greaves, M. W. (1988). Embryopathy in infant conceived one year after termination of maternal etretinate: A reappraisal. *Lancet* 2: 1254.

Grote, W. et al. (1985). Malformation of fetus conceived 4 months after termination of maternal etretinate treatment. *Lancet* 1: 1276.

Guillonneau, M. and Jacqz-Aigrain, E. (1997). [Teratogenic effects of vitamin A and its derivatives]. *Arch. Pediatr.* 4: 867–874.

Happle, R. et al. (1984). Teratogenicity of etretinate in humans. *Dtsch. Med. Wochenschr.* 109: 1476–1480.

Hardman, J. G., Limbird, L. E., and Gilman, A. G., Eds. (2001). *Goodman & Gilman's The Pharmacological Basis of Therapeutics*, Tenth ed., McGraw-Hill, New York, pp. 1776–1777.

Hopf, G. and Mathias, B. (1988). Teratogenicity of isotretinoin and etretinate. *Lancet* 2: 1143.

Hummler, H. and Schuepbach, M. E. (1981). Studies in reproductive toxicology and mutagenicity with Ro 10-9359. In *Retinoids (Proc. Int. Dermatol. Symp.)*, C. E. Orfanos, O. Braun-Falco, and E. M. Farber, Eds., Springer, Berlin, pp. 49–59.

Kietzmann, H. et al. (1986). Fetal malformation after maternal etretinate treatment of Darier's disease. *Dtsch. Med. Wochenschr.* 111: 60–62.

Lambert, D. et al. (1988). Malformations foetales après etretinate. *Nouv. Dermatol.* 7: 448–451.

Lammer, E. J. (1988). Embryopathy in infant conceived one year after termination of maternal etretinate. *Lancet* 2: 1080–1081.

Lofberg, B. et al. (1990). Teratogenicity of steady-state concentrations of etretinate and metabolite acitretin maintained in maternal plasma and embryo by intragastric infusion during organogenesis in the mouse — A possible model for the extended elimination phase in human therapy. *Dev. Pharm.* 15: 45–51.

Martinez-Tallo, M. E. et al. (1989). Agenesia de pene y syndrome polimalformativo asciado con ingestion maternal de etretinato. *An. Esp. Pediatr.* 31: 399–400.

Mitchell, A. A. (1992). Oral retinoids. What should the prescriber know about their teratogenic hazards among women of child-bearing potential. *Drug Saf.* 7: 79–85.

Monga, M. (1997). Vitamin A and its congeners. *Semin. Perinatol.* 21: 135–142.

NRC (National Research Council). (2000). *Scientific Frontiers in Developmental Toxicology and Risk Assessment*, National Academy Press, Washington, D.C., pp. 75–80.

PDR® (*Physicians' Desk Reference®*). (2005). Medical Economics Co., Inc., Montvale, NJ.

Reiners, J. et al. (1988). Transplacental pharmacokinetics of teratogenic doses of etretinate and other aromatic retinoids in mice. *Reprod. Toxicol.* 2: 19–29.

Rinck, G., Gollnick, H., and Orfanos, C. G. (1989). Duration of contraception after etretinate. *Lancet* 1: 845–846.

Rosa, F. (1991). Detecting human retinoid embryopathy. *Teratology* 43: 419.

Vahlquist, A. and Rollman, O. (1990). Etretinate and the risk for teratogenicity. Drug-monitoring in a pregnant woman for 9 months after stopping treatment. *Br. J. Dermatol.* 123: 131.

Verloes, A. et al. (1990). Etretinate embryotoxicity 7 months after discontinuation of treatment. *Am. J. Med. Genet.* 37: 437–438.

Williams, K. J., Ferm, V. H., and Willhite, C. C. (1984). Teratogenic dose-response relationship of etretinate in the golden hamster. *Fundam. Appl. Toxicol.* 4: 977–982.

44 Toluene

Alternate names: Methylbenzene, phenylmethane, toluol

CAS #: 108-88-3

SMILES: c1(ccccc1)C

INTRODUCTION

Toluene is a colorless organic liquid solvent widely used in the paint, lacquer, and resin industries; as a thinner for inks, perfumes, and dyes; as a gasoline additive; and in the manufacture of a number of chemicals, explosives, dyes, and other organic compounds. Toluene can be absorbed through the skin or via inhalation, with target sites of liver, kidney, and blood. Inhalation of airborne toluene is the main source of human exposure, and both occupational and inhalational abuse scenarios exist with the chemical (see below). The threshold limit value (TLV; 8-h time-weighted average) for occupational or environmental exposures is 50 ppm in air to skin (ACGIH, 2005). However, the chemical is abused by acute inhalational exposures in which 500 to 5000 ppm or greater may be experienced (Wilkins-Haug, 1997). Excessively high exposure levels, possibly on the order of 5000 to 30,000 ppm, that produce maternal toxicity have been associated with developmental effects (Ron, 1986; Hathaway and Proctor, 2004; see below). Sources for abuse as a psychotropic agent usually involve sniffing paint, spray-paint, or paint thinner, glue, or gasoline, all substances containing toluene. This effect is of major concern due to the fact that toluene is produced in very large quantities, said to be 927 million pounds produced annually 10 years earlier (Hayes, 2001), and due to its wide availability, low cost, and its ubiquitousness in the environment. Due to these factors, abuse of toluene may be preferred by some over "harder" agents (Davies et al., 1985). In one large city hospital, toluene abuse accounted for 7.5% of all adult admissions for drug abuse (Hershey, 1982). A study in 1994 of eighth grade students disclosed that 19.9% reported that they had used inhalants (Sharp and Rosenberg, 1997). Its presence in the environment is widespread: A 1988 U.S. Environmental Protection Agency (EPA) survey of hazardous waste sites detected toluene levels of 7.5 ppb in surface waters, 21 ppb in groundwaters, and 77 ppb in soil (U.S. EPA, 1988).

DEVELOPMENTAL TOXICOLOGY

ANIMALS

In the laboratory, toluene has been developmentally toxic by the inhalational route in the rat, mouse, and hamster, but was not in the rabbit under the conditions employed. Consistent findings in animal species were not observed. In the rat, postnatal physical development was retarded at maternally toxic levels of 1200 ppm given for 6 h/day for 13 days in gestation (Thiel and Chahoud, 1997).

Developmental neurotoxicity was also demonstrated in this species (Hass et al., 1998). In the mouse, 1000 ppm toluene given for 18 days during gestation was teratogenic, producing rib malformations (Shigeta et al., 1981). In the hamster, 800 mg/m³ elicited postnatal neuromotor alterations when exposures were given on 6 days (6 h/day) in gestation (da-Silva et al., 1990). Rabbit does exposed to toluene for up to 500 ppm for 13 days in pregnancy displayed no developmental toxicity to the resulting bunnies (Klimisch et al., 1992).

HUMANS

In the human, chronic inhalation abuse or occupational exposures of toluene during pregnancy have been associated with teratogenicity and other developmental toxicity in a number of case reports. Due to the wide variation in developmentally toxic effects, the classes of toxicity to development will be separated as follows. It should be mentioned that commercial solvents containing toluene also contain one or more volatile hydrocarbons; thus, some effects attributed to it may be due to concomitant exposure to the others, alone or in combination.

Malformation

Developmental toxicity of toluene in humans was initially reported by Toutant and Lippmann in 1979 in which they described an infant with microcephaly, depressed nasal bridge, hypoplastic mandible, short palpebral fissures, low-set ears, sacral dimple, and sloping forehead; the child also was incoordinate, growth retarded, and mentally disabled. The alcoholic mother had been exposed to toluene and other hydrocarbons for a period of 14 years. An earlier report of two cases was published, but exposure to toluene was not exclusive. Following these reports, a number of publications appeared, reporting a total of at least 73 cases of malformations as shown in Table 1.

TABLE 1
Case Reports of Toluene-Induced Malformations in Humans

Number of Cases	Exposure Characteristics	Ref.
2	Exposed to 298 ppm occupationally in shoemaking factory (also with another chemical)	Euler, 1967
1	Continual abuse by addict over 14 yr to solvents (primarily toluene); also alcoholic	Toutant and Lippmann, 1979
3	Exposed occupationally to solvents, including toluene	Holmberg, 1979; Holmberg and Nurminen, 1980
1	Sniffed paint throughout pregnancy	Streicher et al., 1981
2	Sniffed during pregnancy	Hersh et al., 1985
1	Inhaled spray paint in pregnancy	Medrano, 1988
3	Abuse during pregnancy by addicts	Goodwin, 1988
2	Inhalation of paint over 7 yr (1), sniffing of chemical over 10 yr (1)	Hersh, 1989
35	Sniffing glue and spray paint by abusers	Arnold and Wilkins-Haug, 1990; Wilkins-Haug and Gabow, 1991; Arnold et al., 1994
18	Sniffing spray paint	Seaver et al., 1991; Hoyme et al., 1993; Pearson et al., 1994
2	Sniffing paint thinner	Lindemann, 1991
1	Sniffing paint	Erramouspe et al., 1996
2	Inhaled organic solvents, mainly toluene	Arai et al., 1997

Source: Modified after Schardein, J. L., *Chemically Induced Birth Defects*, Third ed., Marcel Dekker, New York, 2000.

The syndrome of defects, termed the "fetal solvents syndrome" or more appropriately, "toluene embryopathy," was described in a number of reports in the 1980s and 1990s. Basically, there are craniofacial features consistent with fetal alcohol syndrome (FAS), including microcephaly, short palpebral fissures, and poorly developed philtrum with thin upper lip. Hydronephrosis is an occasional internal finding, and renal tubular acidosis is common. This constellation of findings is usually accompanied by intrauterine growth retardation and postnatal growth deficiency in survivors (see below). The frequency of the features comprising the embryopathy is tabulated in Table 2. In

TABLE 2
Major Features of Toluene Embryopathy in 44 Human Cases

Clinical Features[a]	Incidence (%)
Craniofacial features	
Micrognathia	65
Small palpebral fissures	65
Ear anomalies	57
Narrow bifrontal diameter	48
Abnormal scalp hair pattern	43
Thin upper lip	43
Smooth philtrum	35
Small nose	35
Downturned mouth corners	33
Large anterior fontanel	22
Mortality	
Perinatal death	9
Growth and development	
Developmental delay	80
Postnatal microcephaly	67
Small for gestation age	54
Postnatal growth deficiency	52
Prematurity	39
Prenatal microcephaly	33
Other anomalies	
Nail hypoplasia	39
Altered palmar creases	35
Abnormal muscle tone	35
Hemangiomas	28
Renal anomalies	26
Clinodactyly	22
Hirsutism	6

[a] Features not observed in all reports.

Source: After Schardein, J. L., *Chemically Induced Birth Defects*, Third ed., Marcel Dekker, New York, 2000 from Hersh, J. H. et al., *J. Pediatr.*, 106, 922–927, 1985; Hersh, J. H., *J. Med. Genet.*, 25, 333–337, 1989; Arnold, G. and Wilkins-Haug, L., *Am. J. Hum. Genet.*, 47, A46, 1990; Wilkins-Haug, L. and Gabow, P. A., *Obstet. Gynecol.*, 77, 504–509, 1991; Pearson, M. A. et al., *Pediatrics*, 93, 211–215, 1994.

a case-referent study of occupationally exposed women, 301 infants with a congenital deformity were paired with 301 normal infants (McDonald et al., 1987). Toluene was found to be associated with an increased incidence of renal, urinary, gastrointestinal, and cardiac anomalies compared to the controls.

Growth Retardation

As noted above, growth in various forms was affected in high incidence among the cases with malformations (Table 2). Arnold and associates (1994) indicated that maternal toluene abuse of 4 or more years was positively correlated with body weight lower than the fifth percentile and microcephaly in childhood. In another study, this one investigating the pregnancy outcomes of 168 women exposed occupationally to toluene-containing varnishes of electrical insulators in concentrations averaging 55 ppm, there were twice as many babies born with low birth weight (2500 to 3000 g) in the exposed group than in the control group of 201 unexposed women (Syrovadko, 1977).

Death

Perinatal death (Table 2) and spontaneous abortion (Hamill et al., 1982; Axelsson et al., 1984; Lindbohm et al., 1990; Ng et al., 1992; Taskinen et al., 1994) in increased incidence have been associated with excessive toluene exposures.

Functional Deficit

Also in association with the embryopathic features, retarded growth, and increased mortality are signs of central nervous system dysfunction, as this is the primary target of the chemical. There is developmental delay in a high proportion of cases listed in Table 1 and noted in Table 2. Attention deficit disorder and delays in cognition, speech, and motor skills were recorded in toluene-exposed infants (Arai et al., 1997). The neurobehavioral consequences of high concentrations of toluene in the human have been described (Jones and Balster, 1997; Filley et al., 2004).

The central nervous system defects produced by chemical alterations in astrocyte proliferation and maturation may represent the mode of action of these effects (Costa et al., 2002). Interestingly, toluene can cause a persisting motor syndrome in rats that resembles (i.e., a wide-based ataxic gait) the syndrome seen in some heavy abusers of toluene-containing products (Pryor, 1991).

The mechanism of toluene toxicity is not known, but it has been speculated from *in vitro* studies that the chemical causes inhibition of the initiation of DNA synthesis which may result from denaturation of the cell membrane or damage to the translational process required for synthesis of initiator proteins (Winston and Matsushima, 1975). One group of experts placed the magnitude of teratogenic risk for usual occupational exposures as unlikely, but for abuse of the chemical, moderate to high (Friedman and Polifka, 2000).

A number of thorough reviews of toluene-induced developmental toxicity in animals and humans were published (Lawrence et al., 1988; Donald et al., 1991; Wilkins-Haug, 1997; Arnold, 1997; McMartin and Koren, 1999).

CHEMISTRY

Toluene is one of the smaller-sized human developmental toxicants. The structure of toluene contains no heteroatoms and therefore is incapable of hydrogen bonding. It is a nonpolar hydrophobic compound. The calculated physicochemical and topological properties are listed below.

PHYSICOCHEMICAL PROPERTIES

Parameter	Value
Molecular weight	92.140 g/mol
Molecular volume	98.30 A^3
Density	0.871 g/cm^3
Surface area	120.82 A^2
LogP	2.791
HLB	0.000
Solubility parameter	17.641 J$^{(0.5)}$/cm$^{(1.5)}$
Dispersion	17.636 J$^{(0.5)}$/cm$^{(1.5)}$
Polarity	0.428 J$^{(0.5)}$/cm$^{(1.5)}$
Hydrogen bonding	0.000 J$^{(0.5)}$/cm$^{(1.5)}$
H bond acceptor	0.00
H bond donor	0.03
Percent hydrophilic surface	0.00
MR	30.925
Water solubility	0.662 log (mol/M^3)
Hydrophilic surface area	0.00 A^2
Polar surface area	0.00 A^2
HOMO	−9.369 eV
LUMO	0.542 eV
Dipole	0.263 debye

TOPOLOGICAL PROPERTIES (UNITLESS)

Parameter	Value
x0	5.113
x1	3.394
x2	2.743
xp3	1.894
xp4	1.307
xp5	0.901
xp6	0.204
xp7	0.000
xp8	0.000
xp9	0.000
xp10	0.000
xv0	4.387
xv1	2.411
xv2	1.655
xvp3	0.940
xvp4	0.534
xvp5	0.304
xvp6	0.064
xvp7	0.000
xvp8	0.000
xvp9	0.000
xvp10	0.000
k0	4.712
k1	5.143
k2	2.344
k3	1.500
ka1	4.381
ka2	1.783
ka3	1.038

REFERENCES

ACGIH (American Conference of Government Industrial Hygienists). (2005). *TLVs® and BEIs®. Threshold Limit Values for Chemical Substances and Physical Agents & Biological Exposure Indices*, ACGIH, Cincinnati, OH, p. 56.

Arai, H. et al. (1997). [Two cases of toluene embryopathy with severe motor and intellectual disabilities syndrome]. *No To Hattasu* 29: 361–366.

Arnold, G. (1997). Solvent abuse and developmental toxicity. In *Environmental Toxicology and Pharmacology of Human Development*, S. Kacew and G. H. Lambert, Eds., Taylor & Francis, Washington, D.C., pp. 145–151.

Arnold, G. and Wilkins-Haug, L. (1990). Toluene embryopathy syndrome. *Am. J. Hum. Genet.* 47: A46.

Arnold, G. L. et al. (1994). Toluene embryopathy: Clinical delineation and developmental followup. *Pediatrics* 93: 216–220.

Axelsson, G., Lutz, C., and Rylander, R. (1984). Exposure to solvents and outcome of pregnancy in university laboratory employees. *Br. J. Ind. Med.* 41: 305.

Costa, L. G. et al. (2002). Developmental neurotoxicity: Do similar phenotypes indicate a common mode of action? A comparison of fetal alcohol syndrome, toluene embryopathy and maternal phenylketonuria. *Toxicol. Lett.* 127: 197–205.

da-Silva, V. A., Malheiro, L. R., and Bueno, F. M. R. (1990). Effects of toluene exposure during gestation in neurobehavioral development of rats and hamsters. *Braz. J. Med.* 23: 533–537.

Davies, B., Thorley, A., and O'Connor, D. (1985). Progression of addiction careers in young adult solvent misusers. *Br. Med. J.* 290: 109–110.

Donald, J. M., Hooper, K., and Hopenhayn-Rich, C. (1991). Reproductive and developmental toxicity of toluene: A review. *Environ. Perspect.* 94: 237–244.

Erramouspe, J., Galvez, R., and Fischel, D. R. (1996). Newborn renal tubular acidosis associated with prenatal maternal toluene sniffing. *J. Psychoactive Drugs* 28: 201–204.

Euler, H. H. (1967). [Animal experimental studies of an industrial noxa]. *Arch. Gynakol.* 204: 258–259.

Filley, C. M., Halliday, W., and Kleinschmidt-DeMasters, B. K. (2004). The effects of toluene on the central nervous system. *J. Neuropathol. Exp. Neurol.* 63: 1–12.

Friedman, J. M. and Polifka, J. E. (2000). *Teratogenic Effects of Drugs. A Resource for Clinicians (TERIS)*, Second ed., Johns Hopkins University Press, Baltimore, MD.

Goodwin, T. M. (1988). Toluene abuse and renal tubular acidosis in pregnancy. *J. Obstet. Gynecol.* 71: 715–718.

Hamill, P. V. V. et al. (1982). The epidemiologic assessment of male reproductive hazard from occupational exposure to TDA and DNT. *J. Occup. Med.* 24: 985–993.

Hass, U. et al. (1998). Toluene causes developmental neurotoxicity in rats. *Teratology* 58: 23A.

Hathaway, G. J. and Proctor, N. H. (2004). *Proctor and Hughes' Chemical Hazards of the Workplace*, Fifth ed., John Wiley & Sons, Hoboken, NJ, pp. 681–682.

Hayes, A. W., Ed. (2001). *Principles and Methods of Toxicology*, Fourth ed., Taylor & Francis, Philadelphia, PA, p. 532.

Hersh, J. H. (1989). Toluene embryopathy: Two new cases. *J. Med. Genet.* 25: 333–337.

Hersh, J. H. et al. (1985). Toluene embryopathy. *J. Pediatr.* 106: 922–927.

Hershey, C. O. (1982). Solvent abuse: A shift to adults. *Int. J. Addict.* 17: 1085–1089.

Holmberg, P. C. (1979). Central-nervous-system defects in children born to mothers exposed to organic solvents during pregnancy. *Lancet* 2: 177–179.

Holmberg, P. C. and Nurminen, M. (1980). Congenital defects of the central nervous system and occupational factors during pregnancy. A case-referent study. *Am. J. Ind. Health* 1: 167–176.

Hoyme, H. E. et al. (1993). Toluene embryopathy: Elucidation of phenotype and mechanism of teratogenesis in 12 patients. *Reprod. Toxicol.* 7: 158–159.

Jones, H. E. and Balster, R. L. (1997). Neurobehavioral consequences of intermittent prenatal exposure to high concentrations of toluene. *Neurotoxicol. Teratol.* 19: 305–313.

Klimisch, H., Hellwig, J., and Hofman, A. (1992). Studies on the prenatal toxicity of toluene in rabbits following inhalation exposure and proposal of a pregnancy guidance value. *Arch. Toxicol.* 66: 373–381.

Lawrence, K. et al. (1988). Health effects of the alkylbenzenes. I. Toluene. *Toxicol. Ind. Health* 4: 49–76.

Lindbohm, M. L. et al. (1990). Spontaneous abortions among women exposed to organic solvents. *Am. J. Ind. Med.* 17: 449–463.

Lindemann, R. (1991). Congenital renal tubular dysfunction associated with maternal sniffing of organic solvents. *Acta Paediatr. Scand.* 80: 882–884.

McDonald, J. C. et al. (1987). Chemical exposures at work in early pregnancy and congenital defects: A case referent study. *Br. J. Ind. Med.* 44: 527–533.

McMartin, K. I. and Koren, G. (1999). Proactive approach for the evaluation of fetal safety in chemical industries. *Teratology* 60: 130–136.

Medrano, M. (1988). Tracking a syndrome in the barrios of San Antonio. *Nasotros* 1: 1–2.

Ng, T. P., Foo, S. C., and Yoong, T. (1992). Risk of spontaneous abortion in workers exposed to toluene. *Br. J. Ind. Med.* 49: 804–808.

Pearson, M. A. et al. (1994). Toluene embryopathy: Delineation of the phenotype and comparison with fetal alcohol syndrome. *Pediatrics* 93: 211–215.

Pryor, G. T. (1991). A toluene-induced motor syndrome in rats resembling that seen in some human solvent abusers. *Neurotoxicol. Teratol.* 13: 387–400.

Ron, M. A. (1986). Volatile substance abuse: A review of possible long-term neurological, intellectual and psychiatric sequelae. *Int. J. Psychiatry* 148: 235–246.

Schardein, J. L. (2000). *Chemically Induced Birth Defects*, Third ed., Marcel Dekker, New York, pp. 926–927.

Seaver, L. H. et al. (1991). Toluene embryopathy: Elucidation of phenotype and mechanism of teratogenesis in 12 patients. *Am. J. Hum. Genet.* 49 (Suppl. 4): 237.

Sharp, C. W. and Rosenberg, N. L. (1997). Inhalants. In *Substance Abuse. A Comprehensive Textbook*, J. J. Lowinson, P. Ruiz, R. B. Millman, and J. G. Langrod, Eds., Lippincott Williams & Wilkins, Baltimore, MD, pp. 246–264.

Shigeta, S., Aikawa, H., and Misawa, T. (1981). Effects of toluene exposure on mice fetuses. *J. Toxicol. Sci.* 6: 254–255.

Streicher, H. Z., Gabow, P. A., and Moss, A. H. (1981). Syndrome of toluene sniffing in adults. *Ann. Intern. Med.* 94: 758–762.

Syrovadko, O. N. (1977). Working conditions and health status of women handling organosilicon varnishes containing toluene. *Gig. Tr. Prof. Zabol.* 21: 15–19.

Taskinen, H. et al. (1994). Laboratory work and pregnancy outcome. *J. Occup. Med.* 36: 311–319.

Thiel, R. and Chahoud, I. (1997). Postnatal development and behavior of Wistar rats after prenatal toluene exposure. *Arch. Toxicol.* 71: 258–265.

Toutant, C. and Lippmann, S. (1979). Fetal solvents syndrome. *Lancet* 1: 1356.

U.S. Environmental Protection Agency (EPA). (1988). *National Ambient Volatile Organic Compounds (VOCs) Data Base Update* (EPA-600/3-88/010(a)). Atmospheric Sciences Research Laboratory, EPA, RTP, NC.

Wilkins-Haug, L. (1997). Teratogen update: Toluene. *Teratology* 55: 145–151.

Wilkins-Haug, L. and Gabow, P. A. (1991). Toluene abuse during pregnancy: Obstetric complications and perinatal outcomes. *Obstet. Gynecol.* 77: 504–509.

Winston, S. and Matsushima, T. (1975). Permanent loss of chromosome initiation in toluene-treated *Bacillus subtilis* cells. *J. Bacteriol.* 123: 921–927.

45 Ethisterone

Chemical name: 17α-Hydroxypregn-4-en-20-yn-3-one

Alternate names: Anhydrohydroxyprogesterone, 17α-ethinyltestosterone, pregneninolone

CAS #: 434-03-7

SMILES: C12C3C(C4(C(CC3)=CC(CC4)=O)C)CCC1(C(CC2)(C#C)O)C

INTRODUCTION

Ethisterone is a progestational steroid with therapeutic uses similar to those of progesterone — that of treating cases of threatened and habitual abortion and endometriosis. However, it also has estrogenic and androgenic properties, and its usefulness has been recently limited; the drug has largely been replaced in the therapeutic armamentarium. It has been available by prescription under the trade names Pranone®, Ora-Lutin®, Progesteral®, and Lutocylol®, among other names. It has a pregnancy category of D. This is due, presumably, to the causal association of ethisterone to genital malformations in an earlier interval (1950s and 1960s) when the drug was used extensively therapeutically. No significant nongenital malformations were reported with use of the drug, and the restriction that existed for those was lifted by the U.S. Food and Drug Administration (FDA) in 1999 (Brent, 2000).

DEVELOPMENTAL TOXICOLOGY

ANIMALS

In laboratory animals, ethisterone caused masculinization of female fetuses in both rats and rabbits. In rats, oral doses (the route used in humans) of 5 or 10 mg given for 5 days late in gestation were effective in this regard (Kawashima et al., 1977). Rabbits were more sensitive, with doses <1 mg given orally over 20 days in gestation causing virilization (Courrier and Jost, 1942).

HUMANS

In the human, as with some other progestational agents, virilization of female issue were recorded in 78 cases, as tabulated in Table 1. No recent cases have appeared in the published literature, and

TABLE 1
Reports of Virilization Associated with Ethisterone in Humans (Females)

Ref.	Number of Cases
Gross and Meeker, 1955	1
Jones, 1957	1
Wilkins et al., 1958; Wilkins and Jones, 1958	14
Reilly et al., 1958 (Grossman case)	1
Moncrieff, 1958[a]	2
Hillman, 1959	1
Grumbach et al., 1959	8
Jolly, 1959	1
Wilkins, 1960	23
Bongiovanni and McPadden, 1960	2
Jones and Wilkins, 1960	5
Jacobson, 1961	1
Dubowitz, 1962[a]	1
Rawlings, 1962	2
Greenstein, 1962	1
Breibart et al., 1963	1
Erhardt and Money, 1967	5
Serment and Ruf, 1968	8

[a] Includes cases with estrogen (ethinyl estradiol).

no cases of virilization in male issue, in the form of hypospadias, have been apparently recorded. The anomalies appear to be identical to those produced by androgenic agents. They were variously described as virilization, masculinization, and pseudohermaphroditism. The defects were first described almost half a century ago (Jones, 1957; Wilkins et al., 1958), and the descriptions were elaborated on by others more recently (Keith and Berger, 1977; Schardein, 1980, 2000; Wilson and Brent, 1981). Basically, there is phallic (clitoral) and labial enlargement, and usually labioscrotal fusion that may have progressed to the degree that it has resulted in the formation of a urogenital sinus. There is usually a normal vulva, endoscopic evidence of a cervix, and a palpable though sometimes infantile uterus. The anomalies correlated with the timing of drug exposure and the dose of the drug. The time of treatment recorded in the cited cases, when provided, varied from as early as the third or fourth gestational week to as late as pregnancy termination. Doses ranged from 10 to 250 mg/day over the treatment interval. These doses were similar to those producing effects in the two species of laboratory animals.

No other class of developmental toxicity appeared to be associated with the virilization. It clearly is a toxicant limited to hormonal-malforming effects in female issue.

CHEMISTRY

Ethisterone is a larger hydrophobic human developmental toxicant. Structurally it differs from norethindrone by the presence of an additional methyl group. It is of lower polarity. Ethisterone can engage in hydrogen bonding. The calculated physicochemical and topological properties for this compound are shown in the following.

PHYSICOCHEMICAL PROPERTIES

Parameter	Value
Molecular weight	312.452 g/mol
Molecular volume	312.64 A^3
Density	0.959 g/cm^3
Surface area	386.66 A^2
LogP	3.389
HLB	1.321
Solubility parameter	21.694 J$^{(0.5)}$/cm$^{(1.5)}$
Dispersion	19.371 J$^{(0.5)}$/cm$^{(1.5)}$
Polarity	3.477 J$^{(0.5)}$/cm$^{(1.5)}$
Hydrogen bonding	9.128 J$^{(0.5)}$/cm$^{(1.5)}$
H bond acceptor	0.70
H bond donor	0.46
Percent hydrophilic surface	12.09
MR	91.178
Water solubility	−2.968 log (mol/M^3)
Hydrophilic surface area	46.75 A^2
Polar surface area	40.46 A^2
HOMO	−10.063 eV
LUMO	−0.136 eV
Dipole	4.392 debye

TOPOLOGICAL PROPERTIES (UNITLESS)

Parameter	Value
x0	16.458
x1	10.839
x2	10.946
xp3	10.484
xp4	8.415
xp5	6.956
xp6	5.273
xp7	4.058
xp8	3.034
xp9	2.192
xp10	1.430
xv0	14.399
xv1	9.280
xv2	8.968
xvp3	8.352
xvp4	6.812
xvp5	5.356
xvp6	3.988
xvp7	2.940
xvp8	2.090
xvp9	1.368
xvp10	0.819
k0	31.320
k1	16.468
k2	5.247
k3	2.083
ka1	15.457
ka2	4.729
ka3	1.835

REFERENCES

Bongiovanni, A. M. and McPadden, A. J. (1960). Steroids during pregnancy and possible fetal consequences. *Fertil. Steril.* 11: 181–186.

Breibart, S., Bongiovanni, A. M., and Eberlein, W. R. (1963). Progestins and skeletal maturation. *N. Engl. J. Med.* 268: 255.

Brent, R. L. (2000). Nongenital malformations and exposure to progestational drugs during pregnancy; the final chapter of an erroneous allegation. *Teratology* 61: 449.

Courrier, R. and Jost, A. (1942). Fetal intersexuality provoked by pregneninolone administered during pregnancy. *C. R. Soc. Biol. (Paris)* 136: 395–396.

Dubowitz, V. (1962). Virilization and malformation of a female infant. *Lancet* 2: 405–406.

Ehrhardt, A. A. and Money, J. (1967). Progestin-induced hermaphroditism: IQ and psychosexual identity in a study of 10 girls. *J. Sex Res.* 3: 83–100.

Greenstein, N. M. (1962). Iatrogenic female pseudohermaphroditism. *Jewish Mem. Hosp. Bull. (N.Y.)* 7: 191–195.

Gross, R. E. and Meeker, I. A. (1955). Abnormalities of sexual development. Observations from 75 cases. *Pediatrics* 16: 303–324.

Grumbach, M. M., Ducharme, J. R., and Moloshok, R. E. (1959). On the fetal masculinizing action of certain oral progestins. *J. Clin. Endocrinol. Metab.* 19: 1369–1380.

Hillman, D. A. (1959). Fetal masculinization with maternal progesterone therapy. *Can. Med. Assoc. J.* 80: 200–201.

Jacobson, B. D. (1961). Abortion: Its prediction and management. *Fertil. Steril.* 12: 474–485.

Jolly, H. (1959). Non-adrenal female pseudohermaphroditism associated with hormone administration in pregnancy. *Proc. R. Soc. Med.* 52: 300–301.

Jones, H. W. (1957). Female hermaphroditism without virilization. *Obstet. Gynecol. Surv.* 12: 433–460.

Jones, H. W. and Wilkins, L. (1960). The genital anomaly associated with prenatal exposure to progestogens. *Fertil. Steril.* 11: 148–156.

Kawashima, K. et al. (1977). Virilizing activities of various steroids in female rat fetuses. *Endocrinol. Jpn.* 24: 77–81.

Keith, L. and Berger, G. S. (1977). The relationship between congenital defects and the use of exogenous progestational contraceptive hormones during pregnancy: A 20-year review. *Int. J. Gynaecol. Obstet.* 15: 115–124.

Moncrieff, A. (1958). Non-adrenal female pseudohermaphroditism associated with hormone administration in pregnancy. *Lancet* 2: 267–268.

Rawlings, W. J. (1962). Progestogens and the foetus. *Br. Med. J.* 1: 336–337.

Reilly, W. A. et al. (1958). Phallic urethra in female pseudohermaphroditism. *Am. J. Dis. Child.* 95: 9–17.

Schardein, J. L. (1980). Congenital abnormalities and hormones during pregnancy: A clinical review. *Teratology* 22: 251–270.

Schardein, J. L. (2000). *Chemically Induced Birth Defects*, Third ed., Marcel Dekker, New York, pp. 298–299.

Serment, H. and Ruf, H. (1968). Les dangers pour le produit de conception de medicaments administers a la femme enceinte. *Bull. Fed. Soc. Gynecol. Obstet. Lang. Fr.* 20: 69–76.

Wilkins, L. (1960). Masculinization of female fetus due to use of orally given progestins. *JAMA* 172: 1028–1032.

Wilkins, L. and Jones, H. W. (1958). Masculinization of the female fetus. *Obstet. Gynecol.* 11: 355.

Wilkins, L. et al. (1958). Masculinization of female fetus associated with administration of oral and intramuscular progestins during gestation: Nonadrenal pseudohermaphroditism. *J. Clin. Endocrinol. Metab.* 18: 559–585.

Wilson, J. G. and Brent, R. L. (1981). Are female sex hormones teratogenic? *Am. J. Obstet. Gynecol.* 141: 567–580.

46 Acitretin

Chemical name: (all-E)-9-(4-Methoxy-2,3,6-trimethylphenyl)-3,7-dimethyl-2,4,6,8-nonatetraenoic acid

Alternate names: Etretin, Ro-10-1670

CAS #: 55079-83-9

SMILES: c1(c(c(c(c(cc1C)OC)C)C)C)C=CC(=CC=CC(=CC(O)=O)C)C

INTRODUCTION

Acitretin is a retinoid analog of vitamin A and active metabolite of another developmental toxicant, etretinate, which it is gradually replacing in the marketplace. It has therapeutic activity in treating severe psoriasis and other skin (keratinizing) disorders. Its mechanism of action is that of etretinate, by bonding to specific nuclear receptors and modulating gene expression (Hardman et al., 2001). Acitretin is available as a prescription drug under the trade names Neotigason® or Soriatane®, and it has a pregnancy category of X. The package label for the drug contains a "CAUSES BIRTH DEFECTS. DO NOT GET PREGNANT" icon plus a "black box" warning that acitretin must not be used by females who are pregnant or who intend to become pregnant during therapy or at any time during at least the 3 years following discontinuation of therapy (*PDR*, 2005). It also must not be used by females who may not use reliable contraception while undergoing treatment and for at least 3 years following discontinuation of treatment. Further, females of reproductive potential must not be given a prescription for acitretin until pregnancy is excluded and a four-step program is undergone to ensure this condition is followed. The statement on the package label continues with the warning that human fetal abnormalities have been reported with the administration of acitretin (see below). Potentially, any fetus can be affected. Spontaneous abortion and premature birth are also listed as abnormal outcomes of recorded pregnancies.

DEVELOPMENTAL TOXICOLOGY

ANIMALS

In laboratory animals, acitretin is a potent teratogen by the oral route (the route pertinent to human therapy), producing malformations in rabbits, mice, and rats in decreasing order of sensitivity

TABLE 1
Developmental Toxicity Profile of Acitretin in Humans

Case Number	Malformations	Growth Retardation	Death	Functional Deficit	Ref.
1	Embryopathy		✔		Die-Smulders et al., 1995; Sturkenboom, 1995
(?)	None		✔		Geiger et al., 1994 (Manufacturer's data)
2	Embryopathy				Geiger et al., 1994 (Manufacturer's data)
3–6	"Nontypical"				Geiger et al., 1994 (Manufacturer's data)
7–17	None		✔		Maradit and Geiger, 1999
18	Embryopathy	✔		✔	Barbero et al., 2004

related to dosage (Kistler and Hummler, 1985). Effective doses ranged from lower to slightly larger (0.2, 0.3, and 3X, respectively) than used in human subjects (25 to 50 mg/day). At the higher dose of 100 mg/kg/day on gestation day 11, the drug elicited a high incidence of limb defects and cleft palate in the mouse, effects the authors concluded were "model" for those in the human (Lofberg et al., 1990).

HUMANS

In the human, acitretin has, as stated in the package insert, been associated with birth defects in the progeny of women treated during pregnancy. The published cases are provided in Table 1. The recorded malformations resemble those reported for tretinoin, isotretinoin, and etretinate, namely, facial, ear, limb, and heart defects, the "retinoic acid embryopathy" as it has been termed. Only three cases are known at present (cases 1, 2, and 18). The remainder of the cases cited are a significant number of spontaneous abortions, and four cases undescribed or described as "nontypical malformations." A recent study suggested that different retinoids produce only one malformation pattern, but that it has variable phenotypic expression (Barbero et al., 2004). A report published in 1994 related information on 75 women exposed to acitretin in populations both before and during pregnancy and also reviewed pregnancy outcomes from the manufacturer's data over the previous 11 years (Geiger et al., 1994). They indicated one typical embryopathy, a large number of spontaneous and induced abortions, a few nontypical malformations, and at least one normal liveborn. Another study, with one of the same investigators, published 5 years later detailed pregnancy outcomes from 123 cases, again with treatment both prior to and during pregnancy and including both retrospective and prospective exposure data (Maradit and Geiger, 1999). This report also listed different outcomes: abortion was common, but malformations were insignificant. A single case of functional deficits was recorded, that being neurodevelopmental delay and bilateral sensorineural deafness (Barbero et al., 2004). However, the latter does not fit the death/malformation response of other retinoids (excluding isotretinoin, a case in which the drug has been more widely studied). Additionally, growth retardation is not a feature of retinoid therapy.

The half-life of acitretin is shorter (2 to 4 days) than its parent etretinate (120+ days), but it may be converted into it in the body (Katz et al., 1999), explaining the rationale for the long discontinuation process as described on the package label. According to some, an assessment of etretinate concentrations in plasma and fat should be made to clarify the duration necessary for contraception (Maier and Honigsmann, 2001). Concurrent alcohol consumption also permits conversion of acitretin back to etretinate with the longer half-life, so alcohol is contraindicated along with the other restrictions of its use (Gronhoj Larsen et al., 2000).

The mechanism of teratogenicity by the retinoids has been investigated perhaps the most thoroughly of all teratogens, and the reader is referred to the published review article on retinoic acid metabolism by the National Research Council (NRC, 2000). The receptors for retinoids are of two types (RAR and RXR) of the nuclear hormone ligand-dependent, transcription-factor super-family, and in general, the receptor specificities of retinoids correlate with their teratogenic actions. RAR agonists are potent, and RXR agonists are ineffective; mixed agonists have intermediate activity (Kochhar et al., 1996). Further, RAR appears to be essential for the induction of defects of truncation of the posterior axial skeleton and is partially required for neural tube and craniofacial defects (Iulianella and Lohnes, 1997). In contrast, RXR is required for the induction of limb defects (Sucov et al., 1995). In both cases, the receptor, when activated by exogeneously added retinoic acid, is affecting gene expression at abnormal times and sites, as compared with that done by endogeneous retinoid. Further details are available (NRC, 2000).

The magnitude of teratogenic risk by acitretin is considered high according to one group of experts (Friedman and Polifka, 2000). The drug represents not only a significant risk during pregnancy, but also a risk for an unknown duration (perhaps several years) after therapy has ceased (Briggs et al., 2005). Katz and associates (1999) published a review of acitretin and its use in pregnancy.

CHEMISTRY

Acitretin is the hydrolyzed derivative of etretinate. It also includes a conjugated network of double bonds. It is a large molecule of high hydrophobicity that can participate in donor/acceptor hydrogen bonding. Acitretin is of lower polarity in comparison to the other human developmental toxicants. The calculated physicochemical and topological properties are listed below.

PHYSICOCHEMICAL PROPERTIES

Parameter	Value
Molecular weight	326.436 g/mol
Molecular volume	324.80 A^3
Density	0.918 g/cm^3
Surface area	416.09 A^2
LogP	5.740
HLB	2.130
Solubility parameter	20.050 $J^{(0.5)}/cm^{(1.5)}$
Dispersion	18.797 $J^{(0.5)}/cm^{(1.5)}$
Polarity	1.999 $J^{(0.5)}/cm^{(1.5)}$
Hydrogen bonding	6.684 $J^{(0.5)}/cm^{(1.5)}$
H bond acceptor	0.62
H bond donor	0.31
Percent hydrophilic surface	15.61
MR	98.621
Water solubility	–4.288 log (mol/M^3)
Hydrophilic surface area	64.94 A^2
Polar surface area	49.69 A^2
HOMO	–7.714 eV
LUMO	–1.518 eV
Dipole	7.375 debye

TOPOLOGICAL PROPERTIES (UNITLESS)

Parameter	Value
x0	18.276
x1	11.256
x2	10.112
xp3	7.438
xp4	5.649
xp5	3.561
xp6	2.465
xp7	1.330
xp8	0.801
xp9	0.482
xp10	0.308
xv0	15.305
xv1	7.850
xv2	5.889
xvp3	3.911
xvp4	2.399
xvp5	1.347
xvp6	0.793
xvp7	0.346
xvp8	0.157
xvp9	0.086
xvp10	0.048
k0	33.125
k1	22.042
k2	10.871
k3	7.424
ka1	19.816
ka2	9.160
ka3	6.068

REFERENCES

Barbero, P. et al. (2004). Acitretin embryopathy: A case report. *Birth Defects Res. (A)* 70: 831–833.

Briggs, G. G., Freeman, R. K., and Yaffe, S. J. (2005). *Drugs in Pregnancy and Lactation. A Reference Guide to Fetal and Neonatal Risk*, Seventh ed., Lippincott Williams & Wilkins, Philadelphia.

Die-Smulders, C. E. M. et al. (1995). Severe limb defects and craniofacial anomalies in a fetus conceived during acitretin therapy. *Teratology* 52: 215–219.

Friedman, J. M. and Polifka, J. E. (2000). *Teratogenic Effects of Drugs. A Resource for Clinicians (TERIS)*, Second ed., Johns Hopkins University Press, Baltimore, MD.

Geiger, J. M., Boudin, M., and Saurot, J.-H. (1994). Teratogenic risk with etretinate and acitretin treatment. *Dermatology* 189: 109–116.

Gronhoj Larsen, F. et al. (2000). Acitretin is converted to etretinate only during concomitant alcohol intake. *Br. J. Dermatol.* 143: 1164–1169.

Hardman, J. G., Limbird, L. E., and Gilman, A. G., Eds. (2001). *Goodman & Gilman's The Pharmacological Basis of Therapeutics*, Tenth ed., McGraw-Hill, New York, pp. 1776–1777.

Iulianella, A. and Lohnes, D. (1997). Contribution of retinoic acid receptor gamma to retinoid-induced craniofacial and axial defects. *Dev. Dyn.* 209: 92–104.

Katz, W. I., Waalen, J., and Leach, E. E. (1999). Acitretin in psoriasis: An overview of adverse effects. *J. Am. Acad. Dermatol.* 41: S7–S12.

Kistler, A. and Hummler, H. (1985). Teratogenesis and reproductive safety evaluation of the retinoid etretin (Ro 10-1670). *Arch. Toxicol.* 58: 50–56.

Kochhar, D. M. et al. (1996). Differential teratogenic response of mouse embryos to receptor selective analogs of retinoic acid. *Chem. Biol. Interact.* 100: 1–12.

Lofberg, B. et al. (1990). Teratogenicity of the 13-*cis* and all-*trans* isomers of the aromatic retinoid etretin: Correlation to transplacental pharmacokinetics in mice during organogenesis after a single oral dose. *Teratology* 41: 707–718.

Maier, H. and Honigsmann, H. (2001). Assessment of acitretin-treated female patients of childbearing age and subsequent risk of teratogenicity. *Br. J. Dermatol.* 145: 1028–1029.

Maradit, H. and Geiger, J. M. (1999). Potential risk of birth defects after acitretin discontinuation. *Dermatology* 198: 3–4.

NRC (National Research Council). (2000). *Scientific Frontiers in Developmental Toxicology and Risk Assessment*, National Academy Press, Washington, D.C., pp. 75–80.

PDR® (*Physicians' Desk Reference®*). (2005). Medical Economics Co., Inc., Montvale, NJ.

Sturkenboom, M. C. (1995). The "unexpected" teratogenic aspects of acitretin. *Hum. Exp. Toxicol.* 14: 681.

Sucov, H. M. et al. (1995). Mouse embryos lacking RXR alpha are resistant to retinoic acid induced limb defects. *Development* 121: 3997–4003.

47 Valsartan

Chemical name: *N*-(1-Oxopentyl)-*N*-[[2′-(1*H*-tetrazol-5-yl)[1,1′-biphenyl]-4-yl]methyl]-L-valine

CAS #: 137862-53-4

SMILES: n1nc([nH]n1)c2ccccc2c3ccc(cc3)CN(C(C(C)C)C(O)=O)C(CCCC)=O

INTRODUCTION

Valsartan is one of a group of eight presently available nonpeptide orally active angiotensin type 1 (ATI) receptor drugs collectively called "sartans" that cause vasoconstriction and retention of sodium and fluid. They act by binding to the main effector (AII) of the renal-angiotensin system (RAS) as AII receptor antagonists and are thus used in the treatment of essential hypertension and heart failure (Hardman et al., 2001). Valsartan is available by prescription as Diovan®, and it has a pregnancy category ranging from C to D. The package label for the drug contains a "black box" warning stating that when used in pregnancy during the second and third trimesters, drugs that act directly on the renal-angiotensin system can cause injury and even death to the developing fetus (*PDR*, 2005; see below). When pregnancy is detected, the drug should be discontinued as soon as possible. This warning translates into a D pregnancy category. First trimester treatment is designated a C category (as the adverse toxicity has not been reported from treatment early in human pregnancy).

DEVELOPMENTAL TOXICOLOGY

ANIMALS

No laboratory animal studies have been published. The package label refers to studies conducted (apparently by the manufacturer) orally, the route of administration for valsartan in the human, in mice, rats, and rabbits. It caused reduced fetal body weight in all three species, and additionally in rabbits, increased fetal resorption and abortion at maternally toxic dose levels. Of the three species, rabbits were the most sensitive, followed by mice, then rats, at doses of 0.5, 9, and 18 mg/kg/day, respectively, during the organogenesis period of gestation.

TABLE 1
Developmental Toxicity Profile of Valsartan in Humans

Case Number	Malformations	Growth Retardation	Death	Functional Deficit	Ref.
1, 2	Multiple: skull, face, kidneys, digits				Martinovic et al., 2001
3	Lungs, kidneys	✔	✔		Briggs and Nageotte, 2001
4, 5	None		✔		Biswas et al., 2002
6	Multiple: skull, limbs, kidneys				Schaefer, 2003
7	None	✔			Serreau et al., 2005
8	Multiple: kidneys, skull, heart				Serreau et al., 2005

HUMANS

In the human, valsartan has been associated with a few cases of fetopathy late in pregnancy, as shown in Table 1. The fetopathy is characterized by skull hypoplasia, enlarged or dystrophic kidneys with attendant clinical findings of oligohydramnios and neonatal anuria, and occasional pulmonary hypoplasia and facial deformity. Another case reported oligohydramnios amd neonatal anuria only, without fetopathic anomalies (Schaefer, 2003). The cases are similar to those documented with the ACE inhibitors (Sorensen et al., 1998). Fetal growth retardation was recorded in only one of six cases thus far and appears not to represent a consistent parameter of valsartan toxicity. The neonatal renal toxicity must be included as a functional impairment, and the single-term stillborn infant (at week 33) and two miscarriages are also considered treatment-related toxicity. All of the exposed infants were from mothers treated at the low end of the therapeutic dose scale (80 to 320 mg/kg/day orally), and all five fetopathic cases were resultant from treatment over the range of 0–24 to 28–36 gestational weeks (the second or third trimesters of pregnancy). Several cases treated in the first trimester were without effect (Chung et al., 2001; Biswas et al., 2002), and several others were treated in the first trimester and later and had anhydramnios, but these reversed themselves within a short interval (Berkane et al., 2004; Bos-Thompson et al., 2005).

Of the other sartans in clinical use, similar fetopathic findings were observed in cases with losartan, with candesartan, and single cases with telmisartan and irbesartan, to date. The characteristic recurrent pattern of fetal anomalies reported in association with maternal sartan treatment during the second half of pregnancy, the compatibility of these features with the known effects of RAS inhibition produced by AT1 receptor antagonists, and the striking similarity of this pattern with that seen after maternal treatment with ACE inhibitors, a class of therapeutic agents that also block RAS activity (although by a different mechanism), leave no doubt that maternal sartan treatment can cause fetal anomalies and death (Alwan et al., 2005, 2005a).

Several recent reviews of the sartans and their use in late pregnancy were published (Alwan et al., 2005; Bos-Thompson et al., 2005).

CHEMISTRY

Valsartan is a large compound with a high polar surface area. It is slightly hydrophilic and is capable of participating as both a hydrogen bond acceptor and as a hydrogen bond donor. The calculated physicochemical and topological properties of valsartan are as follows.

PHYSICOCHEMICAL PROPERTIES

Parameter	Value
Molecular weight	435.526 g/mol

Continued.

Parameter	Value
Molecular volume	399.54 A^3
Density	1.102 g/cm^3
Surface area	494.98 A^2
LogP	−0.329
HLB	6.883
Solubility parameter	23.458 $J^{(0.5)}/cm^{(1.5)}$
Dispersion	20.706 $J^{(0.5)}/cm^{(1.5)}$
Polarity	5.541 $J^{(0.5)}/cm^{(1.5)}$
Hydrogen bonding	9.532 $J^{(0.5)}/cm^{(1.5)}$
H bond acceptor	0.88
H bond donor	0.61
Percent hydrophilic surface	36.27
MR	127.409
Water solubility	−4.880 log (mol/M^3)
Hydrophilic surface area	179.55 A^2
Polar surface area	118.39 A^2
HOMO	−8.942 eV
LUMO	−0.842 eV
Dipole	6.667 debye

TOPOLOGICAL PROPERTIES (UNITLESS)

Parameter	Value
x0	23.087
x1	15.418
x2	13.522
xp3	10.760
xp4	9.249
xp5	6.798
xp6	4.435
xp7	3.189
xp8	2.292
xp9	1.572
xp10	1.074
xv0	18.654
xv1	10.867
xv2	8.145
xvp3	5.430
xvp4	3.793
xvp5	2.386
xvp6	1.384
xvp7	0.836
xvp8	0.515
xvp9	0.281
xvp10	0.166
k0	46.359
k1	26.602
k2	13.185
k3	7.250
ka1	23.590
ka2	10.991
ka3	5.802

REFERENCES

Alwan, S., Polifka, J. E., and Friedman, J. M. (2005). Angiotensin II receptor antagonist treatment during pregnancy. Teratogen update. *Birth Defects Res. (A)* 70: 123–130.

Alwan, S. et al. (2005a). Addendum: Sartan treatment during pregnancy. *BDR (A)*. 73: 904–905.

Berkane, N. et al. (2004) Fetal toxicity of valsartan and possible reversible adverse side effects. *Birth Defects Res. (A)* 70: 547–549.

Biswas. P. N., Wilton, L. V., and Shakir, S. W. (2002). The safety of valsartan: Results of a postmarketing surveillance study on 12881 patients in England. *J. Hum. Hypertens.* 16: 795–803.

Bos-Thompson, M. A. et al. (2005). Fetal toxic effects of angiotensin II receptor antagonists: Case report and follow-up after birth. *Ann. Pharmacother.* 39: 157–161.

Briggs, G. G. and Nageotte, M. P. (2001). Combined use of valsartan and atenolol. *Ann. Pharmacother.* 35: 859–861.

Chung, N. et al. (2001). Angiotensin-II-receptor inhibitors in pregnancy. *Lancet* 357: 1620–1621.

Hardman, J. G., Limbird, L. E., and Gilman, A. G., Eds. (2001). *Goodman & Gilman's The Pharmacological Basis of Therapeutics*, Tenth ed., McGraw-Hill, New York, pp. 829–833.

Martinovic, J. et al. (2001). Fetal toxic effects and angiotensin-II-receptor antagonists. *Lancet* 358: 241–242.

PDR® (Physicians' Desk Reference®). (2005). Medical Economics Co., Inc., Montvale, NJ.

Schaefer, C. (2003). Angiotensin II-receptor antagonists: Further evidence of fetotoxicity but not teratogenicity. *Birth Defects Res. (A)* 67: 591–594.

Serrau, R. et al. (2005). Developmental toxicity of the angiotensin II type 1 receptor antagonists during human pregnancy: a report of 10 cases. *Br. J. Gynaecol.* 112: 710–712.

Sorensen, A. M. et al. (1998). [Teratogenic effects of ACE-inhibitors and angiotensin II receptor antagonists]. *Ugeskr. Laeger.* 160: 1460–1464.

48 Diethylstilbestrol

Chemical name: 4,4′[(*1E*)-1,2-Diethyl-1,2-ethenediyl]bisphenol

Alternate names: DES, stilbestrol, (others — see below)

CAS #: 56-53-1

SMILES: C(=C(c1ccc(cc1)O)CC)(c2ccc(cc2)O)CC

INTRODUCTION

Diethylstilbestrol (DES) is a nonsteroidal synthetic estrogen presently used in the treatment of ovarian insufficiency, in the palliative treatment of breast malignancy, and as a contraceptive when used postcoitally. It was formerly used to prevent miscarriages, but was found not to be efficacious for this purpose (see below). DES is available by prescription by a large number of generic names (cyren A, domestrol, fonatol, oestromenin, palestrol, and synthoestrin, among others), and by various trade names, including Estrobene® and Stilboestrol DP®, among other names. It has a pregnancy category of X. The package label of an earlier time was a "black box" warning stating that "estrogens should not be used during pregnancy." The statement was continued: It has been reported that females exposed *in utero* to diethylstilbestrol may have an increased risk of developing later in life a rare form of vaginal or cervical cancer. This risk has been established to be 0.14 to 1.4 per 1000 exposures. Furthermore, 30% to 90% of such exposed women have been found to have vaginal adenosis and epithelial changes of the vagina and cervix. Although these changes are histologically benign, it is not known whether they are precursors of malignancy. Stated on the label was the concluding remarks that if diethylstilbestrol is administered during pregnancy, or if the patient becomes pregnant while taking this drug, she should be apprised of the potential risks to the fetus and of the advisability of pregnancy continuation. The National Cancer Institute, at the time, published a list of DES-type drugs introduced under registered trade names that may have been prescribed to pregnant women — it contained 68 names.

DEVELOPMENTAL TOXICOLOGY

ANIMALS

In studies conducted in laboratory animals prior to discovery in the early 1970s of the developmental problems in humans, researchers reported neither developmental toxicity nor teratogenicity in the three species tested. No effects by the oral route (the route used mainly in the human) were recorded in mice or rats given up to 0.4 mg/kg/day for 2 days in two different intervals in gestation (Einer-Jensen, 1968). Similarly, 0.25 mg/kg/day given over the first 9 days in gestation elicited no toxicity in hamsters (Giannina et al., 1971). Animal studies carried out later proved to be more fruitful (see below).

HUMANS

In the human, as pointed out on the package label, DES proved to have significant transplacental developmental effects leading to carcinogenesis in the genital organs of females and adverse developmental effects in male offspring. Precancerous and outright malignancies resulted. Because the events that followed beginning in 1970, the history of this unique chemical needs to be retold in context to the pregnancy outcomes that followed. Much of what is described is taken from the summary provided earlier by Schardein (2000).

History

Synthesized by Dodd in 1938 (Weitzner et al., 1981), DES was introduced to clinical medicine as the first orally active estrogen by Dr. O. W. Smith in 1946 (the report appeared 2 yr later); it was apparently thought to be efficacious in the definitive and preventive treatment of abortion and premature delivery. For this reason, it was given to a large number of pregnant women, being approved for use in pregnancy by the U.S. Food and Drug Administration (FDA) in 1947. Paradoxically, Dieckmann and associates showed as early as 1953 that it was not efficacious, but it remained in wide use, with estimates close to 1 to 2% of pregnant women in the United States taking the drug for various reasons (Fenichell and Charfoos, 1981). Total sales figures, peaking in 1953, estimate a population of between 1 and 10 million women using the drug in the 1938 to 1971 time interval. A mid-range figure of 3 million is probably most reliable (Herbst and Bern, 1981). Whatever the exact number, this corresponds to a peak of vaginal cancer in 1972–1973, precisely a 19-year gap (see the following).

The same year, Greenwald et al. (1971) found five additional cases in females 15- to 19-years of age in their review of the New York State Cancer Registry. Follow-up revealed that they, too, had maternal histories of DES (4) or dienestrol (1) usage, strengthening the latency concept just described. These findings have since been confirmed and expanded on by many others over the past three decades, as shown by the representative reports presented in Table 1. In response to the adverse effects reported, the drug was banned by the U.S. FDA in 1972 for use in humans and in 1979 for use in food animals. Up to 85% of U.S. livestock by the 1960s had been raised on DES to fatten them up for market (Seaman, 2003). Given that the literature is immense on DES effects — the National Library of Medicine has recorded almost 900 citations of DES and its effects in humans over the 35-year history (1970–2005) of the DES saga, it is difficult to select

TABLE 1
Representative Reports of Causal Associations by Diethylstilbestrol and Genital Malformations in Humans Following Discovery

Gilson et al., 1973	Stillman, 1982
Lanier et al., 1973	NCI, 1983
Noller and Fish, 1974	Vessey et al., 1983
Bibbo et al., 1975	Noller et al., 1983
Herbst et al., 1975	Chanen and Pagano, 1984
Robboy et al., 1975	Kaufman et al., 1984
Prins et al., 1976	Robboy et al., 1984
Sonek et al., 1976	Jefferies et al., 1984
Bibbo et al., 1977	Veridiano et al., 1984
Ng et al., 1977	McDonnell et al., 1984
Kaufman et al., 1977	Barter et al., 1986
Herbst et al., 1977	Cunha et al., 1987
Herbst et al., 1978	Bornstein et al., 1988
Bibbo, 1979	Horwitz et al., 1988
Nordquist et al., 1979	Linn et al., 1988
Robboy et al. 1979	Edelman, 1989
O'Brien et al., 1979	Vessey, 1989
Herbst et al., 1980	Sharp and Cole, 1990
Cousins et al., 1980	Gustavson et al., 1991
Ostergard, 1981	Marselos and Tomatis, 1992
Weitzner et al., 1981	Giusti et al., 1995
Robboy et al., 1981	Mittendorf, 1995
Sandberg et al., 1981	Newbold and McLachlan, 1996
Kaufman, 1982	Kaufman et al., 2000
Robboy et al., 1982	Herbst, 2000
Mangan et al., 1982	Hammes and Laitman, 2003
Kaufman et al., 1982	Wise et al., 2005

those representative reports that record most completely the story of this remarkable, and uniquely toxic, chemical.

Herbst and associates, after reporting the association between intake of DES by women during pregnancy and the induction of vaginal cancer in their daughters, established a registry (Registry for Research on Hormonal Transplacental Carcinogenesis) of clear cell adenocarcinoma of the genital tract in young females in 1971. A second registry, the Diethylstilbestrol Adenosis (DESAD) project, was also established to further monitor the effects, beginning in 1981 (Robboy et al., 1981). The cases entered in the registries up to the present time are shown in Table 2.

There is still no explanation for the registry cases occurring in patients whose mothers did not knowingly ingest estrogens, as shown above. The risk is currently cited to be on the order of about 0.14:1000 to 1.4:1000. The risk that a woman whose mother took DES during pregnancy will develop clear cell adenocarcinoma of the vagina or cervix by age 34 is further estimated to be about 1:1000 (Melnick et al., 1987; Vessey, 1989; Herbst and Anderson, 1990). A peak in the age incidence curve of the DES-related cases was observed at about 19 years, with the age range (latency) being 7 to 30 years (Herbst, 1981b). Of the reported cases, 91% occurred before age 27 (Melnick et al., 1987). The 5-year survival rate for the patients in the registry has been 80%. Early timing, long duration of exposure, and high dosage were important determinants of risk for vaginal

TABLE 2
Cases Identified in Diethylstilbestrol (DES) Registries

Interval	Number of Cases	Ref.
Up to 1972	91	Herbst et al., 1972
1972–74	170	Herbst et al., 1974; Poskanzer and Herbst, 1977
1974–June 1980	429[a]	Herbst, 1981b
To 1987	311	Melnick et al., 1987
1980–June, 1997	695[b]	Registry, 1998
To January 1999	705	Herbst, 2000

[a] 243 cases implicated by DES or two related estrogens.
[b] 463 cases implicated by DES.

epithelial changes (Shapiro and Slone, 1979; Jefferies et al., 1984). A review indicated that cervical or vaginal structural changes occurred in a range of incidence of 22 to 58% of exposed women (cited, Briggs et al., 2005).

By far, most of the cases reported have been in the United States, but case histories also came from Africa, Australia, Belgium, Canada, Holland, Great Britain, Czechoslovakia, France, Spain, Israel, and Mexico. Countries where DES was never used (e.g., Denmark and West Germany) did not have affected cases, fortifying the view that DES was the responsible agent. In the United States, the greatest number of cases are in California, Massachusetts, New York, and Pennsylvania.

Some 267 companies were said to market the drug, but most women received the drug from about a dozen pharmaceutical companies licensed by the FDA in various forms: tablets, capsules, suppositories, creams, jellies, and liquids. Eli Lilly was the major manufacturer and supplier (Fenichell and Charfoos, 1981).

Litigation resulting from the uterine lesions is probably the largest product liability case ever brought against U.S. industry, with more than 1000 lawsuits pending at last count. The first was filed in 1974, in New York state (Fenichell and Charfoos, 1981). An unusual circumstance relating to DES cancer induction was publicized in the press in 1990: Third-generation injury claims. In two separate cases, granddaughters won the right in court to sue the manufacturer for their grand-mothers' use of the drug some 40 years or so earlier and whose mothers were also DES victims. Both suits were dismissed. The profile of developmental toxicity of DES is discussed below according to class of developmental toxicity affected.

Malformation

Females

There are many sources of information for descriptions of the developmental defects and tumors associated with prenatal exposure to DES. Much of that which follows is taken from the summary prepared earlier by Schardein (2000).

According to Robboy et al. (1975) and Gunning (1976), the tumors developing in the exposed women occurred as small, reddish, polypoid 3 mm nodules to large friable masses up to more than 10 cm in diameter, filling the vagina. The usual site is on the anterior vaginal wall or fornix in the upper one third of the vagina, with 70% occurring there; 20% are in the posterior vagina or fornix; and 10% occur in the lateral vagina or fornix. There is usually extensive ectopy of the cervix. Histologically, the tumors have solid, tubular, cystic, and papillary patterns composed predominantly of clear, hobnail-shaped and flattened cells, and they may be either poorly or well differentiated. Adenosis may be a precursor to vaginal papillary clear cell adenocarcinoma, accompanying the tumor 30 to 90% of the time. It can be expected when the vaginal mucosa is red or granular, does not stain with iodine, or is colposcopically abnormal. Both adenosis and clear cell adenocarcinoma are associated with gestational bleeding of the vagina (Sharp and Cole, 1990). In addition, gross cervicovaginal abnormalities occur in about 20% of exposed patients. These include transverse cervical or vaginal ridges, and cervical erosions, hoods, or cockscombs. Squamous cell dysplasia and carcinoma *in situ* were also reported (Lanier et al., 1973), and leiomyoma formation is a recent unconfirmed finding (Baird and Newbold, 2005). Hysterosalpingograms of 40 women exposed to DES demonstrated changes that differed significantly from those of nonexposed women (Kaufman et al., 1977). These included "T-shaped appearance," constricting bands, uterine hypoplasia, polypoid defects, synechiae, and unicornate uteri. There are other reproductive and developmental sequelae related to the uterine lesions. Some women have had an increased incidence of premature deliveries associated with increased perinatal mortality (Cousins et al., 1980; Herbst et al., 1980). The incidence of spontaneous abortion and preterm delivery has consistently been greater in exposed women, and 1 of every 30 pregnancies reported has been ectopically located. There is suggestive evidence that these women have a higher incidence of pregnancy loss than those without uterine changes (Sandberg et al., 1981; Veridiano et al., 1984; Horne and Kundsin, 1985; Ankum et al., 1996). Additionally, there are adverse effects on the menstrual cycle (Schechter et al., 1991; Hornsby et al., 1994), on fertility (Barnes et al., 1980; Horne and Kundsin, 1985; Berger and Alper, 1986; Kaufman et al., 1986), and on delivery and labor (Thorp et al., 1990; deHaas et al., 1991; Heffner et al., 1993; Lang et al., 1996).

In all of the patients who had vaginal and cervical carcinoma, maternal ingestion of the hormone occurred before the 18th week (Herbst et al., 1974). In fact, 80% of mothers of patients with carcinoma began DES treatment before the 12th week of pregnancy (Herbst, 1981a). Thus, early first trimester exposure appears to be mandatory for subsequent toxicity. Doses in affected cases have ranged from 1 to 300 mg daily (orally). The duration of treatment ranged from 12 days in the first trimester to the whole gestational period. Although no clear-cut relation has been established, it is possible that the extent of the accompanying adenosis is also related to the time administration of estrogen began (Gunning, 1976).

Males

Because laboratory studies in animals demonstrated genital changes in males as well as females (see below) and because the DES-related changes are mullerian in origin, the analogous reaction, testicular cancer, has been sought in males. With one exception, there have been no reported instances of genital cancer in male offspring, but there have been a host of reports of genital abnormalities, with these reports beginning to appear in 1975, later than reports of incidences in females. In some of the early studies, epididymal cysts, hypoplastic penis, hypotrophic testes, cryptorchidism, and capsular induration of the testes were the most common genital lesions found in males, in a frequency of about 25% or less of DES-exposed subjects (Bibbo et al., 1975, 1977; Gill et al., 1976, 1977, 1979; Mills and Bongiovanni, 1978; Leary et al., 1984; Niculescu, 1985). Problems in passing urine and abnormalities of the penile urethra were also reported (Henderson et al., 1976; Cosgrove et al., 1977; Klip, 2002; Palmer et al., 2005). A testicular tumor, a seminoma, was reported in 1983 (Conley et al.). Fortunately, no further reports have been forthcoming.

Thus, while the malignancy and adverse developmental reactions related to the lesions apparently have not materialized as expected at the onset, there is a wide range of reproductive and developmental genital problems associated with DES intake during pregnancy in both females and males. Nongenital malformations have not been reported in a significant number of publications and are not considered pertinent. It would be negligent, however, not to indicate to the reader that there are critics who have or continue to dispute the relation between DES exposure and vaginal cancer (Lanier et al., 1973; Kinlen et al., 1974; Leary et al., 1984; McFarlane et al., 1986; Bornstein et al., 1988; Edelman, 1989; Meara and Fairweather, 1989; Clark, 1998; Kalter, 2004).

Growth Retardation

There are apparently no significant growth deficits associated with DES exposure.

Death

As mentioned above, there are increased incidences of pregnancy loss, spontaneous abortion, and perinatal mortality associated with DES exposure during pregnancy, as noted in many of the publications listed in Table 1.

Functional Deficit

A variety of functional alterations have been noted in reports of DES-exposed pregnancies, both in infants and in young adult females. These include depression, eating disorders, deficits in cognitive function, altered psychosexual behavior, and general neurological dysfunction (Burke et al., 1980; Vessey et al., 1983; Konicki, 1985; Meyer-Bahlburg and Ehrhardt, 1986; Fried-Cassorla et al., 1987; Katz et al., 1987; Saunders, 1988; Benderly, 1988; Ehrhardt et al., 1989; Lish et al., 1991; Gustavson et al., 1991; Newbold, 1993; Scheirs and Vingerhoets, 1995). Certain postnatal behavioral effects were observed in male subjects as well (Reinisch and Sanders, 1992). It was pointed out that these effects may be related, at least in part, to the perception of being "DES-exposed," rather than to prenatal exposure, *per se* (Briggs et al., 2005).

Mechanisms of Teratogenic and Carcinogenic Action

Mechanistically, investigations implicate the role of gene control and modification by estrogen not only because of their properties, but also because of their pharmacokinetics and metabolism (Miller et al., 1982; Henry and Miller, 1986). As an example, the effects of DES on the developing female mouse reproductive tract and the resulting downregulation of Wnt7a (which causes abnormal smooth muscle proliferation) demonstrates a consistent reaction in knockout mice lacking the gene which have malformed female reproductive tracts (Miller and Sassoon, 1998; Miller et al., 1998). A mechanism for the vaginal lesions has also been theorized (Robboy, 1983; Jefferies et al., 1984), in which the chemical may act to sensitize the proliferating stroma of the lower mullerian duct so that it is incapable of fostering upgrowth of urogenital sinus epithelium to spread over and replace the epithelium covering the vagina and cervical portico by 18 weeks, when this event should occur. The drug may also preferentially affect the stroma of the developing cervix.

DES may be one of a very few agents that can modify not only estrogen receptor activity but also expression of uterine lactoferrin through signal transduction mechanisms (NRC, 2000).

Developing Animal Models

As stated above, developmental studies conducted in animals prior to the 1970s failed to demonstrate developmental toxicity, teratogenicity, or frank carcinogenicity. Whether or not the doses utilized were insufficient or the evaluation of potential affected organ systems was incomplete were factors is not known with certainty. However, beginning in the 1980s, when reports of adverse effects in human subjects became known, a large number of studies were conducted in animals to prove or disprove the existence of models. These studies were recently summarized (Odum et al., 2002).

Animal models have now been described in full for the human lesions in two species following prenatal subcutaneous doses of 100 µg/kg. Vaginal adenosis and adenocarcinoma were reported in CD-1 strain mice followed up for 18 months (Newbold and McLachlan, 1982). Genital lesions, including adenosis, vaginal ridges, and cervical metaplasia, were also described in CD-1 strain mice by others (Walker, 1980, 1983). Dose-related vaginal adenocarcinoma and squamous cell carcinoma at an incidence at least 40 to 90 times higher than observed in humans were reported in Wistar strain rats given 0.1 or 0.5 mg/kg subcutaneously on 3 days late in gestation (Miller et al., 1982). Although the basic processes of uterovaginal development in rodents and humans are similar in some respects, substantial differences exist. For instance, development is entirely prenatal in humans, whereas the process is completed postnatally in rodents. Thus, the validity of animal models in DES-induced lesions has been questioned. Furthermore, squamous cell carcinomas of the vagina and cervix have been the predominant tumors in DES-exposed rodents (Bern et al.,

---

TABLE 3
Frequency of Adverse Findings in Diethylstilbestrol-Exposed Humans and Mice

Finding	Species	
	Human	Mouse
Daily dose (mg/kg)	0.02–5	1–2
Total dose (mg/kg)	2.2–357	1–2
Start of treatment (days postconception)	64–83	15 $^1/_2$ –17 $^1/_2$
Vaginal adenosis	35%	40%
Transverse ridges	22%	20%
Cervical metaplasia	84%	80%
Pregnancy failures	31–33%	32%
Vaginal adenocarcinoma	0.14–1.4/1000	Rare

Source: After Walker, B. E., *J. Natl. Cancer Inst.*, 73, 133–140, 1984. With permission.

1976; McLachlan et al., 1980; Forsberg and Kalland, 1981), whereas it is the clear cell adenocarcinoma that has been linked to DES exposure in humans. For these reasons, an *in vivo* model utilizing athymic BALB/c nude mice was developed; the authors conclude that this model provides a valid approach for examining the dynamics and cytodifferentiation in developing genital tracts under experimentally regulated conditions of DES exposure (Robboy et al. 1982). Confirmatory results have not come forth.

In addition to the genital lesions successfully modeled as described above, studies conducted later in Syrian hamsters (Gilloteaux et al., 1982), ferrets (Baggs and Miller, 1983), and rhesus monkeys (Hendrickx et al., 1988) all confirm urogenital malformations in these three species at relatively low doses. In a publication by Walker (1984), a comparison between the frequency of effects in the mouse and human was presented (Table 3). A close similarity exists.

One group of experts places the magnitude of teratogenic risk by DES for nongenital anomalies as unlikely, but for genital tract anomalies in females as small to moderate, for genital tract anomalies in males as minimal, and for clear cell carcinoma of the cervix in females as minimal to small (Friedman and Polifka, 2000).

A large number of useful reviews and perspectives, both popularized and personalized versions as well as scientific ones, are available to the reader who desires further detailed information on prenatal DES of induced genital malformations and neoplasms (Folkman, 1971; Herbst et al., 1975; Gunning, 1976; Ulfelder, 1976, 1980; Poskanzer and Herbst, 1977; Seaman and Seaman, 1977; Bichler, 1981; Herbst, 1981a, 1981b; Herbst and Bern, 1981; Orenberg, 1981; Weitzner et al., 1981; Fenichell and Charfoos, 1981; Kinch, 1982; Stillman, 1982; Meyers, 1983; Apfel and Fisher, 1984; Glaze, 1984; Coppleson, 1984; Lynch and Reich, 1985; Rock and Schloff, 1985; Barber, 1986; Herbst, 1987; Saunders and Saunders, 1990; Potter, 1991; Palmlund et al., 1993; Mittendorf, 1995; Giusti et al., 1995; Palmlund, 1996; Herbst, 2000; Swan, 2000; Newbold, 2004; Blunt, 2004). Two useful sources of information on DES in the public domain are the DES Action group (www.DES Action.org) and the Centers for Disease Control (www.cdc.gov/DES).

CHEMISTRY

Diethylstilbestrol is an average-sized human developmental toxicant. It is of low polarity and high hydrophobicity. It can participate in hydrogen bonding to a certain extent both as an acceptor and donor. The calculated physicochemical and topological properties of diethylstilbestrol are listed below.

PHYSICOCHEMICAL PROPERTIES

Parameter	Value
Molecular weight	268.355 g/mol
Molecular volume	261.17 A^3
Density	0.990 g/cm^3
Surface area	316.00 A^2
LogP	5.126
HLB	0.354
Solubility parameter	25.502 J$^{(0.5)}$/cm$^{(1.5)}$
Dispersion	21.260 J$^{(0.5)}$/cm$^{(1.5)}$
Polarity	3.325 J$^{(0.5)}$/cm$^{(1.5)}$
Hydrogen bonding	13.686 J$^{(0.5)}$/cm$^{(1.5)}$
H bond acceptor	0.68
H bond donor	0.59
Percent hydrophilic surface	7.89
MR	80.705
Water solubility	−2.092 log (mol/M^3)
Hydrophilic surface area	24.92 A^2
Polar surface area	40.46 A^2
HOMO	−7.756 eV
LUMO	−1.088 eV
Dipole	1.801 debye

TOPOLOGICAL PROPERTIES (UNITLESS)

Parameter	Value
x0	14.535
x1	9.651
x2	8.184
xp3	6.825
xp4	5.399
xp5	4.362
xp6	2.514
xp7	1.649
xp8	0.985
xp9	0.425
xp10	0.191
xv0	11.927
xv1	6.961
xv2	4.758
xvp3	3.503
xvp4	2.509
xvp5	1.641
xvp6	0.744
xvp7	0.426
xvp8	0.210
xvp9	0.071
xvp10	0.024
k0	17.592
k1	16.372
k2	7.852
k3	4.250
ka1	14.508
ka2	6.507
ka3	3.366

REFERENCES

Ankum, W. M. et al. (1996). Risk factors for ectopic pregnancy: A meta-analysis. *Fertil. Steril.* 65: 1093–1099.

Apfel, R. J. and Fisher, S. M. (1984). *To Do No Harm. DES and the Dilemmas of Modern Medicine*, Yale University Press, New Haven.

Baggs, R. B. and Miller, R. K. (1983). Induction of urogenital malformation by diethylstilbestrol in the ferret. *Teratology* 27: 28A.

Baird, D. D. and Newbold, R. (2005). Prenatal diethylstilbestrol (DES) exposure is associated with uterine leiomyoma development. *Reprod. Toxicol.* 20: 81–84.

Barber, H. R. (1986). An update on DES in the field of reproduction. *Int. J. Fertil.* 31: 130–144.

Barnes, A. B. et al. (1980). Fertility and outcome of pregnancy in women exposed *in utero* to diethylstilbestrol. *N. Engl. J. Med.* 302: 609–613.

Barter, J. F. et al. (1986). Diethylstilbestrol in pregnancy: An update. *South Med. J.* 79: 1531–1534.

Benderly, B. L. (1988). Depressed DES daughters. *Psychol. Today* 22: 16.

Berger, M. J. and Alper, M. M. (1986). Intractable primary infertility in women exposed to diethylstilbestrol *in utero*. *J. Reprod. Med.* 31: 231–235.

Bern, H. A., Jones, L. A., and Mills, K. T. (1976). Use of the neonatal mouse in studying long-term effects of early exposure to hormones and other agents. *J. Toxicol. Environ. Health* 1: 103–116.

Bibbo, M. (1979). Transplacental effects of diethylstilbestrol. In *Perinatal Pathology*, E. Grundmann, Ed., Springer-Verlag, New York, pp. 191–211.

Bibbo, M. et a l. (1975). Follow up study of male and female offspring of DES-treated mothers. A preliminary report. *J. Reprod. Med.* 15: 29–32.

Bibbo, M. et al. (1977). Followup study of male and female offspring of DES-exposed mothers. *Obstet. Gynecol.* 49: 1–8.

Bichler, J. (1981). *DES Daughter. The Joyce Bichler Story*, Avon Books, New York.

Blunt, E. (2004). Diethylstilbestrol exposure: It's still an issue. *Holist Nurs. Pract.* 18: 187–191.

Bornstein, L. et al. (1988). Development of cervical and vaginal squamous cell neoplasia as a late consequence of *in utero* exposure to diethylstilbestrol. *Obstet. Gynecol. Surv.* 43: 15–21.

Briggs, G. G., Freeman, R. K., and Yaffe, S. J. (2005). *Drugs in Pregnancy and Lactation. A Reference Guide to Fetal and Neonatal Risk*, Seventh ed., Lippincott Williams & Wilkins, Philadelphia.

Burke, L. et al. (1980). Observations on the psychological impact of diethylstilbestrol exposure and suggestions on management. *J. Reprod. Med.* 24: 99–102.

Chanen, W. and Pagano, R. (1984). Diethylstilbestrol (DES) exposure *in utero*. *Med. J. Aust.* 141: 491–493.

Clark, J. H. (1998). Female reproduction toxicology of estrogen. In *Reproductive and Developmental Toxicology*, K. S. Korach, Ed., Marcel Dekker, New York, pp. 259–275.

Conley, G. R. et al. (1983). Seminoma and epididymal cysts in a young man with known diethylstilbestrol exposure *in utero*. *JAMA* 249: 1325–1326.

Coppleson, M. (1984). The DES story. *Med. J. Aust.* 141: 487–489.

Cosgrove, M. D., Benton, B., and Henderson, B. E. (1977). Male genitourinary abnormalities and maternal diethylstilbestrol. *J. Urol.* 117: 220–222.

Cousins, L. et al. (1980). Reproductive outcome of women exposed to diethylstilbestrol *in utero*. *Obstet. Gynecol.* 56: 70–76.

Cunha, G. R. et al. (1987). Teratogenic effects of clomiphene, tamoxifen, and diethylstilbestrol on the developing human female genital tract. *Hum. Pathol.* 18: 1132–1143.

deHaas, I. et al. (1991). Spontaneous preterm birth: A case-control study. *Am. J. Obstet. Gynecol.* 165: 1290–1296.

Dieckmann, W. J. et al. (1953). Does the administration of diethylstilbestrol during pregnancy have therapeutic value? *Am. J. Obstet. Gynecol.* 66: 1062–1081.

Edelman, D. A. (1989). Diethylstilbestrol exposure and the risk of clear cell cervical and vaginal adenocarcinoma. *Int. J. Fertil.* 34: 251–255.

Ehrhardt, A. A. et al. (1989). The development of gender-related behavior in females following prenatal exposure to diethylstilbestrol (DES). *Horm. Behav.* 23: 526–541.

Einer-Jensen, N. (1968). Antifertility properties of two diphenylethenes. *Acta Pharmacol. (Copenh.)* 26 (Suppl. 1): 1–97.

Fenichell, S. and Charfoos, L. S. (1981). *Daughters at Risk. A Personal D. E. S. History*, Doubleday & Co., Garden City, NY.

Folkman, J. (1971). Transplacental carcinogenesis by stilbestrol. *N. Engl. J. Med.* 285: 404–405.

Forsberg, J. G. and Kalland, T. (1981). Neonatal estrogen treatment and epithelial abnormalities in the cervicovaginal epithelium of adult mice. *Cancer Res.* 41: 721–734.

Fried-Cassorla, M. et al. (1987). Depression and diethylstilbestrol exposure in women. *J. Reprod. Med.* 32: 847–850.

Friedman, J. M. and Polifka, J. E. (2000). *Teratogenic Effects of Drugs. A Resource for Clinicians (TERIS)*, Second ed., Johns Hopkins University Press, Baltimore, MD.

Giannina, T. et al. (1971). Comparative effects of some steroidal and nonsteroidal antifertility agents in rats and hamsters. *Contraception* 3: 347–359.

Gill, W. B., Schumacher, G. F. B., and Bibbo, M. (1976). Structural and functional abnormalities in the sex organs of male offspring of mothers treated with diethylstilbestrol. *J. Reprod. Med.* 16: 147–153.

Gill, W. B., Schumacher, G. F. B., and Bibbo, M. (1977). Pathological semen and anatomic abnormalities of the genital tract in human male subjects exposed to diethylstilbestrol *in utero*. *J. Urol.* 117: 477–480.

Gill, W. B. et al. (1979). Association of diethylstilbestrol exposure *in utero* with cryptorchidism, testicular hypoplasia and semen abnormalities. *J. Urol.* 122: 36–39.

Gilloteaux, J., Paul, R. J., and Steggles, A. W. (1982). Upper genital tract abnormalities in the Syrian hamster as a result of *in utero* exposure to diethylstilbestrol. I. Uterine cystadenomatous papilloma and hypoplasia. *Virchows Arch. [A]* 398: 163–183.

Gilson, M. D., Dibona, D. D., and Knab, D. R. (1973). Clear cell adenocarcinomas in young females. *Obstet. Gynecol.* 41: 494–500.

Giusti, R. M., Iwamoto, K., and Hatch, E. E. (1995). Diethylstilbestrol revisited: A review of the long-term health effects. *Ann. Intern. Med.* 122: 778–788.

Glaze, G. M. (1984). Diethylstilbestrol exposure *in utero*: Review of literature. *J. Am. Osteopath. Assoc.* 83: 435–438.

Greenwald, P. et al. (1971). Vaginal cancer after maternal treatment with synthetic estrogens. *N. Engl. J. Med.* 285: 390–392.

Gunning, J. E. (1976). Supplement: The DES story. *Obstet. Gynecol. Surv.* 31: 827–833.

Gustavson, C. R. et al. (1991). Increased risk of profound weight loss among women exposed to diethylstilbestrol *in utero*. *Behav. Neural Biol.* 55: 307–312.

Hammes, B. and Laitman, C. J. (2003). Diethylstilbestrol (DES) update: Recommendations for the identification of management of DES-exposed individuals. *J. Midwifery Womens Health* 48: 19–29.

Heffner, L. J. et al. (1993). Clinical and environmental predictors of preterm labor. *Obstet. Gynecol.* 81: 750–757.

Henderson, B. E. et al. (1976). Urogenital tract abnormalities in sons of women treated with diethylstilbestrol. *Pediatrics* 58: 505–507.

Hendrickx, A. G., Prahalada, S., and Binkerd, P. E. (1988). Long-term evaluation of the diethylstilbestrol (DES) syndrome in adult female rhesus monkeys (*Macaca mulatta*). *Reprod. Toxicol.* 1: 253–261.

Henry, E. C. and Miller, R. K. (1986). Comparison of the disposition of diethylstilbestrol and estradiol in the fetal rat. Correlation with teratogenic potency. *Biochem. Pharmacol.* 35: 1993–2002.

Herbst, A. L. (1981a). Diethylstilbestrol and other sex hormones during pregnancy. *Obstet. Gynecol.* 58: 35S–40S.

Herbst, A. L. (1981b). Clear cell adenocarcinoma and the current status of DES-exposed females. *Cancer* 48: 484–488.

Herbst, A. L. (1987). The effects in the human of diethylstilbestrol (DES) use during pregnancy. *Princess Takamatsu Symp.* 18: 67–75.

Herbst, A. L. (2000). Behavior of estrogen-associated female genital tract cancer and its relation to neoplasia following intrauterine exposure to diethylstilbestrol (DES). *Gynecol. Oncol.* 76: 147–156.

Herbst, A. L. and Anderson, D. (1990). Clear cell adenocarcinoma of the vagina and cervix secondary to intrauterine exposure to diethylstilbestrol. *Semin. Surg. Oncol.* 6: 343–346.

Herbst, A. L. and Bern, H. A., Eds. (1981). *Developmental Effects of Diethylstilbestrol (DES) in Pregnancy*, Thieme-Stratton, New York.

Herbst, A. L. and Scully, R. E. (1970). Adenocarcinoma of the vagina in adolescence. A report of seven cases including six clear cell carcinomas (so-called mesonephromas). *Cancer* 25: 745–757.

Herbst, A. L., Ulfelder, H., and Poskanzer, D. C. (1971). Adenocarcinoma of the vagina. Association of maternal stilbestrol therapy with tumor appearance in young women. *N. Engl. J. Med.* 284: 878–881.

Herbst, A. L. et al. (1972). Clear cell adenocarcinoma of the genital tract in young females. Registry report. *N. Engl. J. Med.* 287: 1259–1264.

Herbst, A. L. et al. (1974). Clear cell adenocarcinoma of the vagina and cervix in girls. Analysis of 170 registry cases. *Am. J. Obstet. Gynecol.* 119: 713–724.

Herbst, A. L. et al. (1975). Prenatal exposure to stilbestrol. A prospective comparison of exposed female offspring with unexposed controls. *N. Engl. J. Med.* 292: 334–339.

Herbst, A. L. et al. (1977). Age-incidence and risk of diethylstilbestrol-related clear cell adenocarcinoma of the vagina and cervix. *Am. J. Obstet. Gynecol.* 128: 43–50.

Herbst, A. L. et al. (1978). Complications of prenatal therapy with diethylstilbestrol. *Pediatrics* 62: 1151–1159.

Herbst, A. L. et al. (1980). A comparison of pregnancy experience in diethylstilbestrol exposed and diethyl-stilbestrol unexposed daughters. *J. Reprod. Med.* 24: 62–69.

Horne, H. W. and Kundsin, R. B. (1985). Results of infertility studies on 1001 DES-exposed and non DES-exposed consecutive patients. *Int. J. Fertil.* 30: 46–49.

Hornsby, P. P. et al. (1994). Effects on the menstrual cycle of *in utero* exposure to diethylstilbestrol. *Am. J. Obstet. Gynecol.* 170: 709–715.

Horwitz, R. I. et al. (1988). Clear cell adenocarcinoma of the vagina and cervix: Incidence, undetected disease, and diethylstilbestrol. *J. Clin. Epidemiol.* 41: 593–597.

Jefferies, J. A. et al. (1984). Structural anomalies of the cervix and vagina of women enrolled in the Diethylstilbestrol Adenosis (DESAD) Project. *Am. J. Obstet. Gynecol.* 148: 59–65.

Kalter, H. (2004). Teratology in the twentieth century. Congenital malformations in humans and how their environmental causes were established. *Neurotoxicol. Teratol.* 25: 131–282.

Katz, D. L. et al. (1987). Psychosis and prenatal exposure to diethylstilbestrol. *J. Nerv. Ment. Dis.* 175: 306–308.

Kaufman, R. H. (1982). Structural changes of the genital tract associated with *in utero* exposure to diethyl-stilbestrol. *Obstet. Gynecol. Annu.* 11: 187–202.

Kaufman, R. H. et al. (1977). Upper genital tract changes with exposure *in utero* to diethylstilbestrol. *Am. J. Obstet. Gynecol.* 128: 51–59.

Kaufman, R. H. et al. (1982). Development of clear cell adenocarcinoma in DES-exposed offspring under observation. *Obstet. Gynecol.* 59 (Suppl.): 685–725.

Kaufman, R. H. et al. (1984). Upper genital tract abnormalities and pregnancy outcome in DES-exposed progeny. *Am. J. Obstet. Gynecol.* 148: 973–984.

Kaufman, R. H. et al. (1986). Upper genital tract changes and infertility in diethylstilbestrol-exposed women. *Am. J. Obstet. Gynecol.* 154: 1312–1318.

Kaufman, R. H. et al. (2000). Continued follow-up of pregnancy outcomes in diethylstilbestrol-exposed offspring. *Obstet. Gynecol.* 96: 483–489.

Kinch, R. A. (1982). Diethylstilbestrol in pregnancy: An update. *Can. Med. Assoc. J.* 127: 812–813.

Kinlen, L. J. et al. (1974). A survey of the use of oestrogens during pregnancy in the United Kingdom and of the genito-urinary cancer mortality and incidence rates in young people in England and Wales. *J. Obstet. Gynaecol. Br. Commonw.* 81: 849–855.

Klip, J. (2002). Hypospadias in sons of women exposed to diethylstilbestrol *in utero*. *Lancet* 2: 1102–1107.

Konicki, A. M. (1985). Physical and psychological effects of DES on exposed offspring. *Cancer Nurs.* 8: 233–237.

Lang, J. M., Lieberman, E., and Cohen, A. (1996). A comparison of risk factors for preterm labor and term small-for-gestational-age birth. *Epidemiology* 7: 369–376.

Lanier, A. P. et al. (1973). Cancer and stilbestrol. A followup of 1,719 persons exposed to estrogens *in utero* and born 1943–1959. *Mayo Clin. Proc.* 48: 793–799.

Leary F. J. et al. (1984). Males exposed *in utero* to diethylstilbestrol. *JAMA* 252: 2984–2989.

Linn, S. et al. (1988). Adverse outcomes of pregnancy in women exposed to diethylstilbestrol *in utero*. *J. Reprod. Med.* 33: 3–7.

Lish, J. D. et al. (1991). Gender-related behavior development in females exposed to diethylstilbestrol (DES) *in utero*: An attempted replication. *J. Am. Acad. Child Adolesc. Psychiatry* 30: 29–37.

Lynch, H. T. and Reich, J. W. (1985). Diethylstilbestrol, genetics, teratogenesis and tumour spectrum in humans. *Med. Hypotheses* 16: 315–332.

Mangan, C. E. et al. (1982). Pregnancy outcome in 98 women exposed to diethylstilbestrol *in utero*, their mothers, and unexposed siblings. *Obstet. Gynecol.* 59: 315–319.

Marselos, M. and Tomatis, L. (1992). Diethylstilbestrol: I. Pharmacology, toxicology and carcinogenicity in humans. *Eur. J. Cancer* 28A: 1182–1189.

McDonnell, J. M., Emers, J. M., and Jordan, J. A. (1984). The congenital cervicovaginal transformation zone in young women exposed to diethylstilbestrol *in utero*. *Br. J. Obstet. Gynaecol.* 91: 574–579.

McFarlane, M. J., Feinstein, A. R., and Horwitz, R. I. (1986). Diethylstilbestrol and clear cell vaginal carcinoma: Reappraisal of the epidemiologic evidence. *Am. J. Med.* 81: 855–863.

McLachlan, J. A., Newbold, R. R., and Bullock, B. C. (1980). Long-term effects of the female genital tract after prenatal exposure to diethylstilbestrol. *Cancer Res.* 40: 3988–3999.

Meara, J. and Fairweather, D. V. (1989). A randomized double-blind controlled trial of the value of diethyl-stilbestrol therapy in pregnancy: 35-year follow-up of mothers and their offspring. *Br. J. Obstet. Gynaecol.* 96: 620–622.

Melnick, S. et al. (1987). Rates and risks of diethylstilbestrol-related clear-cell adenocarcinoma of the vagina and cervix. *N. Engl. J. Med.* 316: 514–516.

Meyer-Bahlburg, H. F. L. and Ehrhardt, A. A. (1986). Prenatal diethylstilbestrol exposure: Behavior consequences in humans. *Monogr. Neural Sci.* 12: 90–95.

Meyers, R. (1983). *D. E. S. The Bitter Pill*, Seaview/Putnam, New York.

Miller, C. and Sassoon, D. A. (1998). Wnt-7a maintains appropriate uterine patterning during the development of the mouse female reproductive tract. *Development* 125: 3201–3211.

Miller, C., Degenhardt, K., and Sassoon, D. A. (1998). Fetal exposure to DES results in de-regulation of Wnt-7a during uterine morphogenesis. *Nat. Genet.* 20: 228–230.

Miller, R. K. et al. (1982). Transplacental carcinogenicity of diethylstilbestrol (DES): A Wistar rat model. *Teratology* 25: 62A.

Mills, J. L. and Bongiovanni, A. M. (1978). Effects of prenatal estrogen exposure on male genitalia. *Pediatrics* 62: 1160–1165.

Mittendorf, R. (1995). Teratogen update: Carcinogenesis and teratogenesis associated with exposure to diethylstilbestrol (DES) *in utero*. *Teratology* 51: 435–445.

NCI (National Cancer Institute). (1983). DES summary. Prenatal diethylstilbestrol (DES) exposure. *Clin. Pediatr.* 22: 139–143.

Newbold, R. R. (1993). Gender-related behavior in women exposed prenatally to diethylstilbestrol. *Environ. Health Perspect.* 101: 208–213.

Newbold, R. R. (2004). Lessons learned from perinatal exposure to diethylstilbestrol. *Toxicol. Appl. Pharmacol.* 199: 142–150.

Newbold, R. R. and McLachlan, J. A. (1982). Vaginal adenosis and adenocarcinoma in mice exposed prenatally or neonatally to diethylstilbestrol. *Cancer Res.* 42: 2003–2011.

Newbold, R. R. and McLachlan, J. A. (1996). Transplacental hormonal carcinogenesis: Diethylstilbestrol as an example. *Prog. Clin. Biol. Res.* 394: 131–147.

Ng, A. B. P. et al. (1977). Natural history of vaginal adenosis in women exposed to diethylstilbestrol *in utero*. *J. Reprod. Med.* 18: 1–13.

Niculescu, A. M. (1985). Effects of *in utero* exposure to DES on male progeny. *J. Obstet. Gynecol. Neonatal Nurs.* 14: 468–470.

Noller, K. and Fish, C. (1974). Diethylstilbestrol usage. *Med. Clin. North Am.* 58: 793–810.

Noller, K. L. et al. (1983). Maturation of vaginal and cervical epithelium in women exposed *in utero* to diethylstilbestrol (DESAD Project). *Am. J. Obstet. Gynecol.* 146: 279–285.

Nordquist, S. A. B., Medhat, I. A., and Ng, A. B. (1979). Teratogenic effects of intrauterine exposure to DES in female offspring. *Compr. Ther.* 5: 69–74.

NRC (National Research Council). (2000). *Scientific Frontiers in Developmental Toxicology and Risk Assessment*, National Academy Press, Washington, D.C., p. 73.

O'Brien, P. C. et al. (1979). Vaginal epithelial changes in young women enrolled in the National Cooperative Diethylstilbestrol Adenosis (DESAD) Project. *Obstet. Gynecol.* 53: 300–308.

Odum, J. et al. (2002). Comparison of the developmental and reproductive toxicity of diethylstilbestrol administered to rats *in utero*, lactationally, preweaning, or postweaning. *Toxicol. Sci.* 68: 147–163.

Orenberg, C. L. (1981). *DES: The Complete Story*, St. Martin's Press, New York.

Ostergard, D. R. (1981). DES-related vaginal lesions. *Clin. Obstet. Gynecol.* 24: 379–394.

Palmer, J. R. et al. (2005). Hypospadias in sons of women exposed to diethylstilbestrol *in utero*. *Epidemiology* 16: 583–586.

Palmlund, I. (1996). Exposure to a xenoestrogen before birth: The diethylstilbestrol experience. *J. Psychosom. Obstet. Gynaecol.* 17: 71–84.

Palmlund, I. et al. (1993). Effects of diethylstilbestrol (DES) medication during pregnancy: Report from a symposium at the 10th international congress. *J. Psychosom. Obstet. Gynaecol.* 14: 71–89.

Poskanzer, D. C. and Herbst, A. L. (1977). Epidemiology of vaginal adenosis and adenocarcinoma associated with exposure to stilbestrol *in utero. Cancer* 39: 1892–1895.

Potter, E. L. (1991). A historical view: Diethylstilbestrol use during pregnancy. A 30-year historical perspective. *Pediatr. Pathol.* 11: 781–789.

Prins, R. P. et al. (1976). Vaginal embryogenesis, estrogens, and adenosis. *Obstet. Gynecol.* 48: 246–250.

Registry (for Research on Hormonal Transplacental Carcinogenesis). (1998). Clear cell adenocarcinoma collaborative studies. University of Chicago DES (diethylstilbestrol) program.

Reinisch, J. M. and Sanders, S. A. (1992). Effects of prenatal exposure to diethylstilbestrol (DES) on hemispheric laterality and spatial ability in human males. *Horm. Behav.* 26: 62–75.

Robboy, S. J. (1983). A hypothetic mechanism of diethylstilbestrol (DES)-induced anomalies in exposed progeny. *Hum. Pathol.* 14: 831–833.

Robboy, S. J., Scully, R. E., and Herbst, A. L. (1975). Pathology of vaginal and cervical abnormalities associated with prenatal exposure to diethylstilbestrol (DES). *J. Reprod. Med.* 15: 13–18.

Robboy, S. J. et al. (1979). Pathologic findings in young women enrolled in the National Cooperative Diethylstilbestrol Adenosis (DESAD) Project. *Obstet. Gynecol.* 53: 309–317.

Robboy, S. J. et al. (1981). Dysplasia and cytologic findings in 4,589 young women enrolled in diethylstilbestrol-adenosis (DESAD) project. *Am. J. Obstet. Gynecol.* 140: 579–586.

Robboy, S. J., Taguchi, O., and Cunha, G. R. (1982). Normal development of the human female reproductive tract and alterations resulting from experimental exposure to diethylstilbestrol. *Hum. Pathol.* 13: 190–198.

Robboy, S. J. et al. (1984). Increased incidence of cervical and vaginal dysplasia in 3,980 diethylstilbestrol-exposed young women. *JAMA* 252: 2979–2983.

Rock, J. A. and Schloff, W. D. (1985). The obstetric consequences of uterovaginal anomalies. *Fertil. Steril.* 43: 681–692.

Sandberg, E. C. et al. (1981). Pregnancy outcome in women exposed to diethylstilbestrol *in utero. Am. J. Obstet. Gynecol.* 140: 194–205.

Saunders, E. J. (1988). Physical and psychological problems associated with exposure to diethylstilbestrol (DES). *Hosp. Comm. Psychiatry* 39: 73–77.

Saunders, E. J. and Saunders, J. A. (1990). Drug therapy in pregnancy: The lessons of diethylstilbestrol, thalidomide, and bendectin. *Health Care Women Int.* 11: 423–432.

Schardein, J. l. (2000). *Chemically Induced Birth Defects*, Third ed., Marcel Dekker, New York, pp. 292–297.

Schechter, D. et al. (1991). Menstrual-cycle functioning in women with a history of prenatal diethylstilbestrol exposure. *J. Psychosom. Obstet. Gynecol.* 12: 51–66.

Scheirs, J. G. M. and Vingerhoets, A. J. J. M. (1995). Handedness and other laterality indices in women prenatally exposed to DES. *J. Clin. Exp. Neuropsychol.* 17: 725–730.

Seaman, B. (2003). *The Greatest Experiment Ever Performed on Women. Exploding the Estrogen Myth*, Hyperion, New York, p. 214.

Seaman, B. and Seaman, G. (1977). The amazing story of DES. In *Women and the Crisis in Sex Hormones*, B. Seaman and G. Seaman, Eds., Rawson, New York, pp. 1–59.

Shapiro, S. and Slone, D. (1979). The effects of exogenous female hormones on the fetus. *Epidemiol. Rev.* 1: 110–123.

Sharp, G. B. and Cole, P. (1990). Vaginal bleeding and diethylstilbestrol exposure during pregnancy: Relationship to genital tract clear cell adenocarcinoma and vaginal adenosis in daughters. *Am. J. Obstet. Gynecol.* 162: 994–1001.

Smith, O. W. (1948). Diethylstilbestrol in the prevention and treatment of complications of pregnancy. *Am. J. Obstet. Gynecol.* 56: 821–834.

Sonek, M., Bibbo, M., and Wied, G. L. (1976). Colposcopic findings in offspring of DES-treated mothers as related to onset of therapy. *J. Reprod. Med.* 16: 65–71.

Stillman, R. J. (1982). *In utero* exposure to diethylstilbestrol — adverse effects on the reproductive tract and reproductive performance in male and female offspring. *Am. J. Obstet. Gynecol.* 142: 905–921.

Swan, S. H. (2000). Intrauterine exposure to diethylstilbestrol: Long-term effects in humans. *APMIS* 108: 793–804.

Thorp, J. M. et al. (1990). Antepartum and intrapartum events in women exposed *in utero* to diethylstilbestrol. *Obstet. Gynecol.* 76: 828–832.

Ulfelder, H. (1976). DES-transplacental teratogen and possibly also carcinogen. *Teratology* 13: 101–104.

Ulfelder, H. (1980). The stilbestrol disorders in historical perspective. *Cancer* 45: 3008–3011.

Veridiano, N. P., Delke, I., and Tanser, M. L. (1984). Pregnancy wastage in DES-exposed female progeny. In *Spontaneous Abortion*, E. S. E. Hafez, Ed., MTP Press, Lancaster, UK, pp. 183–188.

Vessey, M. P. (1989). Epidemiological studies of the effects of diethylstilbestrol. *IARC Sci. Publ.* 96: 335–348.

Vessey, M. P. et al. (1983). A randomized double-blind controlled trial of the value of stilboestrol therapy in pregnancy: Long-term follow-up of mothers and their offspring. *Br. J. Obstet. Gynaecol.* 90: 1007–1017.

Walker, B. E. (1980). Reproductive tract anomalies in mice after prenatal exposure to DES. *Teratology* 21: 313–321.

Walker, B. E. (1983). Complications of pregnancy in mice exposed prenatally to DES. *Teratology* 27: 73–80.

Walker, B. E. (1984). Tumors of female offspring of mice exposed prenatally to diethylstilbestrol. *J. Natl. Cancer Inst.* 73: 133–140.

Weitzner, K., Candidate, J. D., and Hirsh, H. L. (1981). Diethylstilbestrol medicolegal chronology. *Med. Trial Tech. Q.* 28: 145–170.

Wise, L. A. et al. (2005). Risk of benign gynecologic tumors in relation to prenatal diethylstilbestrol exposure. *Obstet. Gynecol.* 105: 167–173.

49 Pseudoephedrine

Chemical name: [S-(R*,R*)]-α-[1-(methylamino)ethyl]-benzenemethanol

Alternate name: *d*-Isoephedrine

CAS #: 90-82-4

SMILES: C(O)(c1ccccc1)C(NC)C

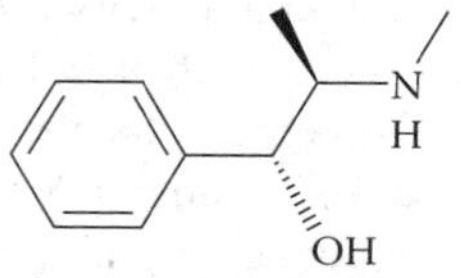

INTRODUCTION

Pseudoephedrine is an adrenergic agonist widely used as a nasal and bronchial decongestant, often in combination with other drugs. It is present in plants of the genus *Ephedra*, known in traditional medicine as Ma Huang. The drug directly stimulates α-adrenergic receptors of respiratory mucosa, causing vasoconstriction, and β-adrenergic receptors, causing bronchial relaxation, increased heart rate, and contractility (Lacy et al., 2004). Pseudoephedrine is available as an over-the-counter (OTC) generic drug and has a large number of trade names, of which Sudafed® and some Dimetapp® formulations are among the most commonly used. The drug is often combined with other agents in the treatment of decongestion, cough, and antihistaminic indications. It has a pregnancy category of C, one in which in the case of pseudoephedrine indicates that animal and human studies have not been done; the drug should be given only if the potential benefit justifies the potential risk to the fetus.

DEVELOPMENTAL TOXICOLOGY

ANIMALS

Laboratory animal studies are meager. The only species for which studies have been published is the rat. In that study, pregnant rats given 240 mg/kg/day orally (the pertinent human route of administration) over 10 days in gestation caused maternal and developmental toxicity manifested in the latter by decreased fetal body weight and reduced ossification (Freeman et al., 1989). This dosage far exceeds the recommended human dose of the drug of 30 to 240 mg/day orally.

HUMANS

In the human, studies have provided conflicting results. In a 1992 case-control study of 76 children with gastroschisis, there was an association (relative risk [RR] = 3.2, 95% confidence interval [CI],

1.3 to 7.7) of first trimester pseudoephedrine use with the defect (Werler et al., 1992). But the group found no association among 416 infants to other congenital anomalies of possible vascular etiology (presumably a cause of gastroschisis). A follow-up study on the risk of maternal medication to gastroschisis also found an association (odds ratio [OR] = 1.8, 95% CI, 1.0 to 3.2) with increased risk for the defect and also to small intestine atresia (OR = 2.0, 95% CI, 1.0 to 4.0; see also Werler et al., 2002). In another recent study by the same group of investigators publishing on potential risk factors for hemifacial microsomia (HFM), another defect of vascular disruption etiology, pseudoephedrine was significantly associated with HFM (OR = 1.7, 95% CI, 1.2 to 3.4; see also Werler et al., 2004). The association of the three vascular disruption defects with small to moderate risks with pseudoephedrine was briefly reviewed recently (Werler, 2005).

Several earlier studies (Heinonen et al., 1977; Jick et al., 1981; Aselton et al., 1985) and several more recent studies (Torfs et al., 1996; Shaw et al., 1998; cited, Briggs et al., 2005), in contrast, found no association between first trimester exposure to pseudoephedrine and malformations of any type among more than 2600 pregnancies. Though data from the above reports provide conflicting information relating to the developmental toxicity potential of pseudoephedrine, one cannot exclude the possibility that this drug has adverse effects when administered during pregnancy with respect to a causal association with malformations of several vascular disruptive types. One group of experts set the magnitude of teratogenic risk to range from none to minimal at this time (Friedman and Polifka, 2000). Nonetheless, the suspicion exists until proven otherwise. If it is a developmental toxicant, it must certainly be a weak one.

CHEMISTRY

Pseudoephedrine is a lower-sized molecule in comparison to the other compounds. It is of low polarity and of average hydrophobicity. Pseudoephedrine can engage in hydrogen bonding both as a hydrogen bond donor and acceptor. The calculated physicochemical and topological properties are shown below.

PHYSICOCHEMICAL PROPERTIES

Parameter	Value
Molecular weight	165.235 g/mol
Molecular volume	168.35 A^3
Density	0.944 g/cm^3
Surface area	215.12 A^2
LogP	1.037
HLB	9.067
Solubility parameter	22.592 J$^{(0.5)}$/cm$^{(1.5)}$
Dispersion	18.555J$^{(0.5)}$/cm$^{(1.5)}$
Polarity	3.606 J$^{(0.5)}$/cm$^{(1.5)}$
Hydrogen bonding	12.373 J$^{(0.5)}$/cm$^{(1.5)}$
H bond acceptor	0.70
H bond donor	0.46
Percent hydrophilic surface	45.77
MR	49.906
Water solubility	1.721 log (mol/M^3)
Hydrophilic surface area	98.46 A^2
Polar surface area	32.26 A^2
HOMO	−9.429 eV
LUMO	0.224 eV
Dipole	2.470 debye

TOPOLOGICAL PROPERTIES (UNITLESS)

Parameter	Value
x0	8.975
x1	5.753
x2	4.643
xp3	3.935
xp4	2.496
xp5	1.672
xp6	0.749
xp7	0.446
xp8	0.184
xp9	0.048
xp10	0.000
xv0	7.489
xv1	4.157
xv2	2.933
xvp3	2.070
xvp4	1.097
xvp5	0.633
xvp6	0.254
xvp7	0.126
xvp8	0.051
xvp9	0.011
xvp10	0.000
k0	11.746
k1	10.083
k2	4.889
k3	2.778
ka1	9.230
ka2	4.237
ka3	2.315

REFERENCES

Aselton, P. et al. (1985). First-trimester drug use and congenital disorders. *Obstet. Gynecol.* 65: 451–455.

Briggs, G. G., Freeman, R. K., and Yaffe, S. J. (2005). *Drugs in Pregnancy and Lactation. A Reference Guide to Fetal and Neonatal Risk*, Seventh ed., Lippincott Williams & Wilkins, Philadelphia.

Freeman, S. J., Irvine, L., and Walker, T. F. (1989). Teratological evaluation in rat of SK&F 93944 administered alone or combined with pseudoephedrine HCl (P). *Toxicologist* 9: 31.

Friedman, J. M. and Polifka, J. E. (2000). *Teratogenic Effects of Drugs. A Resource for Clinicians (TERIS)*, Second ed., Johns Hopkins University Press, Baltimore, MD.

Heinonen, O. P., Slone, D., and Shapiro, S. (1977). *Birth Defects and Drugs in Pregnancy*, Publishing Sciences Group, Littleton, MA.

Jick, H. et al. (1981). First-trimester drug use and congenital disorders. *JAMA* 246: 343–346.

Lacy, C. F. et al. (2004). *Drug Information Handbook (Pocket), 2004–2005*, Lexi-Comp., Inc., Hudson, OH.

Shaw, G. M. et al. (1998). Maternal illness, including fever, and medication use as risk factors for neural tube defects. *Teratology* 57: 1–7.

Torfs, C. P. et al. (1996). Maternal medication and environmental exposures as risk factors for gastroschisis. *Teratology* 54: 84–92.

Werler, M. M. (2005). Teratogen update: Pseudoephedrine. *Birth Defects Res. (A)* 73: 328.

Werler, M. M., Mitchell, A. A., and Shapiro, S. (1992). First trimester maternal medication use in relation to gastroschisis. *Teratology* 45: 361–367.

Werler, M. M., Sheehan, J. E., and Mitchell, A. A. (2002). Maternal medication use and risks of gastroschisis and small intestine atresia. *Am. J. Epidemiol.* 155: 26–31.

Werler, M. M. et al. (2004). Vasoactive exposures, vascular events, and hemifacial microsomia. *Birth Defects Res. (A)* 70: 389–395.

50 Ethanol

Alternate names: Alcohol, ethyl alcohol, ethyl hydrate, ethyl hydroxide

CAS #: 64-17-5

SMILES: C(C)O

HO⌒

INTRODUCTION

Ethanol or "alcohol" as it is better known, is used medicinally as a disinfectant, solvent, and preservative. It is present in many over-the-counter preparations in concentrations ranging from <1% up to 67% in some formulations (personal data, Schardein, 2006). Alcohol is also used universally as a beverage, in which it acts as a central nervous system depressant, with intoxicating properties. The availability of alcohol is ubiquitous throughout virtually all populations. Annual consumption in the United States is estimated at 10.2 l (2.69 gal) per person (Pietrantoni and Knuppel, 1991). While all alcoholic beverages contain container labels stating

> **GOVERNMENT WARNING.** According to the Surgeon General, women should not drink alcoholic beverages during pregnancy because of the risk of birth defects.

it was recently estimated that 14.6% of pregnant women consume alcohol, and 2.1% consume it frequently, according to a large sample of women studied (Ebrahim et al., 1998). This is despite the fact that alcohol is considered, through consumption in pregnancy, the most frequent cause of mental deficiency in the Western world (Clarren and Smith, 1978). It should be apparent from the following discussion that alcohol ranks as the most significant developmental toxicant known.

DEVELOPMENTAL TOXICOLOGY

ANIMALS

Laboratory animal studies clearly demonstrate potent developmental toxicity, including teratogenicity, by all known routes of administration. Because the oral route is that used in human consumption, we will focus on oral administration in the animal studies. Dosage equivalency and procedural differences in administering alcohol in the experimental situation confound interpretation, but an attempt will be made to equivocate these factors. A representative number of experimental studies in nine species of animals administered alcohol orally during gestation and the resultant developmental toxicity profile are shown in Table 1. Notably, the rabbit has been refractory to alcohol developmental toxicity (Blakley, 1988). Most all species cited have reacted to alcohol, usually showing retarded fetal growth, increased mortality, and malformations. Four species — the rat, guinea pig, pig, and pig-tailed monkey — evidenced functional deficits as well, and at least six of the species cited could be designated "models" for the human condition — the mouse, rat,

TABLE 1
Representative Experimental Results in Laboratory Animals Administered Alcohol by the Oral Route

Species	Developmental Toxicity Reported[a]				Dose (days in gestation)	Ref.
	G	D	M	F		
Mouse	✔	✔	✔		15–20%, prior to and throughout	Chernoff, 1977
Rat	✔	✔	✔		30%, prior to and throughout	Tze and Lee, 1975
				✔	4–6 g/kg, throughout	Abel and Dintcheff, 1978
Guinea pig	✔	✔	✔	✔	3 ml/kg 3–4×/wk., throughout	Papara-Nicholson and Telford, 1957
Sinclair mini-pig	✔	✔	✔	✔	20%, prior to and throughout	Dexter and Tumbleson, 1980
Beagle dog	✔	✔	✔		3 mg/kg–4.29 g/kg, 17 days	Ellis and Pick, 1980
Sheep	✔	✔	✔		10%, prior to and throughout	Potter et al., 1981
Ferret		✔	✔		1.5 g/kg, 21 days	McLain and Roe, 1984
Primates:						
Pig-tailed monkey	✔	✔			4.1 g/kg, 145 days	Altschuler and Shippenberg, 1981
				✔	4.1 g/kg/wk., 110 days	Clarren and Bowden, 1982
Cyno monkey	✔	✔			5 g/kg, 130 days	Scott and Fradkin, 1984

[a] G = growth retardation, D = death, M = malformation, F = functional deficit.

pig, dog, sheep, and pig-tailed monkey. In fact, craniofacial development as seen in human fetal alcohol syndrome (FAS) cases (see below) has been studied in animal models (Webster and Ritchie, 1991); Sulik and associates (1981) have drawn a convincing parallel in similarities between mouse and human craniofacial features resulting from FAS.

HUMANS

As suggested by the above introduction, the use of alcohol recreationally during pregnancy can have severe consequences in humans. The outcome of pregnancies of mothers who use alcohol is a distinct syndrome of developmental toxicity, the sum termed the "fetal alcohol syndrome" (FAS). Pregnant women are at high risk. Women of childbearing potential probably constitute about 10% of the 6 million "alcoholics" and 10 million "problem drinkers" in the United States; thus, approximately 65% of embryos or fetuses are exposed to alcohol prenatally according to one report (Pietrantoni and Knuppel, 1991). Further, the pattern of alcohol use among adolescents is of great concern. Translated, the facts illustrate that between 3000 and 6000 babies in the United States will be born mentally retarded each year from maternal (adolescent) alcohol consumption (Mac-Donald, 1987). Because the outcomes of alcohol use are all encompassing, they will be discussed and summarized together below. Much of the discussion that follows is taken in part from a summary published earlier by Schardein (2000).

Pre-FAS History

Observations of toxicity related to alcohol consumption during pregnancy are not new. According to historians, malformations were generally recognized in the offspring of alcoholic women over 250 yr ago (Warner and Rosett, 1975). At the turn of the past century, reports were circulated to indicate that there was increased stillbirth and that "small and sickly" children were born of female drunkards or alcoholics (Sullivan, 1900; Ladrague, 1901). More recently, French studies by Lemarche (1967) and Lemoine et al. (1968) described abnormalities in a number of children born to alcoholic parents. Ulleland (1972) in the United States reported that the offspring of some alcoholic mothers had abnormal appearance.

FAS Discovery

In June of 1973, Jones and Smith, from a total of 11 cases, described a distinct dysmorphic condition associated with maternal, gestational alcoholism (Jones and Smith, 1973; Jones et al., 1973). They termed the condition, which comprised craniofacial, limb, and cardiovascular defects, the "fetal alcohol syndrome" or "FAS." By 1976, these investigators had characterized the syndrome in 41 patients (Jones et al., 1974; Jones and Smith, 1975; Jones et al., 1976; Hanson et al., 1976). Three more patients were added by Palmer and associates (1974). By mid-1978, the number of cases thoroughly studied was about 300 (Mulvihill et al., 1976; Majewski, 1977; Dehaene et al., 1977; Ouellette et al., 1977; Streissguth et al., 1978; Hanson et al., 1978; Rosett et al., 1978; Clarren and Smith, 1978), and by 1980, the number of described cases exceeded 600 (Chua et al., 1979; Pierog et al., 1979; Olegard et al., 1979; Rosett, 1980; Mena et al., 1980; Smith, 1980).

Major or otherwise important reviews and case reports of FAS have appeared regularly since the above reports, confirming the 25 or so associated malformations and the scope of the malformations in offspring of alcoholic women. The published literature on the subject is immense; the National Library of Medicine cites over 5000 published references on alcohol. Thus, only representative reports are provided in Table 2.

TABLE 2
Representative Reports of Fetal Alcohol Syndrome (FAS) in Humans

Kaminski et al., 1981	Ginsburg et al., 1991
Little and Streissguth, 1981	Day and Richardson, 1991
Sokol, 1981	Pietrantoni and Knuppel, 1991
Clarren, 1981	Brien and Smith, 1991
Iosub et al., 1981	Werler et al., 1991
Abel, 1981	Sokol and Abel, 1992
Krous, 1981	Spohr et al., 1994
Neugut, 1981	Niccols, 1994
Ashley, 1981	Abel, 1995
Pratt, 1981	Gladstone et al., 1996
Lamanna, 1982	Koren et al., 1996
Nitowsky, 1982	Kaufman, 1997
Streissguth, 1983	Sampson et al., 1997
Lipson et al., 1983	Larkby and Day, 1997
Rosett et al., 1983a	Thomas and Riley, 1998
Grisso et al., 1984	Jones and Chambers, 1998
Streissguth et al., 1985	Makarechian et al., 1998
Graham, 1986	Polygenis et al., 1998
Jones, 1986	Nulman et al., 1998
Ernhardt et al., 1987	Abel, 1998
Abel and Sokol, 1987	Stoler, 1999
Streissguth and LaDue, 1987	Abel, 1999
Leonard, 1988	Chaudhury, 2000
Abel, 1989	Warren and Foudin, 2001
Ernhardt et al., 1989	Polygenis et al., 2001
Burd and Martsoff, 1989	Mattson et al., 2001
Hill et al., 1989	Chinboga, 2003
Schenker et al., 1990	Olney, 2004
Michaelis, 1990	Huggins et al., 2004
Walpole et al., 1990	Sulik, 2005
Russell, 1991	

TABLE 3
Original Features Associated with Fetal Alcohol Syndrome (FAS) in Humans

Central Nervous System (CNS)
Mild to moderate mental retardation, developmental delay
Fine motor dysfunction, irritability (infancy), hyperactivity (school age)
Holoprosencephaly

Craniofacial
Microcephaly
Short palpebral fissures, ptosis, epicanthal folds, microphthalmia, strabismus, myopia
Maxillary hypoplasia
Protruberant posteriorly rotated ears
Short upturned nose
Cleft lip and palate, small hypoplastic teeth with abnormal enamel, long flat philtrum, thin upper vermillon border
Retrognathia in infancy, micrognathia or relative prognathia in adolescence

Cardiac
Atrial septal defect
Ventricular septal defect

Skeletal
Limited joint mobility, especially fingers and toes
Hypoplasia of the fingers and toe nails, especially fifth
Radioulnar synostosis
Pectus excavatum and carinatum, bifid sternum
Klippel-Feil anomaly
Scoliosis
Abnormal palmar creases
Limb reduction defects

Renal
Labial hypoplasia
Hypospadias
Hydronephrosis
Small rotated kidneys

Other
Hemangiomas
Hursutism in infancy
Diaphragmatic, umbilical, or inguinal hernias; diastasis recti

Source: From Robin, N. H. and Zackai, E. H., *Teratology*, 50, 160–164, 1994. With permission.

Malformation

According to Clarren and Smith (1978), the abnormalities most typically associated with alcohol teratogenesis can be grouped into four categories: (1) central nervous system dysfunctions, (2) growth deficiencies, (3) a characteristic cluster of facial abnormalities, and (4) variable minor and major malformations. A tabulation of the features originally described as being associated with FAS is provided in Table 3.

There is a rather typical facial appearance in individuals with FAS. In fact, it is the craniofacial similarities, rather than the mental and growth deficiencies, among children with the syndrome that unite them into a discernible entity. The facies are characterized by short palpebral fissues, hypoplastic upper lip with thinned vermillon, and diminished or absent philtrum. The face in general has a drawn appearance produced primarily by the hypoplastic lip and philtrum and further accentuated by the frequent additional feature of mid-facial hypoplasia. Eye growth is usually deficient, on rare occasions resulting in frank microphthalmia. Strabismus and myopia are frequent problems, and ptosis and blepharophimosis are reported frequently. The nose is frequently short, with a low bridge and associated epicanthal folds and anteverted nostrils. Cleft lip/palate have occasionally been observed. The ears are involved in some patients; posterior rotation of the helix is common, and alteration in conchal shape occurs occasionally. The mandible is generally small at birth; in some, growth of the jaw is greater than the mid-facial structures with aging, and apparent prognathism may therefore be observed in adolescence.

Although there is an increased frequency of malformations in children with FAS, no one particular type of major malformation occurs in most cases. Associated features not mentioned in the foregoing and that occurred in up to 25% incidence in the large series of cases analyzed by Clarren and Smith (1978) included great vessel anomalies and tetralogy of Fallot and numerous skeletal defects, including polydactyly and bifid xiphoid. Observed more frequently (26 to 50% of cases) were prominent lateral palatine ridges in the mouth and cardiac murmurs. The major skeletal defects (Van Rensburg, 1981) and cardiac anomalies (Sandor et al., 1981) in FAS were described in detail. A number of cases of neural tube defects were reported independently of FAS from maternal alcohol ingestion (Uhlig, 1957; Friedman, 1982; Ronen and Andrews, 1991). A high percentage of placentas from infants with FAS had villitis, raising the suspicion that some of the manifestations of the syndrome might be due to intrauterine virus infection (Baldwin et al., 1982). Placenta abruption has also been associated with high intake levels of alcohol (Marbury et al., 1983). Several cases of neuroblastoma associated with FAS were reported (Seeler et al., 1979; Kinney et al., 1980), as was Hodgkin's disease (Bostrom and Nesbit, 1983), and hepatic cancer or abnormalities (Khan et al., 1979; Habbick et al., 1979). Other associated malformations that may be related to FAS include clubfoot (Halmesmaki et al., 1985), gastroschisis (Sarda and Bard, 1984), malignant tumors (Cohen, 1981; Kiess et al., 1984), skin lesions (Linneberg et al., 2004), and optic nerve hypoplasia (Pinazo-Duran et al., 1997).

Stoler (1999), in a recent reassessment of FAS, indicated that (1) not all alcohol-abusing women will have children with FAS, (2) not every type of birth defect associated with exposure to alcohol is a causal connection, (3) not all cardiac defects are attributable to alcohol exposure, and (4) the facial features associated with FAS are not specific. In 1996, the Institute of Medicine (IOM) compiled a list of new criteria for FAS identification. This is provided in outline form in Table 4. There are two other separate categories that may co-occur in addition to FAS. These are termed "alcohol-related birth defects" (ARBD), discussed in Table 5, and "alcohol-related neurodevelopmental disorders" (ARND), discussed in a later section. Earlier, children who have only some of the characteristics of FAS (i.e., not enough for a full diagnosis) were often said to have "fetal alcohol effects" (FAEs) (Streissguth, 1997).

ARBD

The congenital anomalies, including malformations and dysplasias, are shown in Table 5. In this classification, these are clinical conditions in which there is a history of maternal alcohol exposure and where clinical or animal research has linked maternal alcohol ingestion to an observed outcome. There are two categories that may co-occur (see ARND section). If both diagnoses are present, then both diagnoses should be rendered.

Growth Retardation

Most infants with FAS are growth deficient at birth for both length and weight. In general, they remain more than two standard deviations below the mean, with weight being more severely limited, according to reports presented in Table 2. Decreased adipose tissue is a nearly constant feature. Growth hormone, cortisol, and gonadotropin levels in the children are within normal ranges; diminished prenatal cell proliferation may be responsible for the growth deficiency. Further, the children are unresponsive to growth-promoting hormonal therapy (Castells et al., 1981). Growth retardation of infants of "heavy" drinkers was twice that of abstinent or moderate-drinking mothers

TABLE 4
Current Diagnostic Criteria for Fetal Alcohol Syndrome (FAS)

I. FAS *with* confirmed maternal alcohol exposure[a]
   A. Confirmed maternal alcohol exposure[a]
   B. Evidence of characteristic pattern of facial anomalies, including features such as short palpebral fissures and abnormalities in the premaxillary zone (e.g., flat upper lip, flattened philtrum, and flat midface)
   C. Evidence of growth retardation as in at least one of the following:
     1. Low birth weight for gestational age
     2. Decelerating weight over time not due to nutrition
     3. Disproportional low weight to height
   D. Evidence of central nervous system (CNS) neurodevelopmental abnormalities as in at least one of the following:
     1. Decreased cranial size at birth
     2. Structural brain abnormalities (e.g., microcephaly, partial or complete agenesis of the corpus callosum, cerebellar hypoplasia)
     3. Neurological hard or soft signs (as age appropriate) such as impaired fine motor skills, neurosensory hearing loss, poor tandem gait, poor eye–hand coordination
II. FAS *without* confirmed maternal alcohol exposure
   B, C, and D above
III. Partial FAS *with* confirmed maternal alcohol exposure[a]
   A. Confirmed maternal alcohol exposure[a]
   B. Evidence of some components of pattern of characteristic facial anomalies
   Either C, D, or E
   C. Evidence of growth retardation as in I, C (above)
   D. Evidence of CNS neurodevelopmental abnormalities as in I, D (above)
   E. Evidence of a complex pattern of behavior or cognitive abnormalities that are inconsistent with developmental level and cannot be explained by familial background or environment alone, such as learning difficulties, deficits in school performance, poor impulse control, problems in social perception, deficits in higher-level receptive and expressive language, poor capacity for abstraction or metacognition, specific deficits in mathematical skills, or problems in memory, attention, or judgment

[a] A pattern of excessive intake characterized by substantial, regular intake or heavy episodic drinking. Evidence of this pattern may include frequent episodes of intoxication, development of tolerance or withdrawal, social problems related to drinking, legal problems related to drinking, engaging in physical hazardous behavior while drinking, or alcohol-related medical problems such as hepatic disease.

Source: Modified after IOM (Institute of Medicine), Fetal alcohol syndrome: Diagnosis, epidemiology, prevention, and treatment. Division of Biobehavioral Sciences and Mental Disorders, Committee to study fetal alcohol syndrome, K. R. Stratton, C. J. Howe, and F. C. Battaglia, Eds., National Academy Press, Washington, D.C., 1996. With permission.

(Rosett et al., 1978). A prospective analysis of 31,604 pregnancies demonstrated that newborns below the tenth percentile of weight for gestational age increased as maternal alcohol increased (Mills et al., 1984). Mean birth weight was reduced 14 g in newborns whose mothers drank <1 drink per day and 165 g in those whose mothers drank three to five drinks per day. Several other reports also relate other aspects of the growth deficiency associated with FAS (Little, 1977; Wright et al., 1983; Leichter et al., 1989; Rostland et al., 1990). Researchers conducting studies in rats suggest that prenatal alcohol exposure can also interfere with the development of normal sucking behavior, which might influence normal growth (Chen et al., 1982).

Death

Spontaneous abortion, stillbirth, and premature birth appear to be associated with FAS (Makarechian et al., 1998), and many reports listed in Table 2 reference these endpoints. In one study, perinatal mortality was found in 17% of a small number of cases of FAS examined (Jones et al., 1974). In another larger study of 616 drinking women, spontaneous abortion occurred in ~25% of drinkers

TABLE 5
Malformation Criteria Associated with Alcohol-Related Birth Defects (ARBD)[a,b]

Cardiac

Atrial and ventricular septal defects
Aberrant great vessels
Tetralogy of Fallot

Skeletal

Hypoplastic nails
Shortened fifth digits
Radioulnar synostosis
Flexion contractures
Campto- or clinodactyly
Pectus excavatum and carinatum
Klippel-Feil syndrome
Hemivertebrae
Scoliosis

Renal

Aplastic, dysplastic, or hypoplastic kidneys
Horseshoe kidneys
Hydronephrosis

Ocular

Strabismus
Retinal vascular anomalies
Refractive problems secondary to small globes

Auditory

Conductive or neurosensory hearing loss

Other

Virtually every malformation has been described in some patient with fetal alcohol syndrome (FAS). The etiologic specificity of most of these anomalies to alcohol teratogenesis remains uncertain.

[a] See Footnote a in Table 4.
[b] As further research is completed and as, or if, lower quantities or variable patterns of alcohol use are associated with ARBD or alcohol-related neurodevelopmental disorders (ARND), these patterns of alcohol use should be incorporated into the diagnostic criteria.

Source: Modified after IOM (Institute of Medicine), Fetal alcohol syndrome: Diagnosis, epidemiology, prevention, and treatment. Division of Biobehavioral Sciences and Mental Disorders, Committee to study fetal alcohol syndrome, K. R. Stratton, C. J. Howe, and F. C. Battaglia, Eds., National Academy Press, Washington, D.C., 1996. With permission.

compared to 14% of mothers who drank less than two times per week (Kline et al., 1980). Increased spontaneous abortion and stillbirths were also described in association with FAS in other reports (Harlap and Shiono, 1980; Marbury et al., 1983; Ginsburg et al., 1991; Abel, 1997). Moderate drinking may actually increase the risk of miscarriage by two- to fourfold.

Functional Deficit

Mental retardation is one of the most common and serious problems of the teratogenic syndrome. Although not all affected persons are retarded, rarely have any displayed average or better mental ability. Mental deficiency was considered the most common problem in FAS, occurring in 44% of the 23 cases examined (Jones et al., 1974). In fact, the frequency of functional abnormality among those born to 42 "heavy" drinkers was twice that of those born to abstinent or moderate-drinking mothers in one study (Rosett et al., 1978). Studies by Streissguth et al. (1978) of 20 cases indicated that 60% of the patients had IQs more than two standard deviations below the mean. The severity of the dysmorphic features was related to the degree of mental deficiency. Later studies by these investigators confirmed similar effects on IQs (Streissguth et al., 1989) and on learning disabilities (Streissguth, 1986), but the effect on intelligence was not replicated by others (Greene et al., 1991). Identifiable deficits in sequential memory processes and specific academic skills were reported among fetuses exposed to alcohol throughout pregnancy (Coles et al., 1991). Effects on sustained attention performance could not be demonstrated in alcohol-exposed preschoolers in one study (Boyd et al., 1991), but deficits in the ability to sustain attention were identified as showing attentional and behavioral problems in another study (Brown et al., 1991). Evaluation of neonatal behavior assessment scales of alcohol-exposed neonates revealed few effects of alcohol on neonatal behavior in still another study (Richardson et al., 1989). Another investigator found no deficits in indices of child development at 18 or 42 months of age (Olsen, 1994). However, one documented effect is poor motor performance in 4-year-old children whose mothers had prenatal exposure to alcohol (Barr et al., 1990). Another effect is language difficulty, which was recognized as an associated FAS finding among 63 cases in one study (Iosub et al., 1981), as were language and speech problems in another (Sparks, 1984), although language development was said not to be a sensitive indicator of alcohol exposure by others (Greene et al., 1990).

Other functional abnormalities recorded in the FAS literature include hearing disorders (Church and Gerkin, 1988), effects on social behavior (Roebuck et al., 1999; Kelly et al., 2000), deficits in a variety of test performances (Becker et al., 1990), and functional alterations in a variety of neurobehavioral assessments (Wisniewski and Lupin, 1979; Olson et al., 1998; Steinhausen and Spohr, 1998; Mattson and Riley, 2000; Willford et al., 2004; Nulman et al., 2004; Bailey et al., 2004; Lee et al., 2004; Burden et al., 2005). Limited neuropathological studies performed to date indicate cerebellar dysplasia and heterotopic cell clusters as consistent anomalies. A quantification of neuroanatomical structure was described recently that may be useful in diagnosing fetal alcohol damage more effectively (Bookstein et al., 2001). Microcephaly has also been an important feature of the syndrome, and hydrocephaly may be an occasional variant; neurological abnormalities may be present from birth, as discussed earlier. Such findings convinced Abel (1981) that alcohol is a behavioral teratogen in humans. In fact, there is convincing evidence that the most devastating effects of alcohol are on the developing brain (West and Goodlett, 1990; Konovalov et al., 1997; Nulman et al., 1998; Guerri, 1998; Eckardt et al., 1998). With substantiating studies in laboratory animals, evidence indicates that *in utero* alcohol exposure produces a developmental delay in the maturation of response inhibition mechanisms in the brain rather than an irreversible effect, but other studies show that some of these effects may be long lasting (Abel and Berman, 1994). Newborns are usually irritable and temulous, have a poor suck, and apparently possess hyperacusis; these abnormalities usually persist for several weeks or months. Hyperactivity is a frequent component of FAS in young children. Withdrawal symptoms in the infants, similar to those in adults, have been reported, and may be a reason for the irritability and other clinical signs (Pierog et al., 1977). Older children have also frequently shown mild alterations in cerebellar function and hypotonicity. Neonatal seizures have been observed occasionally, but rarely beyond the neonatal period. Many aspects of neurological factors in alcohol-exposed infants were reviewed (Becker et al., 1990).

As mentioned above, there have been recent diagnostic criteria promulgated by the IOM (1996) in the study of FAS that have resulted in further delineation of findings related to the syndrome, relating in part as to whether maternal consumption has been confirmed. The one that refers to CNS and functional findings was termed "alcohol-related neurodevelopmental disorders" (ARND). The categories of the disorders identified in this classification are shown in Table 6.

ARND

There are two categories (see A and B in Table 6) of neurodevelopmental disorders identified under this classification, as follows. These are clinical conditions in which there is a history of maternal

TABLE 6

Current Diagnostic Disorders of Neurodevelopment Associated with Alcohol-Related Neurodevelopmental Disorders (ARND)[a,b]

A. Evidence of central nervous system (CNS) neurodevelopmental abnormalities as in any one of the following:
1. Decreased cranial size at birth
2. Structural brain abnormalities
   a. Microcephaly
   b. Partial or complete absence of corpus callosum
   c. Cerebellar hypoplasia
2. Neurological hard or soft signs (as age appropriate)
   a. Impaired fine motor skills
   b. Neurosensory hearing loss
   c. Poor tandem gait
   d. Poor eye–hand coordination

and/or

B. Evidence of a complex pattern of behavior or cognitive abnormalities that are inconsistent with developmental level and cannot be explained by familial background or environment alone, such as the following:
1. Learning difficulties
2. Deficits in school performance
3. Poor impulse control
4. Problems in social perception
5. Deficits in higher-level receptive and expressive language
6. Poor capacity for abstraction or metacognition
7. Specific deficits in mathematical skills
8. Problems in memory, attention, or judgment

[a] See Footnote a in Table 4.
[b] See Footnote b in Table 5.

Source: Modified after IOM (Institute of Medicine), Fetal alcohol syndrome: Diagnosis, epidemiology, prevention, and treatment. Division of Biobehavioral Sciences and Mental Disorders, Committee to study fetal alcohol syndrome, K. R. Stratton, C. J. Howe, and F. C. Battaglia, Eds., National Academy Press, Washington, D.C., 1996. With permission.

alcohol exposure and where, through clinical or animal research, maternal alcohol ingestion was linked to an observed outcome. There are two categories that may co-occur (see ARBD above). If both diagnoses are present, then both diagnoses should be rendered.

Characterization of the Syndrome

All affected children recognized to date have been the offspring of chronic alcoholic women who drank heavily during pregnancy (Jones and Smith, 1975). The susceptibility factors in subjects developing FAS were identified: maternal age >30 years, from a low socioeconomic group, and from Native American or African American ancestries, who had a previous child with FAS, were undernourished, and who had specific genetic backgrounds (Jones, 2003). Paternal origin of FAS was described (Bartoshesky et al., 1979; Abel, 1992) but not seriously considered etiologically. Poor nutrition, pyridoxine deficiency, contaminants in alcohol, dehydration, or genetic predisposition were considered to play a role in the production of the syndrome by some, but this is unlikely (Green, 1974; Shepard, 1974; Fisher et al., 1982; Leichter and Lee, 1982). The major metabolite of alcohol, acetaldehyde, was considered the culprit in one study (Dunn et al., 1979). Nonetheless, it has now been established with certainty that ethanol is the etiological agent. As we have seen, animal models demonstrated many of the features of the syndrome in common with humans (see above).

The syndrome is greatly underreported or unrecognized, even in infants of known alcohol-abusing women (Little et al., 1990; Stoler and Holmes, 1999). Fetal alcohol syndrome is generally estimated to occur in the United States in 0.97 cases per 1000 live births in the general obstetric population and in 4.3% of infants of heavy drinkers (Abel, 1995). The incidence of partial expression is perhaps 3 to 5 per 1000 (Clarren and Smith, 1978). Of FAS and ARND (see above) combined, the incidence is considered to be 9.1:1000 (Sampson et al., 1997). Prevalence rates for FAS reported from case registries range from 0.03 to 2.99 per 1000 (Hymbaugh et al., 2002). The frequency of FAS varies widely geographically, being estimated at 1:100 in northern France (Dehaene et al., 1977), 1:600 in Sweden (Olegard et al., 1979), and approximately 1.9:1000 worldwide (Abel and Sokol, 1987). The incidence is about 20 times higher in the United States than in other countries (Abel, 1995). In this country, the frequency is highest in Native Americans (19.5:1000) and lowest in the White, middle socioeconomic stratum (2.6:1000; see Abel, 1989). Males may be more vulnerable to the effect than females (Qazi and Masakawa, 1976). Unfortunately, we are not certain at what gestational stage the fetus is most vulnerable to the effects of alcohol: The critical period may be close to the time of conception according to one scientist (Ernhart et al., 1987), or according to another, the first 85 days of gestation (the period of most rapid neuromigration) is the window of susceptibility for development of FAS (Koren, 1997). Alcohol is known to cross the placenta and distribute in the fetus and is eliminated slower than in the mother (Obe and Ristow, 1979).

How much can a woman drink during pregnancy without having an effect on her child? Both moderate and high levels of alcohol may result in alterations of growth and morphogenesis (Hanson et al., 1978), and there appears to be a definite risk with six drinks (of 90 ml) per day (Morrison and Maykut, 1979). Another team of investigators place the risk at 5.6% for FAS when the quantities consumed are greater than 3 oz (~90 ml) per day, there being no clear threshold (Ernhart et al., 1987). However, Rosett et al. (1983a) found no difference between rare and moderate drinkers with respect to safety. Beyond these pronouncements, there is disagreement. One statement emerging from studies thus far is that no safe drinking level has been established for pregnant women; in fact, it may never be known with certainty.

Alcohol intake is normally expressed as an average amount of absolute alcohol consumed per day. Servings of beverages are assumed to be of constant size, typically: beer, 12 oz; wine 5 oz; hard liquor, 1.25 oz; and to constant proportion of ethanol by volume, 4, 12, and 45% respectively. Thus, 1 drink of beer, wine or liquor would contain about 0.5, 0.6 and 0.6 oz (12, 14, and 14 g) absolute alcohol respectively. As a rough approximation, 1 drink is 0.5 oz absolute alcohol and 5 drinks per day for a 60 kg person is about 1 g/kg per day.

A study by Mau (1980) analyzing data from 7525 pregnancies indicated that moderate consumption of alcohol had no significant effect on later development. This is in agreement with a meta-analysis performed on seven studies examining this question recently (Polygenis et al., 1998). They found that moderate alcohol consumption (more than two drinks per week to two drinks per day) during the first trimester of pregnancy was not associated with increased risk (relative risk [RR] = 1.01, 95% confidence interval [CI], 0.94 to 1.08) of fetal malformations. On the other hand, even low sporadic doses of alcohol during pregnancy may increase the risk of congenital anomalies, and this risk increases with increasing levels of alcohol exposure (Martinez-Frias et al., 2004). The U.S. Department of Health, Education and Welfare proposed that women limit their daily alcohol intake to 28.5 ml (1 oz) of pure ethanol (two mixed drinks, two beers, or two glasses of wine; see *Medical World News*, June 27, 1997). Above that frequency, there is increased risk of fetal abnormality. The FDA took an even tougher stance. They first issued a government advisory on alcohol and pregnancy in 1981 (*Science* 214: 642 passim 645, 1981); later, they planned to propose federal legislation requiring cautionary labels on alcohol-containing products, including all alcoholic beverages, but this plan apparently fell through from lack of support (*Science* 233: 517-518, 1986).

However, it was implemented initially in California, beginning in November 1989, on alcoholic beverages and is now in effect throughout the U.S.

Scientifically, there appears to be a dose relation between alcohol consumption and FAS. Ouellette et al. (1977) showed that infants born to heavy drinkers have twice the risk of abnormality of those born of abstinent or moderate drinkers, 32% compared to 9% in abstinent and 14% in the moderate drinkers group. Or put another way, the frequency of developmental toxicity (malformations, growth retardation, death, or functional alterations) in offspring of heavy drinkers was twice that of infants born to abstinent or moderate drinking mothers (Rosett et al., 1978). One group of experts places the magnitude of teratogenic risk for heavy drinking (defined as >6 drinks per day) as moderate to high, and minimal to moderate for <2 drinks per day (Friedman and Polifka, 2000). Unrecognized in any consideration of FAS is that many women ingest alcohol in forms other than in the usual alcoholic beverage; one FAS case was reported in which the mother abused cough syrup containing 9.5% alcohol (Chasnoff et al., 1981). There are countless other medications containing alcohol in concentrations ranging up to 67% (e.g., tincture of belladonna) that are easily obtainable over the counter (*Patient Care*, February 28, 1979). FAS cases have been recorded even after drinking has ceased (Scheiner et al., 1979; Veghelyi and Osztovics, 1979), but benefits to offspring have been noted when heavy drinkers stopped before the third trimester (Rosett et al., 1983b). In addition to abnormalities, as such, other effects may be related to lesser consumption of alcohol. Bark (1979) compared pregnancy outcomes of 40 alcoholic women with 40 matched controls and found no significant differences between the two groups relative to fertility, pregnancy outcome, or state of the children. In contrast, a large cohort study of some 9236 pregnancies in France indicated a significantly higher incidence of premature placental separation, stillbirth, and low birth weight among infants of mothers who drank more than 44.4 ml (1.5 oz) of absolute alcohol per day compared with a group whose mothers drank less than this amount or none (Kaminski et al., 1976).

The pathogenesis of FAS remains undefined at present. One group of investigators, however, proposed three main, nonexclusive mechanisms that may explain the genesis of FAS (Schenker et al., 1990). These were impaired placental or fetal blood flow, deranged prostaglandin balance, and direct effects of alcohol (or acetaldehyde) on cellular processes. The cellular toxicity and molecular events involved in FAS were discussed (Michaelis, 1990); the metabolic basis was also described (Luke, 1990). A review of mechanisms was recently published (Goodlet et al., 2005). The frequency of adverse outcomes of pregnancy for chronic alcoholic women, said to be 43%, led several investigators to suggest that serious consideration be given to early termination of pregnancy in severely chronic alcoholic women (Jones and Smith, 1975). Blood alcohol analyses conducted on expectant mothers was recommended in at least one report as a means of identifying drinking mothers at risk so that they may be advised of the potential teratogenic risk to their babies (Erb and Anderson, 1978). Recently, blood markers of alcohol use — acetaldehyde, carbohydrate-deficient transferin, γ-glutamyl transpeptidase, and mean red blood cell volume — were proposed for identifying alcohol abusers and predicting infant outcome (Stoler et al., 1998; Chang et al., 1998). Under this scheme, women with two or more positive markers had infants with significantly smaller birth weights, lengths, and head circumferences than infants with negative maternal screens. These markers could lead to better efforts at detection and prevention of alcohol-induced fetal damage by identifying women at risk (Jones and Chambers, 1998). An estimate of risk for developmental toxicity associated with alcohol consumption by pregnant women has been tabulated in Table 7. The factors constituting this risk were detailed in a review by Sokol and Abel (1992).

Minimum criteria for identifying FAS to simplify, clarify, and standardize the diagnosis were initially proposed by the fetal alcohol study group of the Research Society on Alcoholism (Rosett, 1980). While the diagnosis of FAS has changed little since the mid-1970s (Sampson et al., 1997), FAS is not easily identified, often recognizable only to expert clinicians (Hymbaugh et al., 2002). The slow head growth after birth in affected infants may explain why FAS is not diagnosed in

---

TABLE 7
Estimates of Risk for Developmental Toxicity Associated
with Alcohol Consumption by Mothers during Pregnancy

Outcome	Risk Ratio	Incidence (%)
Growth retardation		
Low birth weight	2	25
Intrauterine growth retardation (IUGR)	2.5	10
Death		
Spontaneous abortion	2	30
Malformation		
Congenital anomalies	4	40
Fetal alcohol syndrome (FAS)	—	2.5
Functional deficits	?	?

Source: Modified from Ernhart, C. B. et al., *Am. J. Obstet. Gynecol.*, 156, 33–39, 1987. Data uncorrected for concomitant risk factors.

---

some cases until 9 to 12 months of age (Streissguth et al., 1985). As we have seen, however, these criteria were defined further by the IOM in 1996 (Table 4, Table 5, and Table 6). Even more recently, the Fetal Alcohol Syndrome Surveillance Network (FASSNet), a four-state network initiated in 1997, proposed a multisource methodology surveillance scheme, based on the IOM framework, for determining its case definition. It is shown in Table 8 as a further example of the diagnostic methodology currently available for FAS, and for the sake of completeness.

Successful rehabilitation programs have been described with subsequent reduction in FAS with reduced drinking (Rosett et al., 1978, 1981; Little et al., 1980; Rosett and Weiner, 1981; Little and Streissguth, 1981; Waterson, 1990; Streissguth, 1997; NIAAA, 2003). Other workers believe that counseling in these cases is useless (Pierog et al., 1979). Even the existence of FAS has been disputed by some under certain conditions (Tennes and Blackard, 1980; Miller, 1982; Marbury et al., 1983; Tolo and Little, 1993; Olsen, 1994). In sum, in dealing with alcohol use in pregnancy, FAS, or *fetal alcoholism syndrome* (Krous, 1981), *fetal alcohol abuse syndrome* (Abel, 1999), or *fetal alcohol spectrum disorder* (FASD) (Streissguth and O'Malley, 2000), as it has more recently been termed, the most conservative advice to render is that mothers should abstain from all alcohol consumption from conception through delivery and lactation. It appears that daily intake of more than 28.5 ml (1 oz) of absolute alcohol presents a risk to the fetus, and this risk rises progressively with increasing intake during pregnancy (Newman and Correy, 1980). However, the risk from light drinking (<1 oz absolute alcohol daily) has not been demonstrated and should not be overstated, because exaggeration could decrease credibility about the adverse effects of heavy drinking and may cause parents of children with abnormalities to feel guilty that small amounts of alcoholic beverages caused abnormalities that were actually due to other factors (Rosett, 1980).

A large number of reviews of alcohol consumption during pregnancy and FAS, including a personalized version (Dorris, 1989) and a popularized article (Steinmetz, 1992), were published from early on in the history up to the present (Hanson et al., 1976; Witti, 1978; Morrison and Maykut, 1979; Chernoff, 1980; Beagle, 1981; Rosett et al., 1981; Sandor et al., 1981; Sokol, 1981; Krous, 1981; Neugut, 1981; Pratt, 1982; Little et al., 1982; Streissguth, 1983, 1986; Ernhart et al., 1987; Blakley, 1988; Hoyseth and Jones, 1989; Wiedemann et al., 1989; Driscoll et al., 1990; Tatha, 1990; Pietrantoni and Knuppel, 1991; McCance-Katz, 1991; Brien and Smith, 1991; Gladstone et al., 1996; Abel, 1998; May and Gossage, 2001; Golden, 2005; Briggs et al., 2005). The Web sites of the National Organization on Fetal Alcohol Syndrome (www.nofas.org) and the Canadian Fetal

TABLE 8
Diagnostic Criteria for Assessing Fetal Alcohol Syndrome (FASSNet)

Diagnostic Category	FACE	Phenotype Positive Central Nervous System (CNS)	GROWTH
Confirmed fetal alcohol syndrome (FAS) phenotype with or without documentation[a] of *in utero* alcohol exposure	Abnormal facial features consistent with FAS as reported by a physician	At least one structural or functional anomaly	Growth delay indicated in at least one of the following:
	or	Structural — head circumference 10th centile at birth or any age	Intrauterine — weight or height corrected for gestational age 10th centile
	Two of the following: short palpebral fissures, abnormal philtrum, thin upper lip	or	or
		Functional — standardized measure of intellectual function 1 S.D. below the mean	Postnatal — weight or height 10th centile for age
		or	or
		Standardized measure of developmental delay 1 S.D. below the mean	Weight or height 10th centile
		or	
		Developmental delay or mental retardation diagnosed by psychologist or physician	
		or	
		Attention-deficit/hyperactivity disorder (ADHD) diagnosed by qualified examiner	
Probable FAS phenotype with or without documentation[a] of *in utero* alcohol exposure	Required; same as confirmed FAS phenotype above	Must meet either CNS or GROWTH criteria as outlined in the confirmed FAS phenotype above	
Suspect	All children referred into the surveillance system, including all children with ICD-9 Codes 760.71, provider referrals, children identified by abstractors who meet predetermined criteria from the specific referral source, newborn nursery logs, etc.		

[a] Determined from the availability of documentation in the records of some level of maternal alcohol use during the index pregnancy.

Source: From Hymbaugh, K. et al., *Teratology*, 66, S41–S49, 2002. With permission.

Alcohol Spectrum Disorders FASlink (www.acbr.com/fas/index.htm) contain considerable information on fetal alcohol syndrome.

CHEMISTRY

Ethanol is one of the smallest human developmental toxicants. It is hydrophilic and can participate
in hydrogen bonding. The calculated physicochemical and topological properties of ethanol are
listed below.

PHYSICOCHEMICAL PROPERTIES

Parameter	Value
Molecular weight	46.069 g/mol
Molecular volume	52.19 A^3
Density	0.822 g/cm^3
Surface area	80.00 A^2
LogP	−0.235
HLB	5.075
Solubility parameter	25.095 J$^{(0.5)}$/cm$^{(1.5)}$
Dispersion	15.032 J$^{(0.5)}$/cm$^{(1.5)}$
Polarity	8.351 J$^{(0.5)}$/cm$^{(1.5)}$
Hydrogen bonding	18.277 J$^{(0.5)}$/cm$^{(1.5)}$
H bond acceptor	0.33
H bond donor	0.23
Percent hydrophilic surface	28.41
MR	12.961
Water solubility	3.880 log (mol/M^3)
Hydrophilic surface area	22.73 A^2
Polar surface area	20.23 A^2
HOMO	−10.916 eV
LUMO	3.478 eV
Dipole	1.599 debye

TOPOLOGICAL PROPERTIES (UNITLESS)

Parameter	Value
x0	2.707
x1	1.414
x2	0.707
xp3	0.000
xp4	0.000
xp5	0.000
xp6	0.000
xp7	0.000
xp8	0.000
xp9	0.000
xp10	0.000
xv0	2.154
xv1	1.023
xv2	0.316
xvp3	0.000

Continued.

Parameter	Value
xvp4	0.000
xvp5	0.000
xvp6	0.000
xvp7	0.000
xvp8	0.000
xvp9	0.000
xvp10	0.000
k0	1.431
k1	3.000
k2	2.000
k3	0.000
ka1	2.960
ka2	1.960
ka3	0.000

REFERENCES

Abel, E. L. (1981). Behavioral teratology of alcohol. *Psychol. Bull.* 90: 564–581.

Abel, E. L. (1989). *Fetal Alcohol Syndrome: Fetal Alcohol Effects*, Plenum Press, New York.

Abel, E. L. (1992). Paternal exposure to alcohol. In *Perinatal Substance Abuse*, T. B. Snoderegger, Ed., Johns Hopkins University Press, Baltimore, MD, pp. 132–160.

Abel, E. L. (1995). An update on incidence of FAS: FAS is not an equal opportunity birth defect. *Neurotoxicol. Teratol.* 17: 437–443.

Abel, E. L. (1997). Maternal alcohol consumption and spontaneous abortion. *Alcohol Alcohol.* 32: 211–219.

Abel, E. L. (1998). Fetal alcohol syndrome: The "American paradox." *Alcohol Alcohol.* 33: 195–201.

Abel, E. L. (1999). What really causes FAS? *Teratology* 59: 4–6.

Abel, E. L. and Berman, R. F. (1994). Long-term behavioral effects of prenatal alcohol exposure in rats. *Neurotoxicol. Teratol.* 16: 467–470.

Abel, E. L. and Dintcheff, B. A. (1978). Effects of prenatal alcohol exposure on growth and development in rats. *J. Pharmacol. Exp. Ther.* 207: 916–921.

Abel, E. L. and Sokol, R. J. (1987). Incidence of fetal alcohol syndrome and economic impact of FAS related anomalies. *Drug Alcohol Depend.* 19: 51–70.

Altschuler, H. L. and Shippenberg, T. S. (1981). A subhuman primate model for fetal alcohol syndrome research. *Neurobehav. Toxicol. Teratol.* 3: 121–126.

Ashley, M. J. (1981). Alcohol use during pregnancy: A challenge for the '80s. *Can. Med. Assoc. J.* 125: 141–143.

Bailey, B. N. et al. (2004). Prenatal exposure to binge drinking and cognitive and behavioral outcomes at age 7 years. *Am. J. Obstet. Gynecol.* 191: 1037–1043.

Baldwin, V. J., MacLeod, P. M., and Benirschke, K. (1982). Placental findings in alcohol abuse in pregnancy. *Birth Defects* 18: 89–94.

Bark, N. (1979). Fertility and offspring of alcoholic women: An unsuccessful search for the fetal alcohol syndrome. *Br. J. Addict.* 74: 43–49.

Barr, H. M. et al. (1990). Prenatal exposure to alcohol, caffeine, tobacco, and aspirin: Effects on fine and gross motor performance in 4-year old children. *Dev. Psychol.* 26: 339–348.

Bartoshesky, L. E. et al. (1979). A paternal fetal alcohol syndrome and fetal alcohol syndrome in a child whose alcoholic parents had stopped drinking. *Birth Defects Conf. Absts.*

Beagle, W. S. (1981). Fetal alcohol syndrome: A review. *J. Am. Diet. Assoc.* 79: 274–276.

Becker, M., Warrleeper, G. A., and Leeper, H. A. (1990). Fetal alcohol syndrome — a description of oral motor, articulatory, short-term memory, grammatical and semantic abilities. *J. Commun. Dis.* 23: 97–124.

Blakley, P. M. (1988). Experimental teratology of ethanol. In *Issues and Reviews in Teratology, Vol. 4*, H. Kalter, Ed., Plenum Press, New York, pp. 237–282.

Bookstein, F. L. et al (2001). Geometric morphometrics of corpus callosum and subcortical structures in the fetal-alcohol-affected brain. *Teratology* 64: 4–32.

Bostrom, B. and Nesbit, M. E., Jr. (1983). Hodgkin disease in a child with fetal alcohol-hydantoin syndrome. *J. Pediatr.* 103: 760–762.

Boyd, T. A. et al. (1991). Prenatal alcohol exposure and sustained attention in the preschool years. *Neurotoxicol. Teratol.* 13: 49–55.

Brien, J. F. and Smith, G. N. (1991). Effects of alcohol (ethanol) on the fetus. *J. Dev. Physiol.* 15: 21–32.

Briggs, G. G., Freeman, R. K., and Yaffe, S. J. (2005). *Drugs in Pregnancy and Lactation. A Reference Guide to Fetal and Neonatal Risk*, Seventh ed., Lippincott Williams & Wilkins, Philadelphia.

Brown, R. T. et al. (1991). Effects of alcohol exposure at school age. II. Attention and behavior. *Neurotoxicol. Teratol.* 13: 369–376.

Burd, L. and Martsoff, J. T. (1989). Fetal alcohol syndrome: Diagnosis and syndromal variability. *Physiol. Behav.* 46: 39–43.

Burden, M. J. et al. (2005). Effects of prenatal alcohol exposure on attention and working memory at 7.5 years of age. *Alcohol Clin. Exp. Res.* 29: 443–452.

Castells, S. et al. (1981). Growth retardation in fetal alcohol syndrome. Unresponsiveness to growth-promoting hormones. *Dev. Pharmacol. Ther.* 3: 232–241.

Chang, G. et al. (1998). Pregnant women with negative alcohol screens do drink less. A prospective study. *Am. J. Addict.* 7: 299–304.

Chasnoff, I. J., Diggs, G., and Schnoll, S. H. (1981). Fetal alcohol effects and maternal cough syrup abuse. *Am. J. Dis. Child.* 135: 968.

Chaudhury, J. D. (2000). An analysis of the teratogenic effects that could possibly be due to alcohol consumption by pregnant mothers. *Indian J. Med. Sci.* 54: 425–431.

Chen, J. S., Driscoll, C. D., and Riley, E. P. (1982). Ontogeny of suckling behavior in rats prenatally exposed to alcohol. *Teratology* 26: 145–153.

Chernoff, G. F. (1977). The fetal alcohol syndrome in mice: An animal model. *Teratology* 15: 223–229.

Chernoff, G. F. (1980). Introduction: A teratologist's view of the fetal alcohol syndrome. *Curr. Alcohol* 7: 7–13.

Chinboga, C. A. (2003). Fetal alcohol and drug effects. *Neurologist* 9: 267–279.

Chua, A. et al. (1979). Maternal drinking and the outcome of pregnancy. *Pediatr. Res.* 13: 485.

Church, M. W. and Gerkin, K. P. (1988). Hearing disorders in children with fetal alcohol syndrome — findings from case reports. *Pediatrics* 82: 147–154.

Clarren, S. K. (1981). Recognition of the fetal alcohol syndrome. *JAMA* 245: 2436–2445.

Clarren, S. K. and Bowden, D. M. (1982). A new primate model for binge drinking and its relevance to human ethanol teratogenesis. *Teratology* 25: 35A–36A.

Clarren, S. K. and Smith, D. W. (1978). The fetal alcohol syndrome. *N. Engl. J. Med.* 298: 1063–1067.

Cohen, M. M. (1981). Neoplasia and the fetal alcohol and hydantoin syndromes. *Neurobehav. Toxicol. Teratol.* 3: 161–162.

Coles, C. D. et al. (1991). Effects of prenatal alcohol exposure at school age. I. Physical and cognitive development. *Neurotoxicol. Teratol.* 13: 351–367.

Day, N. L. and Richardson, G. A. (1991). Prenatal alcohol exposure: A continuum of effects. *Semin. Perinatal.* 15: 271–279.

Dehaene, P., Samaille-Villette, C., and Samaille, P. (1977). The fetal alcohol syndrome in the north of France. *Rev. Alcohol.* 23: 145–158.

Dexter, J. D. and Tumbleson, M. E. (1980). Fetal alcohol syndrome in Sinclair (S-1) miniature swine. *Teratology* 21: 35A–36A.

Dorris, M. (1989). *The Broken Cord*, Harper & Row, New York.

Driscoll, C. D., Streissguth, A. P., and Riley, E. P. (1990). Prenatal alcohol exposure: Comparability of effects in humans and animal models. *Neurotoxicol. Teratol.* 12: 231–237.

Dunn, P. M., Stewart-Brown, S., and Peel, R. (1979). Metronidazole and the fetal alcohol syndrome. *Lancet* 2: 144.

Ebrahim, S. H. et al. (1998). Alcohol consumption by pregnant women in the United States during 1988–1995. *Obstet. Gynecol.* 92: 187–192.

Eckardt, M. J. et al. (1998). Effects of moderate alcohol consumption on the central nervous system. *Alcohol Clin. Exp. Res.* 22: 998–1040.

Ellis, F. W. and Pick, J. R. (1980). An animal model of the fetal alcohol syndrome in beagles. *Alcohol Clin. Exp. Res.* 4: 123–134.

Erb, L. and Anderson, B. D. (1978). The fetal alcohol syndrome. *Clin. Pediatr.* 17: 644–649.

Ernhart, C. B. et al. (1987). Alcohol teratogenicity in the human: A detailed assessment of specificity, critical period, and threshold. *Am. J. Obstet. Gynecol.* 156: 33–39.

Ernhart, C. B. et al. (1989). Alcohol-related birth defects: Assessing risks. *Ann. N.Y. Acad. Sci.* 562: 159–172.

Fisher, S. E. et al. (1982). Ethanol-associated selective fetal malnutrition: A contributing factor in the fetal alcohol syndrome. *Alcohol Clin. Exp. Res.* 6: 197–201.

Friedman, J. M. (1982). Can maternal alcohol ingestion cause neural tube defects? *J. Pediatr.* 101: 232–234.

Friedman, J. M. and Polifka, J. E. (2000). *Teratogenic Effects of Drugs. A Resource for Clinicians (TERIS)*, Second ed., Johns Hopkins University Press, Baltimore, MD.

Ginsburg, K. A. et al. (1991). Fetal alcohol exposure and adverse pregnancy outcomes. *Contrib. Gynecol. Obstet.* 18: 115–129.

Gladstone, J., Nulman, I., and Koren, G. (1996). Reproductive risks of binge drinking during pregnancy. *Reprod. Toxicol.* 10: 3–13.

Golden, J. (2005). *Message in a Bottle. The Making of Fetal Alcohol Syndrome*, Harvard University Press, Cambridge.

Goodlet, C. R., Horn, K. H., and Zhou, F. C. (2005). Alcohol teratogenesis: Mechanisms of damage and strategies for intervention. *Exp. Biol. Med.* 230: 394–406.

Graham, J. M. (1986). Current issues in alcohol teratogenesis. In *The Intrauterine Life. Management and Therapy (Proc. Second Int. Symp., The Fetus as a Patient — Diagnosis and Treatment*, Jerusalem, 1985), J. D. Schenker and D. Weinstein, Eds., Excerpta Medica, New York, pp. 383–388.

Green, H. G. (1974). Infants of alcoholic mothers. *Am. J. Obstet. Gynecol.* 118: 713–716.

Greene, T. et al. (1990). Prenatal alcohol exposure and language development. *Alcohol Clin. Exp.Res.* 14: 937–945.

Greene, T. et al. (1991). Prenatal alcohol exposure and cognitive development in the preschool years. *Neurotoxicol. Teratol.* 13: 57–68.

Grisso, J. A. et al. (1984). Alcohol consumption and outcome of pregnancy. *J. Epidemiol. Comm. Health* 38: 232–235.

Guerri, C. (1998). Neuroanatomical and neurophysiological mechanisms involved in central nervous system dysfunctions induced by prenatal alcohol exposure. *Alcohol Clin. Exp. Res.* 22: 304–312.

Habbick, B. F. et al. (1979). Liver abnormalities in three patients with fetal alcohol syndrome. *Lancet* 1: 580–581.

Halmesmaki, E., Raivio, K., and Ylikorkala, O. (1985). A possible association between maternal drinking and fetal clubfoot. *N. Engl. J. Med.* 312: 790.

Hanson, J. W., Jones, K. L., and Smith, D. W. (1976). Fetal alcohol syndrome. Experience with 41 patients. *JAMA* 235: 1458–1460.

Hanson, J. W., Streissguth, A. P., and Smith, D. W. (1978). The effects of moderate alcohol consumption during pregnancy on fetal growth and morphogenesis. *J. Pediatr.* 92: 457–460.

Harlap, S. and Shiono, P. H. (1980). Alcohol, smoking, and incidence of spontaneous abortions in the first and second trimester. *Lancet* 2: 173–176.

Hill, R. M., Hegemier, S., and Tennyson, L. M. (1989). The fetal alcohol syndrome: A multihandicapped child. *Neurotoxicology* 10: 585–596.

Hoyseth, K. S. and Jones, P. J. H. (1989). Minireview — ethanol induced teratogenesis: Characterization, mechanisms and diagnostic approaches. *Life Sci.* 44: 643–649.

Huggins, J. E. et al. (2004). Screening for FASD at delivery. *Birth Defects Res. (A)* 70: 289.

Hymbaugh, K. et al. (2002). A multiple source methodology for the surveillance of fetal alcohol syndrome — the fetal alcohol syndrome surveillance network (FASSNet). *Teratology* 66: S41–S49.

IOM (Institute of Medicine). (1996). Fetal alcohol syndrome: Diagnosis, epidemiology, prevention, and treatment. Division of Biobehavioral Sciences and Mental Disorders, Committee to study fetal alcohol syndrome, K. R. Stratton, C. J. Howe, and F. C. Battaglia, Eds., National Academy Press, Washington, D.C.

Iosub, S. et al. (1981). Fetal alcohol syndrome revisited. *Pediatrics* 68: 475–479.

Jones, K. L. (1986). Fetal alcohol syndrome. *Pediatr. Rev.* 8: 122–126.

Jones, K. L. (2003). From recognition to responsibility: Josef Warkany, David Smith, and the fetal alcohol syndrome in the 21st century. *Birth Defects Res. (A)* 67: 13–20.

Jones, K. L. and Chambers, C. (1998). Biomarkers of fetal exposure to alcohol: Identification of at-risk pregnancies. *J. Pediatr.* 133: 316–317.

Jones, K. L. and Smith, D. W. (1973). Recognition of the fetal alcohol syndrome in early infancy. *Lancet* 2: 999–1001.

Jones, K. L. and Smith, D. W. (1975). The fetal alcohol syndrome. *Teratology* 12: 1–10.

Jones, K. L., Smith, D. W., and Ulleland, C. J. (1973). Pattern of malformation in offspring of chronic alcoholic mothers. *Lancet* 1: 1267–1271.

Jones, K. L. et al. (1974). Outcome in offspring of chronic alcoholic women. *Lancet* 1: 1076–1078.

Jones, K. L., Smith, D. W., and Hanson, J. W. (1976). The fetal alcohol syndrome: Clinical delineation. *Ann. N.Y. Acad. Sci.* 273: 130–139.

Kaminsky, M., Rumeau-Rouquette, C., and Schwartz, D. (1976). Consommation d'alcool chez les femmes enceintes et issue de la grossese. *Rev. Epidemiol. Med. Sante Publique* 24: 27–40.

Kaminsky, M. et al. (1981). Moderate alcohol use and pregnancy outcome. *Neurobehav. Toxicol. Teratol.* 3: 173–182.

Kaufman, M. H. (1997). The teratogenic effects of alcohol following exposure during pregnancy, and its influence on the chromosome constitution of the pre-ovulatory egg. *Alcohol Alcohol.* 32: 113–128.

Kelly, S. J., Day, N., and Streissguth, A. P. (2000). Effects of prenatal alcohol exposure on social behavior in humans and other species. *Neurotoxicol. Teratol.* 22: 143–149.

Khan, A. et al. (1979). Hepatoblastoma in child with fetal alcohol syndrome. *Lancet* 1: 1403–1404.

Kiess, W. et al. (1984). Fetal alcohol syndrome and malignant disease. *Eur. J. Pediatr.* 143: 160–161.

Kinney, H., Faix, R., and Brazy, J. (1980). The fetal alcohol syndrome and neuroblastoma. *Pediatrics* 66: 130–132.

Kline, J. et al. (1980). Drinking during pregnancy and spontaneous abortion. *Lancet* 2: 176–180.

Konovalov, H. V. et al. (1997). Disorders of brain development in the progeny of mothers who used alcohol during pregnancy. *Early Hum. Dev.* 48: 153–166.

Koren, G. (1997). *The Children of Neverland. The Silent Human Disaster*, The Kid in Us®, Toronto, pp. 50–51.

Koren, G., Koren, T., and Gladstone, J. (1996). Moderate alcohol drinking in pregnancy. Perceived vs. true risk. *Clin. Chim. Acta* 246: 155–162.

Krous, H. F. (1981). Fetal alcohol syndrome: A dilemma of maternal alcoholism. *Pathol. Annu.* 16 (Part 1): 295–311.

Ladrague, P. (1901). *Alcoolisme et Enfants,* Steinheil, Paris.

LaDue, R. A., Stresissguth, A. P., and Randels, S. P. (1992). Clinical considerations pertaining to adolescents and adults with fetal alcohol syndrome. In *Perinatal Substance Abuse: Research Findings and Clinical Implications,* T. B. Sonderegger, Ed., Johns Hopkins University Press, Baltimore, MD, pp. 104–131.

Lamanna, M. (1982). Alcohol related birth defects: Implications for education. *J. Drug Educ.* 12: 113–122.

Larkby, C. and Day, N. (1997). The effects of prenatal alcohol exposure. *Alcohol Health Res. World* 21: 192–198.

Lee, K. T., Mattson, S. N., and Riley, E. P. (2004). Classifying children with heavy prenatal alcohol exposure using measures of attention. *J. Int. Neuropsychol. Soc.* 10: 271–277.

Leichter, J. and Lee, M. (1982). Method of ethanol administration as a confounding factor in studies of fetal alcohol syndrome. *Life Sci.* 31: 221–228.

Leichter, J., Lee, M., and Delparte, S. L. (1989). Possible mechanisms linking maternal alcohol-consumption and fetal growth-retardation. *Biochem. Arch.* 5: 279–287.

Lemarche, M. A. (1967). Reflexions sur la descendenace des alcooliques. *Bull. Acad. Nat. Med.* 151: 517–521.

Lemoine, P. et al. (1968). Les enfants de parents alcooliques: Anomalies observees a propos de 127 cas. *Quest Med.* 25: 476–482.

Leonard, B. E. (1988). Alcohol as a social teratogen. *Biochem. Basis Funct. Neuroteratol.* 73: 305–317.

Linneberg, A. et al. (2004). Alcohol during pregnancy and atropic dermatitis in the offspring. *Clin. Exp. Allergy* 34: 1678–1683.

Lipson, A. H., Walsh, D. A., and Webster, W. S. (1983). Fetal alcohol syndrome. A great paediatric imitator. *Med. J. Aust.* 1: 266–269.

Little, B. B. et al. (1990) Failure to recognize fetal alcohol syndrome in newborn infants. *Am. J. Dis. Child.* 144: 1142–1146.

Little, R. E. (1977). Moderate alcohol use during pregnancy and decreased infant birth weight. *Am. J. Public Health* 67: 1154–1156.

Little, R. E., Streissguth, A. P., and Guzinski, G. M. (1980). Prevention of fetal alcohol syndrome: A model program. *Alcohol Clin. Exp. Res.* 4: 185–189.

Little, R. E., Graham, J. M., and Samson, H. H. (1982). Fetal alcohol effects in humans and animals. *Adv. Alcohol Subst. Abuse* 1: 103–125.

Little, R. R. and Streissguth, A. P. (1981). Effects of alcohol on the fetus: Impact and prevention. *Can. Med. Assoc. J.* 125: 159–166.

Luke, B. (1990). The metabolic basis of the fetal alcohol syndrome. *Int. J. Fertil.* 35: 333–337.

MacDonald, D. I. (1987). Patterns in alcohol and drug use among adolescents. *Pediatr. Clin. North Am.* 34: 275–288

Majewski, F. (1977). [On certain embryopathies induced by teratogenic agents]. *Monatsschr. Kinderheilkd.* 125: 609–620.

Makarechian, N. et al. (1998). The association between moderate alcohol consumption during pregnancy and spontaneous abortion, stillbirth and premature birth: A meta-analysis. *Can. J. Clin. Pharmacol.* 5: 169–176.

Marbury, M. C. et al. (1983). The association of alcohol consumption with outcome of pregnancy. *Am. J. Public Health* 73: 1165–1168.

Martinez-Frias, M. L. et al. (2004). Risk for congenital anomalies associated with different sporadic and daily doses of alcohol consumption during pregnancy: A case-control study. *Birth Defects Res. (A)* 70: 194–200.

Mattson, S. N. and Riley, E. P. (2000). Parent ratings of behavior in children with heavy prenatal alcohol exposure and IQ-matched controls. *Alcohol Clin. Exp. Res.* 24: 226–231.

Mattson, S. N., Schoenfeld, A. M., and Riley, E. P. (2001). Teratogenic effects of alcohol on brain and behavior. *Alcohol Res. Health* 25: 175–184.

Mau, G. (1980). Moderate alcohol consumption during pregnancy and child development. *Eur. J. Pediatr.* 133: 233–237.

May, P. A. and Gossage, J. P. (2001). Estimating the prevalence of fetal alcohol syndrome. A summary. *Alcohol Res. Health* 25: 159–167.

McCance-Katz, E. F. (1991). The consequences of maternal substance abuse for the child exposed *in utero*. *Psychosomatics* 32: 268–274.

McLain, D. E. and Roe, D. A. (1984). Fetal alcohol syndrome in the ferret *(Mustela putorius)*. *Teratology* 30: 203–210.

Mena, M. et al. (1980). [Fetal alcohol syndrome, a study of 19 clinical cases]. *Rev. Chil. Pediatr.* 51: 414–423.

Michaelis, E. K. (1990). Fetal alcohol exposure: Cellular toxicity and molecular events involved in toxicity. *Alcohol Clin. Exp. Res.* 14: 819–826.

Miller, M. (1982). Prenatal alcohol effect disputed. *Pediatrics* 70: 322–323.

Mills, J. L. et al. (1984). Maternal alcohol consumption and birth weight. How much drinking during pregnancy is safe? *JAMA* 252: 1875–1879.

Morrison, A. B. and Maykut, M. O. (1979). Potential adverse effects of maternal alcohol ingestion on the developing fetus and their sequelae in the infant and child. *Can. Med. Assoc. J.* 120: 826–828.

Mulvihill, J. J. et al. (1976). Fetal alcohol syndrome: Seven new cases. *Am. J. Obstet. Gynecol.* 125: 937–941.

Neugut, R. H. (1981). Epidemiological appraisal of the literature on the fetal alcohol syndrome in humans. *Early Hum. Dev.* 5: 411–429.

Newman, N. M. and Correy, J. F. (1980). Effects of alcohol in pregnancy. *Med. J. Aust.* 2: 5–10.

NIAAA (National Institute on Alcohol Abuse and Alcoholism). (2003). Assessing Alcohol Problems. A Guide for Clinicians and Researchers, Second ed., J. P. Allen, Ed., NIH Publication No. 03-3745. Public Health Service USDHHS, Washington, DC.

Niccols, G. A. (1994). Fetal alcohol syndrome: Implications for psychologists. *Clin. Psychol. Rev.* 14: 91–111.

Nitowsky, H. M. (1982). Fetal alcohol syndrome and alcohol-related birth defects. *N.Y. State J. Med.* 82: 1214–1217.

Nulman, I. et al. (1998). The effects of alcohol on the fetal brain. The central nervous system tragedy. In *Handbook of Developmental Neurotoxicity*, W. Slikker, Jr. and L. W. Chang, Eds., Academic Press, New York, pp. 567–586.

Nulman, I. et al. (2004). Binge alcohol consumption by non-alcohol-dependent women during pregnancy affects child behavior, but not general intellectual functioning: A prospective controlled study. *Arch. Women Ment. Health* 7: 173–181.

Obe, G. and Ristow, H. (1979). Mutagenic, cancerogenic and teratogenic effects of alcohol. *Mutat. Res.* 65: 229–259.

Olegard, R. et al. (1979). Effects on the child of alcoholic abuse during pregnancy. *Acta Paediatr. Scan.* Suppl. 275: 112–121.

Olney, J. W. (2004). Fetal alcohol syndrome at the cellular level. *Addict. Biol.* 9: 137–149.

Olsen, J. (1994). Effects of moderate alcohol consumption during pregnancy on child development at 18 and 42 months. *Alcohol Clin. Exp. Res.* 18: 1109–1113.

Olson, H. C. et al. (1998). Neuropsychological deficits in adolescents with fetal alcohol syndrome: Clinical findings. *Alcohol Clin. Exp. Res.* 22: 1998–2012.

Ouellette, E. M. et al. (1977). Adverse effects on offspring of maternal alcohol abuse during pregnancy. *N. Engl. J. Med.* 297: 528–530.

Palmer, R. H. et al. (1974). Congenital malformations in offspring of a chronic alcoholic mother. *Pediatrics* 53: 490–494.

Papara-Nicholson, D. and Telford, I. R. (1957). Effects of alcohol on reproduction and fetal development in the guinea pig. *Anat. Rec.* 127: 438–439.

Pierog, S., Chandavasu, O., and Wexler, I. (1977). Withdrawal symptoms in infants with the fetal alcohol syndrome. *J. Pediatr.* 90: 630–633.

Pierog, S., Chandavasu, O., and Wexler, I. (1979). The fetal alcohol syndrome: Some maternal characteristics. *Int. J. Gynaecol. Obstet.* 16: 412–415.

Pietrantoni, M. and Knuppel, R. A. (1991). Alcohol use in pregnancy. *Clin. Perinatol.* 18: 93–111.

Pinazo-Duran, M. D. et al. (1997). Optic nerve hypoplasia in fetal alcohol syndrome: An update. *Eur. J. Ophthalmol.* 7: 262–270.

Polygenis, D. et al. (1998). Moderate alcohol consumption during pregnancy and the incidence of fetal malformations: A meta-analysis. *Neurotoxicol. Teratol.* 20: 61–67.

Polygenis, D. et al. (2001). Moderate alcohol consumption during pregnancy and the incidence of fetal malformations: A meta-analysis. In *Maternal and Fetal Toxicology. A Clinician's Guide*, Third ed., G. Koren, Ed., Marcel Dekker, New York, pp. 495–505.

Potter, B. J. et al. (1981). Teratogenic effects of ethanol in pregnant sheep: A model for the fetal alcohol syndrome. In *Man, Drugs, Society Current Perspective (Proc. Pan-Pac. Conf. Drugs Alcohol, 1st)*, Australian Foundation Alcohol Drug Dependency, Canberra, pp. 303–306.

Pratt, O. (1981). Alcohol and the women of childbearing age a public health problem. *Br. J. Addict.* 76: 383–390.

Pratt, O. E. (1982). Alcohol and the developing fetus. *Br. Med. Bull.* 38: 48–52.

Qazi, Q. H. and Masakawa, A. (1976). Altered sex ratio in fetal alcohol syndrome. *Lancet* 2: 42.

Richardson, G. A., Day, N. L., and Taylor, P. M. (1989). The effect of prenatal alcohol, marijuana, and tobacco exposure on neonatal behavior. *Infant Behav. Dev.* 12: 199–209.

Robin, N. H. and Zackai, E. H. (1994). Unusual craniofacial dysmorphism due to prenatal alcohol and cocaine exposure. *Teratology* 50: 160–164.

Roebuck, T. M., Mattson, S. N., and Riley, E. P. (1999). Behavioral and psychosocial profiles of alcohol-exposed children. *Alcohol Clin. Exp. Res.* 23: 1070–1076.

Ronen, G. M. and Andrews, W. L. (1991). Holoprosencephaly as a possible embryonic alcohol effect. *Am. J. Med. Genet.* 40: 151–154.

Rosett, H. L. (1980). A clinical perspective of the fetal alcohol syndrome. *Alcohol Clin. Exp. Res.* 4: 119–122.

Rosett, H. L. and Weiner, L. (1981). Identifying and treating pregnant patients at risk from alcohol. *Can. Med. Assoc. J.* 125: 149–158.

Rosett, H. L. et al. (1978). Therapy of heavy drinking during pregnancy. *Obstet. Gynecol.* 51: 41–46.

Rosett, H. L., Weiner, L., and Edelin, K. C. (1981). Strategies for prevention of fetal alcohol effects. *Obstet. Gynecol.* 57: 1–7.

Rosett, H. L. et al. (1983a). Alcohol consumption and fetal development. *Obstet. Gynecol.* 61: 539–546.

Rosett, H. L., Weiner, L., and Edelin, K. C. (1983b). Treatment experience with pregnant problem drinkers. *JAMA* 249: 2029–2033.

Rostland, A. et al. (1990). Alcohol use in pregnancy, craniofacial feature, and fetal growth. *J. Epidemiol. C.* 44: 302–306.

Russell, M. (1991). Clinical implications of recent research on the fetal alcohol syndrome. *B. N.Y. Acad. Med.* 67: 207–222.

Sampson, P. D. et al. (1997). Incidence of fetal alcohol syndrome and prevalence of alcohol-related neurodevelopmental disorder. *Teratology* 56: 317–326.

Sandor, G. G. S., Smith, D. W., and MacLeod, P. M. (1981). Cardiac malformations in the fetal alcohol syndrome. *J. Pediatr.* 98: 771–773.

Sarda, P. and Bard, H. (1984). Gastroschisis in a case of dizygotic twins — the possible role of maternal alcohol consumption. *Pediatrics* 74: 94–96.

Schardein, J. L. (2000). *Chemically Induced Birth Defects*, Third ed., Marcel Dekker, New York, pp. 735–743.

Scheiner, A. P., Donovan, C. M., and Bartoshesky, L. E. (1979). Fetal alcohol syndrome in child whose parent had stopped drinking. *Lancet* 1: 1077–1078.

Schenker, S. et al. (1990). Fetal alcohol syndrome — current status of pathogenesis. *Alcohol Clin. Exp. Res.* 14: 635–647.

Scott, W. J. and Fradkin, R. (1984). The effects of prenatal ethanol in cynomolgus monkeys. *Teratology* 29: 49–56.

Seeler, R. A. et al. (1979). Ganglioneuroblastoma and fetal hydantoin-alcohol syndromes. *Pediatrics* 63: 524–527.

Shepard, T. H. (1974). Teratogenicity from drugs an increasing problem. *Dis. Month.*, 3–32.

Smith, D. W. (1980). Alcohol effects on the fetus. In *Drug and Chemical Risks to the Fetus and Newborn*, R. H. Schwarz and S. J. Yaffe, Eds., Alan R. Liss, New York, pp. 73–82.

Sokol, R. J. (1981). Alcohol and abnormal outcomes of pregnancy. *Can. Med. Assoc. J.* 125: 143–148.

Sokol, R. J. and Abel, E. L. (1992). Risk factors for alcohol-related birth defects: Threshold, susceptibility, and prevention. In *Perinatal Substance Abuse*, T. B. Sondregger, Ed., Johns Hopkins University Press, Baltimore, MD, pp. 90–103.

Sparks, S. N. (1984). Speech and language in fetal alcohol syndrome. *ASHA* 26: 27–31.

Spohr, H.-L., Willims, J., and Steinhausen, H.-C. (1994). The fetal alcohol syndrome in adolescence. *Acta Paediatr.* Suppl. 404: 19–26.

Steinhausen, H.-C. and Spohr, H.-L. (1998). Long-term outcome of children with fetal alcohol syndrome: Psychopathology, behavior, and intelligence. *Alcohol Clin. Exp. Res.* 22: 334–338.

Steinmetz, G. (1992). The preventable tragedy. Fetal alcohol syndrome. *National Geographic*, February, pp. 37–39.

Stoler, J. M. (1999). Reassessment of patients with the diagnosis of fetal alcohol syndrome. *Pediatrics* 103: 1313–1315.

Stoler, J. M. and Holmes, L. B. (1999). Under-recognition of prenatal alcohol effects in infants of known alcohol abusing women. *J. Pediatr.* 135: 430–436.

Stoler, J. M. et al. (1998). The prenatal detection of significant alcohol exposure with maternal blood markers. *J. Pediatr.* 133: 346–352.

Streissguth, A. P. (1983). Alcohol and pregnancy: An overview and an update. *Subst. Alcohol Actions Misuse* 4: 149–173.

Streissguth, A. P. (1986). Smoking and drinking during pregnancy and offspring learning disabilities. A review of the literature and development of a research strategy. In *Learning Disabilities and Prenatal Risk*, M. Lewis, Ed., University of Illinois Press, Urbana, pp. 28–67.

Streissguth, A. P. (1997). *Fetal Alcohol Syndrome. A Guide for Families and Communities*, P. H. Brookes, Baltimore, MD.

Streissguth, A. P. and LaDue, R. A. (1987). Fetal alcohol. Teratogenic causes of developmental disabilities. *Monogr. Am. Assoc. Ment. Defic.* 8: 1–32.

Streissguth, A. P. and O'Malley, K. (2000). Neuropsychiatric implications and long-term consequences of fetal alcohol spectrum disorders. *Semin. Clin. Neuropsych.* 5: 179–190.

Streissguth, A. P., Herman, C. S., and Smith, D. W. (1978). Intelligence, behavior, and dysmorphogenesis in the fetal alcohol syndrome. A report on 20 patients. *J. Pediatr.* 92: 363–367.

Streissguth, A. P., Clarren, S. K., and Jones, K. L. (1985). Natural history of the fetal alcohol syndrome. A 10 year follow-up of eleven patients. *Lancet* 2: 85–91.

Streissguth, A. P. et al. (1989). IQ at age 4 in relation to maternal alcohol use and smoking during pregnancy. *Dev. Psychol.* 25: 3–11.

Sulik, K. K. (2005). Genesis of alcohol-induced craniofacial dysmorphism. *Exp. Biol. Med.* 230: 366–375.

Sulik, K. K., Johnston, M. C., and Webb, M. A. (1981). Fetal alcohol syndrome: Embryogenesis in a mouse model. *Science* 214: 936–938.

Sullivan, W. C. (1900). The children of the female drunkard. *Med. Temp. Rev.* 1: 72–79.

Tatha, C. (1990). [Alcohol and pregnancy. A review and risk assessment]. *J. Toxicol. Clin. Exp.* 10: 105–114.

Tennes, K. and Blackard, C. (1980). Maternal alcohol consumption, birth weight, and minor physical anomalies. *Am. J. Obstet. Gynecol.* 138: 774–780.

Thomas, J. D. and Riley, E. P. (1998). Fetal alcohol syndrome: Does alcohol withdrawal play a role? *Alcohol Health Res. World* 22: 47–53.

Tolo, K. A. and Little, R. E. (1993). Occasional binges by moderate drinkers: Implications for birth outcomes. *Epidemiology* 4: 415–420.

Tze, W. J. and Lee, M. (1975). Adverse effects of maternal alcohol consumption on pregnancy and foetal growth in rats. *Nature* 257: 479–480.

Uhlig, H. (1957). [Abnormalities in undesired children]. *Aerztl. Wochenschr.* 12: 61–64.

Ulleland, C. N. (1972). The offspring of alcoholic mothers. *Ann. N.Y. Acad. Sci.* 197: 167–169.

Van Rensburg, L. J. (1981). Major skeletal defects in the fetal alcohol syndrome. A case report. *S. Afr. Med. J.* 59: 687–688.

Veghelyi, P. V. and Osztovics, M. (1979). Fetal-alcohol syndrome in child whose parents had stopped drinking. *Lancet* 2: 35–36.

Walpole, I., Zubrick, S., and Pontre, J. (1990). Is there a fetal effect with low to moderate alcohol use before or during pregnancy. *J. Epidemiol. C.* 44: 297–301.

Warner, R. H. and Rosett, H. L. (1975). The effects of drinking on offspring: An historical survey of the American and British literature. *J. Stud. Alcohol* 36: 1395–1420.

Warren, K. R. and Foudin, L. L. (2001). Alcohol-related birth defects — the past, present, and future. *Alcohol Res. Health* 25: 153–158.

Waterson, E. J. (1990). Preventing alcohol related birth damage a review. *Social Sci. Med.* 30: 349–364.

Webster, W. S. and Ritchie, H. E. (1991). Teratogenic effects of alcohol and isotretinoin on craniofacial development — an analysis of animal models. *J. Cran. Genet. Dev. Biol.* 11: 296–302.

Werler, M. M. et al. (1991). Maternal alcohol use in relation to selected birth defects. *Am. J. Epidemiol.* 134: 691–698.

West, J. R. and Goodlett, C. R. (1990). Teratogenic effects of alcohol on brain development. *Ann. Med.* 22: 319–325.

Wiedemann, H.-R., Kunze, J., and Dibbern, H. (1989). *Atlas of Clinical Syndromes. A Visual Aid to Diagnosis*, Second ed., Mosby Year Book, St. Louis, pp. 480–483.

Willford, J. A. et al. (2004). Verbal and visuospatial learning and memory function in children with moderate prenatal alcohol exposure. *Alcohol Clin. Exp. Res.* 28: 497–507.

Wisniewski, K. and Lupin, R. (1979). Fetal alcohol syndrome and related CNS problems. *Neurology* 29: 1429–1430.

Witti, F. P. (1978). Alcohol and birth defects. *FDA Consumer* 12: 20–23.

Wright, J. T. et al. (1983). Alcohol consumption, pregnancy, and low birth weight. *Lancet* 1: 663–665.

51 Discussion and Summary

As set forth in the Preface, we have identified approximately 70 developmental toxicants affecting humans. That is, agents that adversely affect one or more of the four classes of developmental toxicity (i.e., growth, viability, structural malformation or terata, and function). We selected 50 for characterization in this text, based on considerations laid out in the Preface.

TOXICOLOGICAL CHARACTERIZATION OF HUMAN DEVELOPMENTAL TOXICANTS

RESULTS OF EVALUATION

The results of developmental toxicity reported for the 50 developmental toxicants by class are tabulated in Table 1. There does not appear to be a typical pattern of response by the toxicants (Table 2): 70% exhibited two or more classes of developmental toxicity, but the remaining 30% elicited only one class of toxicity. Twelve developmental toxicants (24%) exhibited all four classes and would generally be considered by observers to be among the most hazardous agents of the group. These were the potent cancer chemotherapeutic drugs aminopterin, cyclophosphamide, and methotrexate; the anticonvulsant paramethadione, now largely replaced as a medicament due to its toxicity; several ubiquitous environmental toxicants, carbon monoxide and toluene; the ACE inhibitor captopril; the antithyroid agent methimazole; the widely used coumarin warfarin; a useful therapeutic drug ergotamine; and two widely used (and abused) agents with medicinal and social indications, cocaine and alcohol. The hazard posed by alcohol is certainly substantial, where the incidence of its induced syndrome of effects is considered to be in the range of up to 4.3% among offspring of heavy drinkers. With regard to those agents affecting a single endpoint, malformation was the most common solitary class of developmental toxicity encountered.

The societal impact of developmental toxicants based on the number of recorded cases involved is shown in Table 3. From this perspective, it is shown that 14 (28%) of the toxicants studied have significant effects on public health, with hundreds and even thousands of victims of disabling perinatal disease and death. Among the most involved group, in the case of ethanol, labeling restrictions on bottles of liquor and for thalidomide, strict labeling (category X) and registration for prescription plus massive educational programs for limiting their use already exist and, presumably, aid, in part, in controlling widespread indiscriminate use. This is not so with phenytoin. Here, the pregnancy category is D (an equivocal warning regarding its association to birth defects), and there has been no notable further effort to educate the user to its realistic risk, calculated as two- to threefold the normal risk for malformation and significant growth-retarding and neurological dysfunction properties in offspring of women administered the drug during pregnancy. However, in this case, phenytoin has risk:benefit considerations with regard to controlling epilepsy in users that permit its widespread use. However, additional efforts to evaluate the current therapeutic indications of this drug are clearly warranted.

Of the other two groups of major concern, those in which recorded numbers of cases total 100 or more, two agents were unclassified (being chemicals), one was category B (caffeine), one was category C (propranolol), three were category D, and the remaining three were category X. The high-rated agents, caffeine and propranolol, were presumably classed as more innocuous, because

TABLE 1
Developmental Toxicity Reported with 50 Human Developmental Toxicants

Chapter and Agent	Developmental Toxicity Reported			
	Growth Retardation	Death	Malformation	Functional Deficit
1. Aminopterin	X	X	X	X
2. Busulfan	X	X	X	
3. Cyclophosphamide	X	X	X	X
4. Methotrexate	X	X	X	X
5. Chlorambucil		X	X	
6. Mechlorethamine		X	X	
7. Cytarabine		X	X	
8. Tretinoin		X	X	
9. Propranolol	X			
10. Penicillamine		X	X	
11. Vitamin A			X	
12. Carbamazepine	X		X	X
13. Danazol			X	
14. Paramethadione	X	X	X	X
15. Carbon monoxide	X	X	X	X
16. Formaldehyde		X		
17. Isotretinoin		X	X	X
18. Captopril	X	X	X	X
19. Misoprostol		X	X	X
20. Streptomycin			X	X
21. Methimazole	X	X	X	X
22. Ethylene oxide		X		
23. Tetracycline			X	
24. Caffeine	X			
25. Thalidomide		X	X	X
26. Primidone	X		X	
27. Fluconazole	X	X	X	
28. Ergotamine	X	X	X	X
29. Propylthiouracil				X
30. Medroxyprogesterone			X	
31. Cocaine	X	X	X	X
32. Quinine		X	X	X
33. Methylene blue		X	X	X
34. Warfarin	X	X	X	X
35. Phenobarbital	X		X	X
36. Trimethoprim			X	
37. Methyltestosterone			X	
38. Disulfiram			X	
39. Valproic acid	X		X	X
40. Carbon disulfide		X	X	X
41. Norethindrone			X	
42. Phenytoin	X		X	X
43. Etretinate		X	X	
44. Toluene	X	X	X	X
45. Ethisterone			X	
46. Acitretin		X	X	
47. Valsartan		X	X	X
48. Diethylstilbestrol		X	X	X
49. Pseudoephedrine			X	
50. Ethanol	X	X	X	X

TABLE 2
Summary of Endpoints Associated with
50 Human Developmental Toxicants

Only 1 class	15
2 classes	9
3 classes	14
All 4 classes	12

TABLE 3
Recorded Cases with 50 Developmental Toxicants

Estimated Number of Cases of Developmental Toxicity[a]	Developmental Toxicant
≤ 10	Busulfan, cyclophosphamide, chlorambucil, disulfiram, medroxyprogesterone, methyltestosterone, paramethadione, penicillamine, tretinoin (topical), valsartan
≤ 100	Acitretin, aminopterin, captopril, carbon disulfide[c], carbon monoxide[c], cytarabine, danazol, ergotamine, ethisterone, ethylene oxide[c], etretinate, fluconazole, mechlorethamine, methimazole, methotrexate, methylene blue[c], norethindrone, phenobarbital[c], primidone, propylthiouracil, pseudoephedrine[c], quinine[c], streptomycin[c], trimethoprim[c], vitamin A (excess), warfarin
Hundreds	Carbamazepine[c], diethylstilbestrol, formaldehyde[c], misoprostol, propranolol[c,d], toluene[c], valproic acid
Thousands	Caffeine[c,d], cocaine[c], isotretinoin[c], tetracycline[b,c]
Tens of thousands	Ethanol[c], phenytoin[c], thalidomide

[a] Cases not cumulative.
[b] Dental staining only.
[c] Exact number unknown.
[d] Growth retardation only.

(1) caffeine is really used as an additive, mainly in beverages, in uncontrolled quantities; and (2) propranolol's effects are somewhat controversial and consist only of questionable, if not probable, retardation of fetal weight. As with all anticonvulsants, carbamazepine and valproic acid have the "D" warning. The more potent agents, diethylstilbestrol, isotretinoin, and cocaine, carry the category "X", as expected.

The distribution of the four classes represented by the developmental toxicants evaluated with respect to response is shown in Table 4. Malformation again was the most common class observed, being present in 90% of the toxicants evaluated. Distribution of the other three classes was variable

TABLE 4
Summary of Reactions Associated with Classes of 50
Developmental Toxicants in Humans

Endpoint	Number of Positives	Number of Negatives
Growth retardation	21	29
Death	31	19
Malformation	45	5
Functional deficit	26	24

with respect to positives:negatives, with functional deficit fairly equivalent (52 versus 48%), with the remaining two skewed 42 versus 58% (growth retardation) and 62 versus 38% (death). This result is contrary to the popular but unproven perception that in animals, fetal growth retardation is the most sensitive indicator of the classes in the developmental toxicity continuum. Growth retardation in the human appears to be the least sensitive endpoint of the four classes, at least in the toxicants evaluated here. Intrauterine growth retardation (IUGR) is said to complicate up to 10% of pregnancies in the United States (Seeds, 1984). Further, the association of IUGR or low birth weight to the other classes of developmental toxicity in the human has been amply demonstrated with respect to perinatal mortality (Low and Galbraith, 1974), spontaneous abortion (Nelson et al., 1971), severe congenital anomalies (Christianson et al., 1981), and neurological dysfunction (Hill et al., 1971; Miller, 1981).

The high proportion evidenced by those agents associated with lethality confirms, in part, that spontaneous abortion is said to be the most likely outcome of exposure to toxicants, the ability to detect an effect of exposure being much greater for this endpoint than for other adverse outcomes of pregnancy (Stellman, 1979; Sever and Hessol, 1984), an association also observed in this series but only to some extent. Lethality has also been linked to congenital malformation (Shepard and Fantel, 1979; Haas and Schottenfeld, 1979; Poland et al., 1981). This is not too surprising, as malformation accounts for approximately 14% of infant deaths (Warkany, 1957). Perinatal deaths are also associated with IUGR in a high percentage of cases (Callan and Witter, 1990). The association of functional changes to other adverse developmental effects was noted. Up to 61% of infants with mental retardation have displayed associated congenital anomalies, according to several investigators (Illingworth, 1959; Malamud, 1964; Smith and Bostian, 1964).

ANIMAL AND HUMAN RELATIONSHIPS

GENERAL CONCORDANCE AND "MODELS"

Relationships between agents considered developmental toxicants in humans and in laboratory animals used in nonclinical testing to demonstrate safety in humans show a number of interesting associations. While it has been repeatedly stated by various investigators that all putative human teratogens have demonstrated teratogenicity in animals, there appears to be one exception, at least among the representative developmental toxicants selected here. The exception is misoprostol. While it is shown here (Chapter 19) that the drug affects three classes of developmental toxicity in humans, no class has been replicated in laboratory species, even at very large dosage multiples. It may be, however, that the limited toxicity testing conducted with this drug in the laboratory simply may not have been broad enough to demonstrate adverse effects. Animal studies with the remaining 49 agents evaluated all demonstrated one or more adverse effects on development. These are tabulated in Table 5. In fact, animal "models" have been designated in the original reports for 14 of the 50 agents. We added another (propranolol).

COMPLETE CONCORDANCE AND SPECIES SENSITIVITY

Another way of looking at these results is to evaluate those laboratory animals that showed complete concordance to the effects shown in humans. This information is provided in Table 6. In 27 (54%) of the agents evaluated, one or more species exhibited the identical class of developmental toxicity as reported for the human. Reactivity of these species is shown in Table 7, with the rat being the most sensitive. These data are similar to, but not identical to, results published by other investigators in studies of "teratogens" (Brown and Fabro, 1983; Schardein and Keller, 1989; Schardein, 1998). These data relate in part to those species most often used in testing, but the mouse versus rabbit results do not conform to this impression. Further, those results are different than those from animal

TABLE 5

Laboratory Animal Species Developmental Toxicity Response to 50 Human Developmental Toxicants

Agent	Developmental Toxicity[a]				Termed "Model"
	GR	D	M	FD	
Aminopterin	R,S	M, S, Mk, C, Rb	R, S, D, P, Rb		
Busulfan	R, M	R, M	R, M		
Cyclophosphamide	R, M	R, M	R, Rb, M, Mk		
Methotrexate		R, Rb, M, C, Mk	R, Rb, M, C		
Chlorambucil		M, R	M, R		
Mechlorethamine	M, R	M, R	Rb, M, R, F		
Cytarabine	M	M, R	M, R		
Tretinoin	Rb	R, Rb, Mk, P	R, M, Ha, F, P, Rb	R	M, Rb
Propranolol	R				R[b]
Penicillamine	R, Ha, M	R, Ha, M	R, Ha, M		R
Vitamin A	R	G, R, Rb, C, Mk	R, M, G, Ha, P, D, Rb, C, Mk	R	
Carbamazepine	R	R	M, R		
Danazol	Rb				
Paramethadione	R	R			
Carbon monoxide	M, Rb	R, M, Rb, P	R, M, G, Mk	R, M	
Formaldehyde	D	Ha	R, M	R	
Isotretinoin	R	M, R, Mk	Ha, Rb, M, R, Mk		Mk
Captopril	R	S, Rb, R			
Misoprostol					
Streptomycin			M, R		
Methimazole			R		
Ethylene oxide	M, R	M, R, Rb	M		
Tetracycline	R	M, R	R, G		
Caffeine	M, R, Mk	M, R, Mk	M, R, Rb	R	
Thalidomide	R, Rb, M, Mk	R, Rb, M, D, Mk	R[c], Rb, M[c], Ha[c], C[c], D[c], Arm[c], F, P, Mk	R	Rb, Mk
Primidone		M, R	M	M, R	
Fluconazole		Rb, R	R		
Ergotamine	M, R	R			
Propylthiouracil				G, Rb, R, M	
Medroxyprogesterone	R, Rb, M	Rb, Mk	R, Rb, Mk		
Cocaine	R, M	R	R, M, Mk	R	R
Quinine	M	R, M, Rb, D	G, Rb	G, Rb	
Methylene blue		R, M	M		
Warfarin		Rb, M, R	M, R		R
Phenobarbital		Rb	M, R, Rb	R	R
Trimethoprim		Rb	R		
Methyltestosterone	R		R, Rb, D		
Disulfiram		R, M			
Valproic acid	M, R, Mk	R, Rb, Mk	M, R, Ha, Rb, Mk	R	M
Carbon disulfide	Rb	R, Rb	R, Rb	R	
Norethindrone	Mk	R, Mk	M, R, Mk		
Phenytoin	M, R	M, R, Mk, C	M, R, Mk[c]	R, Mk	M

Continued.

TABLE 5 *(Continued)*
Laboratory Animal Species Developmental Toxicity Response to 50 Human Developmental Toxicants

	Developmental Toxicity[a]				
Agent	GR	D	M	FD	Termed "Model"
Etretinate	Ha		Ha, R, Rb, M		M(?)
Toluene	R, M	R, M	R, M	R, Ha	R
Ethisterone			Rb, R		
Acitretin		M, Rb, R	M, Rb, R		M
Valsartan	M, R, Rb	Rb			
Diethylstilbestrol		Mk	R, M, Ha, F, Mk		M
Pseudoephedrine	R				
Ethanol	R, M, G, P, D, S	R, G, P, D, S, F, Mk, M	R, M, Rb, P, D, S, F, Mk	R, G	R, M, P, D, S, Mk[c]

[a] In species cited, by any route; M = mouse, R = rat, Rb = rabbit, G = guinea pig, D = dog, C = cat, Mk = monkey, F = ferret, P = pig, S = sheep, Ha = hamster, Arm = armadillo.
[b] Assigned by authors.
[c] Variable/questionable reaction.

studies conducted by others in which the parameter was "nonteratogens," where a rank order of primate ≈ rabbit > rat > mouse was the usual scenario (Brown and Fabro, 1983; Schardein et al., 1985).

CHEMICAL CHARACTERIZATION OF HUMAN DEVELOPMENTAL TOXICANTS

Both physicochemical and topological parameters were calculated for all of the 50 human developmental toxicants discussed in this book. Physicochemical parameters characterize the steric, transport, and electronic properties of the respective compounds, as well as their ability to engage in intermolecular interactions (e.g., a biological receptor site). Topological parameters encode the connectivity between atoms comprising a molecular entity, as well as the size and degree of branching. Table 8 and Table 9 list the mean, standard deviation, minimum, and maximum values for the respective individual calculated physicochemical and topological parameters for the chemicals discussed in the preceding chapters. Each of the calculated physicochemical and topological parameters was analyzed by means of histograms (utilizing the "project" command within MiniTAB version 11.12; www.minitab.com). The histograms are provided in Appendix I (physicochemical) and Appendix II (topological), and plotted is the frequency of compound distribution as a function of calculated parameter values. The initial histogram pertains to the distribution of the respective parameter across all 50 of the human developmental toxicants. Subsequent histograms plot the respective calculated parameters of chemicals according to their positive and negative (eliciting the respective response) effects within each of the four major classes (endpoints) of developmental toxicity, respectively, growth retardation, death, malformation, and functional deficit.

OVERALL RESULTS AND STRUCTURE–ACTIVITY RELATIONSHIPS (SARS)

Investigation of the positive versus negative compound distribution with respect to each of the calculated parameters indicates that there are a limited number of parameters that can adequately distinguish between chemicals that can or cannot elicit human growth retardation, death, malformation, and functional deficit. Statistical analysis (*t*-test) of positives versus negatives within each

TABLE 6
Complete Concordance by Animals to 50 Human Developmental Toxicants

Agent	Species Concordance
Aminopterin	—
Busulfan	Mouse, rat
Cyclophosphamide	—
Methotrexate	—
Chlorambucil	Mouse, rat
Mechlorethamine	Mouse, rat
Cytarabine	Mouse, rat
Tretinoin	Rat, pig
Propranolol	Rat
Penicillamine	Mouse, rat, hamster
Vitamin A	Mouse, rat, guinea pig, hamster, pig, dog, rabbit, cat, monkey
Carbamazepine	—
Danazol	—
Paramethadione	—
Carbon monoxide	Mouse
Formaldehyde	Hamster
Isotretinoin	—
Captopril	—
Misoprostol	—
Streptomycin	—
Methimazole	—
Ethylene oxide	Mouse, rat, rabbit
Tetracycline	Rat, guinea pig
Caffeine	Mouse, rat, monkey
Thalidomide	—
Primidone	—
Fluconazole	—
Ergotamine	—
Propylthiouracil	Mouse, rat, guinea pig, rabbit
Medroxyprogesterone	Rat, rabbit, monkey
Cocaine	Rat
Quinine	Rabbit
Methylene blue	—
Warfarin	—
Phenobarbital	—
Trimethoprim	Rat
Methyltestosterone	Rat, rabbit, dog
Disulfiram	—
Valproic acid	Rat
Carbon disulfide	Rat
Norethindrone	Mouse, rat, monkey
Phenytoin	Mouse, rat
Etretinate	—
Toluene	Rat
Ethisterone	Rat, rabbit
Acitretin	Mouse, rat, rabbit
Valsartan	—
Diethylstilbestrol	—
Pseudoephedrine	—
Ethanol	Rat

TABLE 7
Laboratory Animal Response to Class of
Developmental Toxicity Shown in Humans

Species	Percent (%) Response
Rat	48
Mouse	26
Rabbit	16
Primate	8
Hamster	6
Guinea pig	6
Pig	4
Dog	4
Cat	2

TABLE 8
Statistical Measures for Calculated Physicochemical Properties

Parameter	Mean	Standard Deviation	Min	Max
Molecular weight	263.2	124.9	28.0	581.7
Molecular volume	240.1	112.2	28.1	502.8
Density	1.1127	0.1922	0.8133	1.5319
Surface area	301.7	136.7	45.5	640.4
LogP	1.054	3.546	−12.158	6.855
HLB	9.15	7.08	0.00	21.54
Solubility parameter	24.245	4.649	14.493	36.455
Dispersion	20.466	3.091	14.493	27.189
Polarity	5.973	3.780	0.000	15.748
Hydrogen bonding	10.173	5.305	0.000	25.626
H bond acceptor	1.040	1.124	0.000	6.640
H bond donor	0.518	0.730	0.000	4.370
Percent hydrophilic surface area	45.88	31.12	0.00	100.00
MR	73.84	34.04	7.03	161.02
Water solubility	−0.287	3.007	−5.080	8.874
Hydrophilic surface area	125.8	114.3	0.0	605.9
Polar surface area	67.25	63.16	0.00	334.59
HOMO	−9.419	1.174	−12.362	−7.453
LUMO	−0.325	1.397	−5.801	3.478
Dipole	3.528	1.938	0.000	7.375

of the individual endpoints indicates a significant difference ($p < 0.05$) between the mean values of xvp6 through xvp10 for growth retardation; xvp3 through xvp10 for death; and dispersion for functional deficit. Results for malformation are questionable due to extreme bias (45 positives vs. 5 negatives). It remains to be seen in further investigations if a combination of physicochemical and topological parameters can be utilized in a SAR study to discern the effect of chemical properties upon human developmental toxicity.

Further statistical analyses were performed on three data sets utilizing the 50 human developmental toxicants and their respective calculated physicochemical and topological parameters. The first data set consisted of the 12 agents that elicited all 4 classes of human developmental toxicity vs. the remaining 38 compounds that exhibited only 1, 2, or 3 classes of human developmental

TABLE 9
Statistical Measures for Calculated Topological Properties

Parameter	Mean	Standard Deviation	Min	Max
x0	13.656	6.635	2.000	30.102
x1	8.796	4.443	1.000	20.676
x2	8.043	4.443	0.000	20.338
xp3	6.528	4.053	0.000	18.595
xp4	5.281	3.564	0.000	17.028
xp5	3.939	3.061	0.000	14.492
xp6	2.699	2.314	0.000	10.745
xp7	1.831	1.825	0.000	8.578
xp8	1.258	1.372	0.000	6.681
xp9	0.845	0.996	0.000	4.770
xp10	0.5354	0.6892	0.0000	3.3773
xv0	11.213	5.230	0.908	24.247
xv1	6.525	3.213	0.204	15.118
xv2	5.201	2.879	0.000	12.784
xvp3	3.797	2.537	0.000	10.132
xvp4	2.748	2.099	0.000	8.123
xvp5	1.860	1.674	0.000	6.020
xvp6	1.167	1.249	0.000	4.538
xvp7	0.724	0.932	0.000	3.205
xvp8	0.4577	0.6662	0.0000	2.3733
xvp9	0.2779	0.4394	0.0000	1.5214
xvp10	0.1570	0.2701	0.0000	0.9356
k0	23.10	15.84	0.60	69.03
k1	15.12	7.51	1.33	34.49
k2	6.566	3.577	0.000	14.189
k3	3.935	2.837	0.000	13.091
ka1	13.872	6.811	1.298	32.879
ka2	5.757	3.129	0.000	13.125
ka3	3.406	2.610	0.000	12.912

toxicity as shown in Tables 1 and 2. In the second comparison, the animal/human model data sets (based on Table 5) referred to those 15 compounds that exhibited identical results (in terms of adverse effects elicited) in both humans and animals (irregardless of species), vs. those 35 compounds that did not. The third comparison was in the animal/human concordance data set (based on Table 6) that consisted of 27 compounds exhibiting the identical class of developmental toxicity as reported for the human in one or more species vs. the remaining 23 agents that did not. All 49 calculated parameters (20 physicochemical and 29 topological) were submitted for comparison of mean values of positive vs. negative chemicals for each respective data set utilizing the Student's t-test. Data sets that contained outliers (a data point that is more than 1.5 times the interquartile range) were resubmitted for analysis with the outliers removed in order to determine if the outliers have an influence on the results. Only the parameters that were statistically significant even after deletion of outliers are discussed below.

For the 12 toxicants that exhibited all 4 classes of human developmental toxicity vs. the 38 toxicants that did not, only the logarithm of the partition coefficient (logP) was statistically significant at a p-value less than 0.05. It is interesting to note that the compounds exhibiting all 4 classes of human developmental toxicity had a lower mean logP value (-0.48) from the remaining 38 compounds (1.54) by a factor of 100. This may be indicative of the major importance that transport phenomena play within developmental toxicity. It is surprising that additional properties are not significant in separating the two groups of agents.

Significant calculated parameters for the 15 human/animal model compounds that shared the same characteristic adverse effects in humans and specific animal species comprised density, hydrophilic–lipophilic balance (HLB), %hydrophilic surface area, xp9, xp10, xvp4, xvp5, and vp6. All of the model compounds had lower mean values than the chemicals that did not. The relatively high number of parameters that were operative in terms of animal/human modeling may be indicative of the complexity of factors which must be satisfied. The majority of these factors consisted of parameters related to molecular size and hydrophilic properties. Significant parameters for animal/human concordance consisted of molecular weight, molecular volume, surface area, molar refractivity (MR), hydrophilic surface area, dipole, x0, x1, x2, xp3, xp4, xv0, xv1, k0, k1, k2, ka1, and ka2. As in the previous comparisons, the concordant compounds all had mean values that are lower in value than the non-concordant chemicals. The majority of the parameters refer to molecular size. The above preliminary univariate statistical analyses represent the first steps in future studies that will compare and contrast the physicochemical and topological properties that are operative within and across species. Future investigations should employ multivariate methods to discern QSAR relationships within human and animal species. Interested investigators may utilize the properties that were calculated by the authors (and present on the accompanying CD), or utilize their own pertinent algorithms on the provided three-dimensional chemical structures (also present on the CD as individual MOL files) to further explore the relationships between chemical structure and developmental toxicity.

There have been numerous reports in the scientific literature regarding the SARs of developmental toxicants (Schumacher, 1975; Kolb-Meyers and Beyler, 1981; Brown et al., 1982; Abramovici and Rachnuth-Roizman, 1983; Enslein et al., 1983; Johnson, 1984; Wassom, 1985; Gaffield, 1986; Willhite, 1986; Otsby et al., 1987; Howard et al., 1987, 1988; Enslein, 1989; DiCarlo, 1990; Kavlock, 1990, 1993; Oglesby et al., 1992; Nau, 1994; Rosenkranz et al., 1998; Gomez et al., 1999; Macina et al., 2001; Arena et al., 2004). These investigations have mainly utilized animal data as the basis for their investigations, with varying degrees of success. The application of SAR models derived from animal data to predictions of human developmental toxicity risk is tenuous. Some of the studies are further limited in application due to their focus on congeneric series (i.e., a collection of chemicals possessing a basic parent structure with variations occurring in the type and placement of substituents). Furthermore, some of the SAR modeling studies utilize a narrow range of descriptors and specific modeling methodologies. Very few studies (Ghanooni et al., 1997; Rosenkranz et al., 1998) concentrated on human developmental toxicity and SAR. These studies, however, utilized a chemical descriptor set and a modeling technique that is not widely available for independent verification. The use of chemical structure in predicting *reproductive* toxicants identified in the human was not successful in a recent report (Maslankiewicz et al., 2005).

To the best of the authors' knowledge, a comprehensive analysis using a broad range of descriptors and SAR modeling techniques has not been performed for human developmental toxicity. As stated by Richard (1998), the goal of accurate and reliable toxicity prediction for any chemical which is based solely on structural information remains elusive.

It is the hope of the authors that the data presented herein will provide further impetus for the investigation of the chemical requirements that induce human developmental toxicity. Human developmental toxicity is composed of complex phenomena that have not been adequately addressed by SAR studies up to the present time. It was stated that one of the biggest limitations in the development of predictive systems is the lack of reliable and consistent data available (Mirkes, 1996; Greene, 2002). We hope that careful scrutiny of developmental data in the present case might provide direction in predictive value of identifying toxicants, especially because it was done in the ultimate target species, the human. With an increasing number of pharmaceuticals and industrial chemicals being marketed, it is imperative that a comprehensive understanding of the factors responsible for human developmental toxicity be established for use in risk assessment.

REFERENCES

Abramovici, A. and Rachnuth-Roizman, P. (1983). Molecular structure-teratogenicity relationships of some fragrance additives. *Toxicology* 29: 143–156.

Arena, V. C. et al. (2004). The utility of structure-activity relationship (SAR) models for prediction and covariate selection in developmental toxicity: Comparative analysis of logistic regression and decisions tree models. *SAR QSAR Environ. Res.* 15: 1–18.

Brown, N. A. and Fabro, S. (1983). The value of animal teratogenicity testing for predicting human risk. *Clin. Obstet. Gynecol.* 26: 467–477

Brown, N. A. et al. (1982). Teratogenicity and lethality of hydantoin derivatives in the mouse: Structure-toxicity relationships. *Toxicol. Appl. Pharmacol.* 64: 271–288.

Callan, N. A. and Witter, E. R. (1990). Intrauterine growth retardation: Characteristics, risk factors and gestational age. *Int. J. Gynecol. Obstet.* 33: 215–220.

Christianson, R. E. et al. (1981). Incidence of congenital anomalies among white and black live births with long-term followup. *Am. J. Public Health* 71: 1333–1341.

DiCarlo, F. J. (1990). Structure-activity relationships (SAR) and structure-metabolism relationships (SMR) affecting the teratogenicity of carboxylic acids. *Drug Metab. Rev.* 22: 411–449.

Enslein, K. (1989). New teratogenesis model. *HDJ Toxicol. Newslett.* 9 (February).

Enslein, K. et al. (1983). Teratogenesis: A statistical structure-activity model. *Teratog. Carcinog. Mutagen.* 3: 289–309.

Gaffield, W. (1986). The significance of isomerism and stereospecificity to the chemistry of plant teratogens. *J. Toxicol.* 5: 229–240.

Ghanooni, M. et al. (1997). Structural determinants associated with risk of human developmental toxicity. *J. Am. Gyn. Obst. Soc.* 176: 799–806.

Gomez, J. et al. (1999). Structural determinants of developmental toxicity in hamsters. *Teratology* 60: 190–205.

Greene, N. (2002). Computer systems for the prediction of toxicity: An update. *Adv. Drug Delivery Rev.* 54: 417–431.

Haas, J. F. and Schottenfeld, D. (1979). Risks to the offspring from occupational exposures. *J. Occup. Med.* 21: 607–613.

Hill, R. M. et al. (1971). The impact of intrauterine malnutrition on the developmental potential of the human infant. A 14 year progressive study. In *Behavioral Effects of Energy and Protein Deficits*, J. Brozek, Ed., Publ. 79–1906, U.S. Department of Health, Education and Welfare, Washington, D.C.

Howard, W.B. et al. (1987). Structure toxicity relationships of the tetramethylated tetralin and indane analogs of retinoic acid. *Teratology* 36: 303–311.

Howard, W.B. et al. (1988). Structure-activity relationships of retinoids in developmental toxicology. III. Contribution of the vitamin A beta-cyclogeranylidene ring. *Toxicol. Appl. Pharmacol.* 95: 122–138.

Illingworth, R. S. (1959). Congenital anomalies associated with cerebral palsy and mental retardation. *Arch. Dis. Child.* 34: 228.

Johnson, E. M. (1984). A prioritization and biological decision tree for developmental toxicity evaluations. *J. Am. Coll. Toxicol.* 3: 141–147.

Kavlock, R. J. (1990). Structure-activity relationships in the developmental toxicity of substituted phenols: *In vivo* effects. *Teratology* 41: 43–59.

Kavlock, R. J. (1993). Structure-activity approaches in the screening of environmental agents for developmental toxicity. *Reprod. Toxicol.* 7: 113–116.

Kolb-Meyers, V. and Beyler, R. E. (1981). How to make an "educated guess" about the teratogenicity of chemical compounds. In *Environmental Toxicology Principles and Policies*, S. M. Somane and F. L. Cavender, Eds., Thomas, Springfield, IL, pp. 124–161.

Low, J. A. and Galbraith, R. S. (1974). Pregnancy characteristics of intrauterine growth retardation. *Obstet. Gynecol.* 44: 122–126.

Macina, O. T. et al. (2001). Physicochemical and graph theoretical descriptors in developmental toxicity SAR: A comparative study. *SAR QSAR Environ. Res.* 11: 345–362.

Malamud, N. (1964). Neuropathology. In *Mental Retardation: A Review of Research*, H. A. Stevens and R. Heber, Eds., University of Chicago Press, Chicago, pp. 126–135.

Maslankiewicz, L. et al. (2005). Can chemical structure predict reproductive toxicity? *Reprod. Toxicol.* 20: 470.

Miller, H. C. (1981). Intrauterine growth retardation. An unmet challenge. *Am. J. Dis. Child.* 135: 944–948.

Mirkes, P. E. (1996). Prospects for the development of validated screening tests that measure developmental toxicity potential: View of one skeptic. *Teratology* 53: 334–338.

Nau, H. (1994). Toxicokinetics and structure-activity relationships in retinoid teratogenesis. *Ann. Oncol.* 5 (Suppl. 9): S39–S49.

Nelson, T. et al. (1971). Collection of human embryos and fetuses. II. Clarification and tabulation of conceptual wastage with observations on type of malformation, sex ratio and chromosome studies, In *Monitoring Birth Defects and Environment*, E. B. Hook, D. T. Janerich, and I. H. Porter, Eds., Academic Press, New York, pp. 45–64.

Oglesby, L. A. et al. (1992). *In vitro* embryotoxicity of a series of para-substituted phenols: Structure, activity, and correlation with *in vivo* data. *Teratology* 45: 11–33.

Otsby, J. S. et al. (1987). The structure activity relationships of azo dyes derived from benzidine (B), dimethylbenzidine (DMB) or dimethoxybenzidine (DMOB) and their teratogenic effects on the testes of the mouse. *Toxicologist* 7: 146.

Poland, B. J. et al. (1981). Spontaneous abortions. A study of 1,961 women and their conceptuses. *Acta Obstet. Gynecol. Scand. Suppl.* 102: 5–32.

Richard, A. M. (1998). Structure-based methods for predicting mutagenicity and carcinogenicity: Are we there yet? *Mutat. Res.* 400: 493–507.

Rosenkranz, H. S. et al. (1998). Human developmental toxicity and mutagenesis. *Mutag. Res.* 422: 347–350.

Schardein, J. L. (1998). Animal/human concordance. In *Handbook of Developmental Neurotoxicology*, W. Slikker, Jr. and M. C. Chang, Eds., Academic Press, New York, pp. 687–708.

Schardein, J. L. and Keller, K. A. (1989). Potential human developmental toxicants and the role of animal testing in their identification and characterization. *CRC Crit. Rev. Toxicol.* 19: 251–339.

Schardein, J. L. et al. (1985). Species sensitivities and prediction of teratogenic potential. *Environ. Health Perspect.* 61: 55–67.

Schumacher, H. J. (1975). Chemical structure and teratogenic properties. In *Methods for Detection of Environmental Agents That Produce Congenital Defects*, T. H. Shepard, J. R. Miller, and M. Marois, Eds., American Elsevier, New York, pp. 65–77.

Seeds, J. W. (1984). Impaired fetal growth: Definition and clinical diagnosis. *Obstet. Gynecol.* 64: 303.

Sever, L. E. and Hessol, N. A. (1984). Overall design considerations in male and female occupational reproductive studies. In *Reproduction: The New Frontier in Occupational and Environmental Health Research*, J. E. Lockey, G. K. Lemasters, and W. R. Keye, Eds., Alan R. Liss, New York, pp. 15–47.

Shepard, T. H. and Fantel, A. G. (1979). Embryonic and early fetal loss. *Clin. Perinatol.* 6: 219–243.

Smith, D. W. and Bostian, K. E. (1964). Congenital anomalies associated with idiopathic mental retardation. *J. Pediatr.* 65: 189.

Stellman, J. M. (1979). The effect of toxic agents on reproduction. *Occup. Health Saf.* 48: 36–43.

Warkany, J. (1957). Congenital malformations and pediatrics. *Pediatrics* 19: 725–733.

Wassom, J. S. (1985). Use of selected toxicology information resources assessing relationships between chemical structure and biological activity. *Environ. Health Perspect.* 61: 287–294.

Willhite, C. C. (1986). Structure-activity relationships of retinoids in developmental toxicology. II. Influence of the polygene chain of the vitamin A molecule. *Toxicol. Appl. Pharmacol.* 83: 563–575.

Appendix I
Physicochemical Parameter Histograms

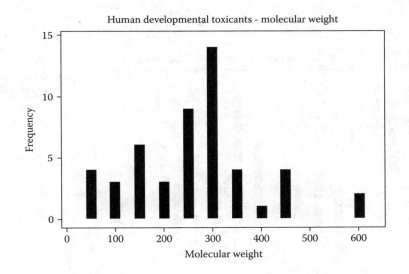

Human developmental toxicants - molecular weight

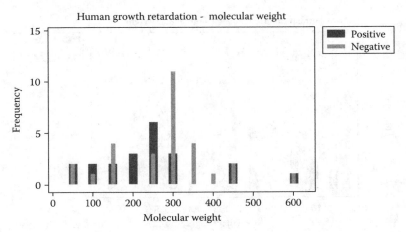

Human growth retardation - molecular weight

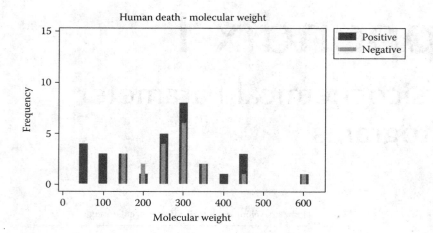

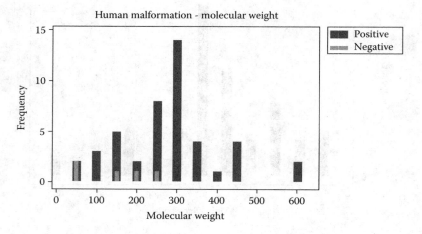

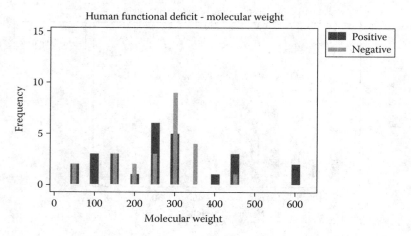

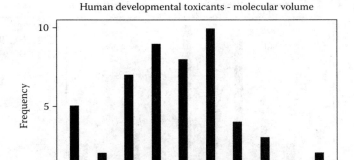

Human developmental toxicants - molecular volume

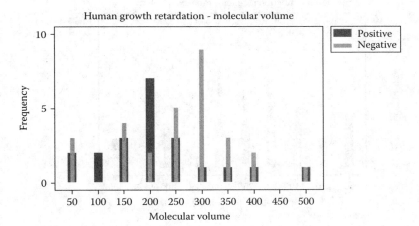

Human growth retardation - molecular volume

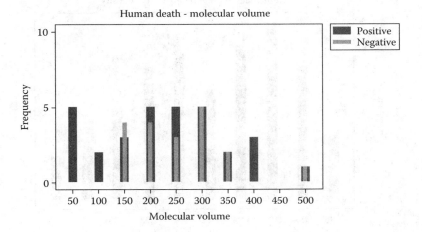

Human death - molecular volume

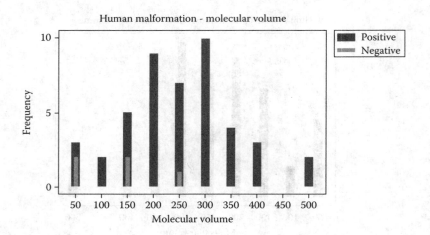

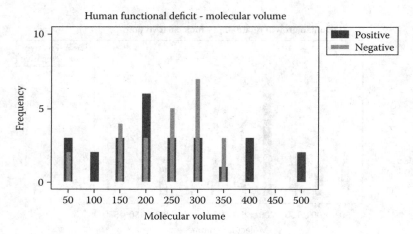

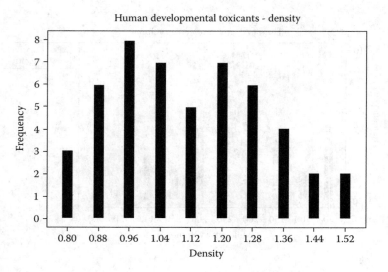

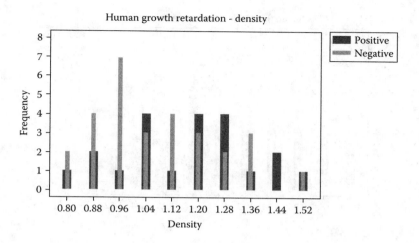

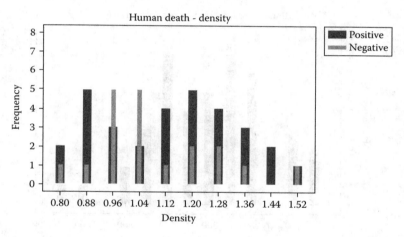

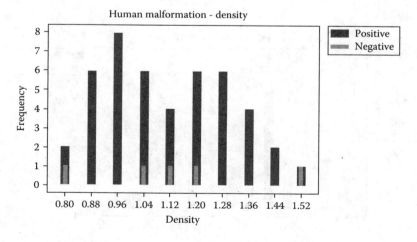

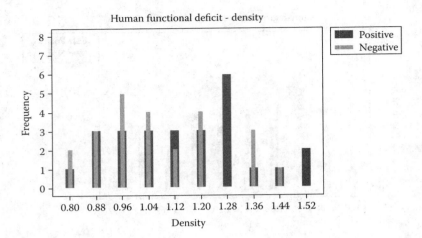

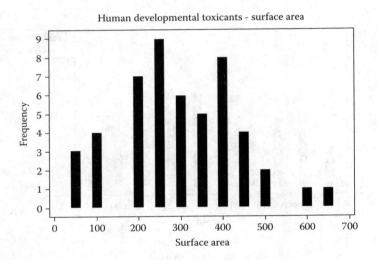

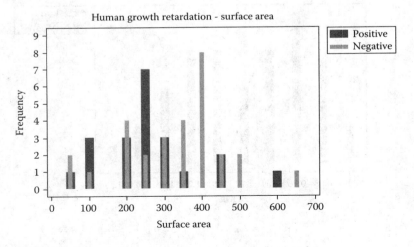

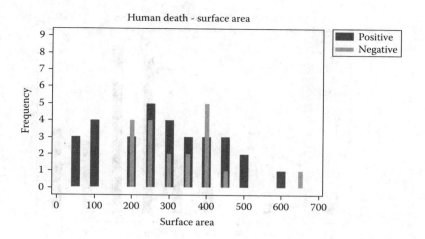

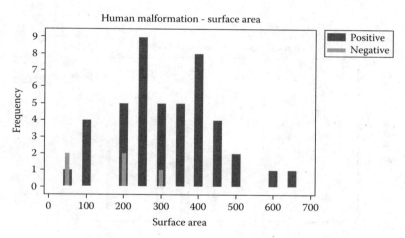

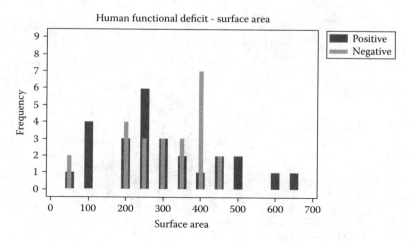

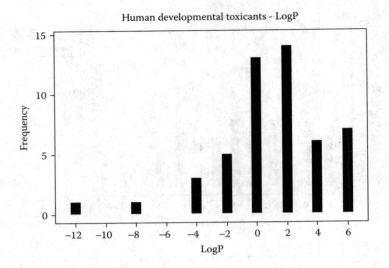

Human developmental toxicants - LogP

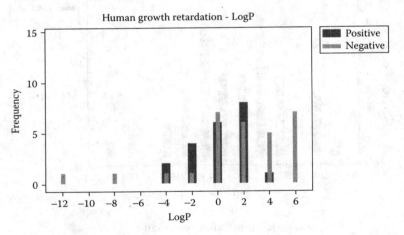

Human growth retardation - LogP

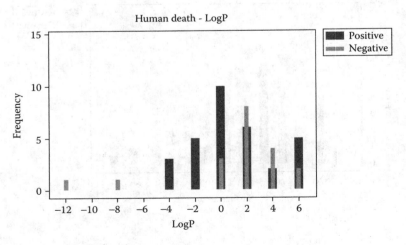

Human death - LogP

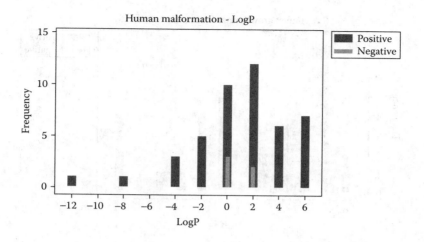

Human malformation - LogP

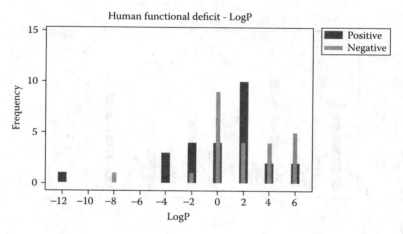

Human functional deficit - LogP

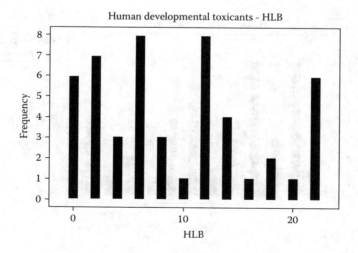

Human developmental toxicants - HLB

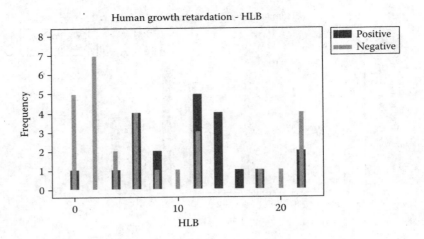

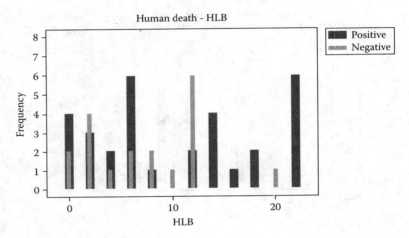

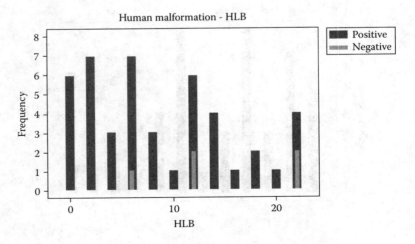

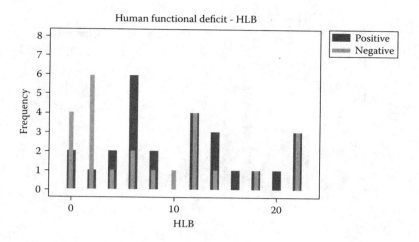

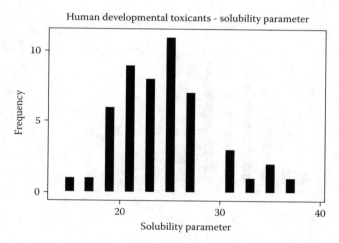

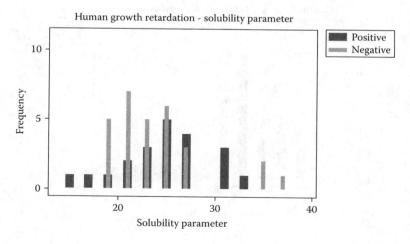

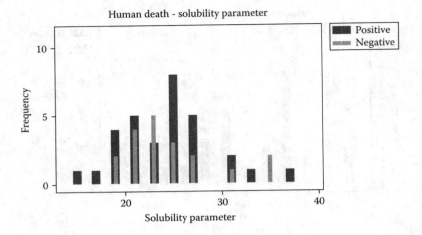

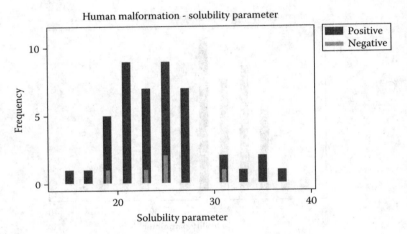

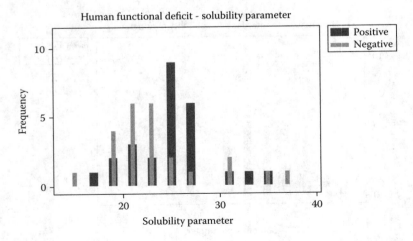

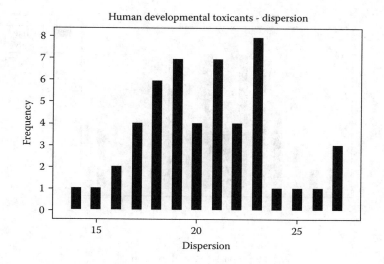

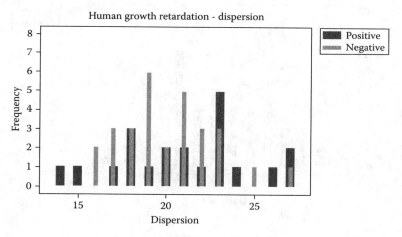

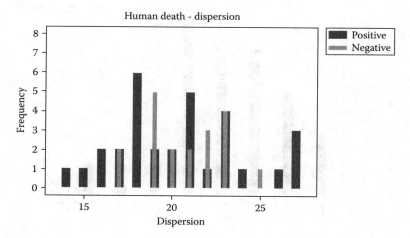

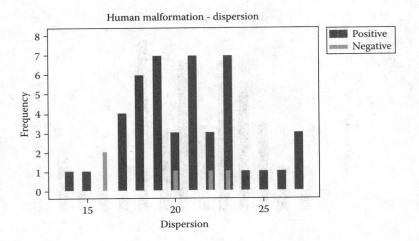

Human malformation - dispersion

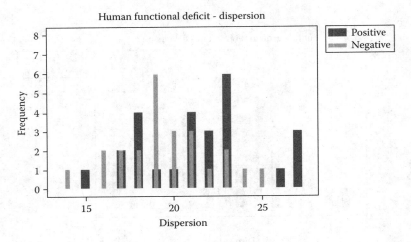

Human functional deficit - dispersion

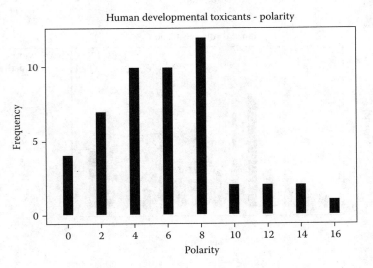

Human developmental toxicants - polarity

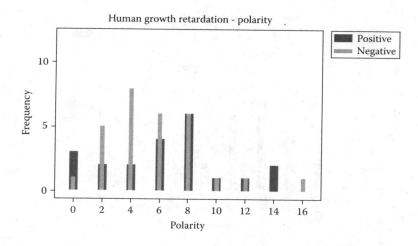

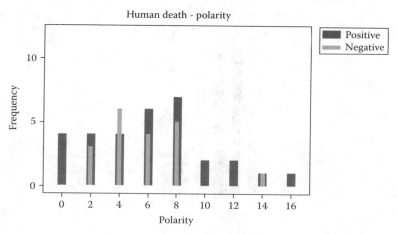

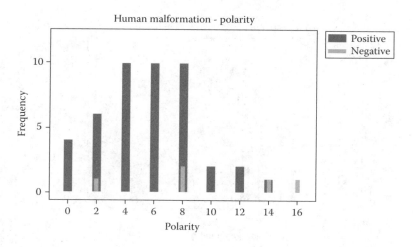

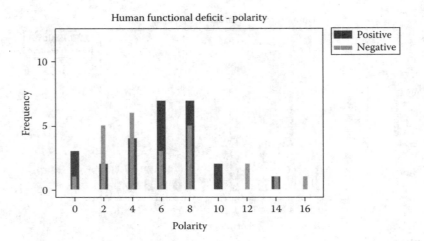

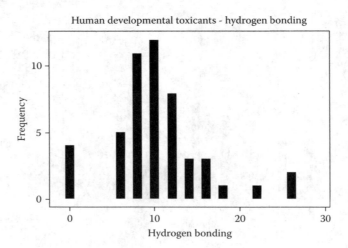

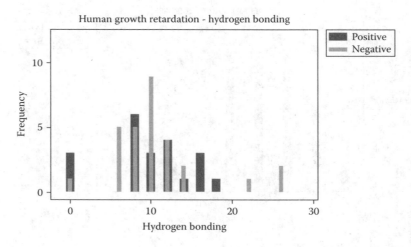

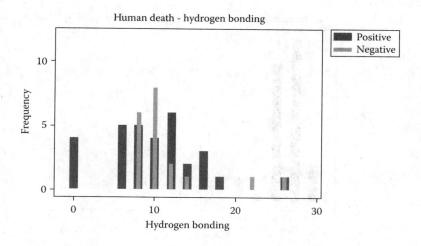

Human death - hydrogen bonding

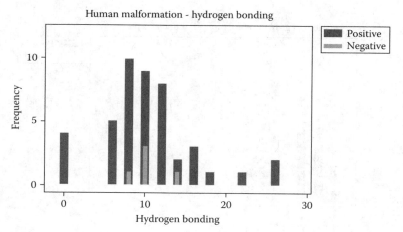

Human malformation - hydrogen bonding

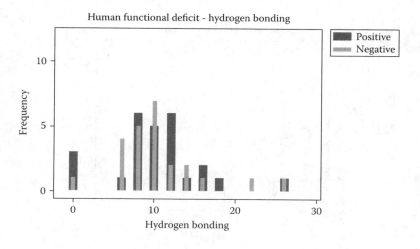

Human functional deficit - hydrogen bonding

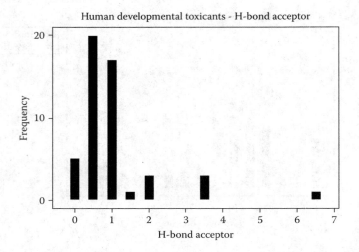

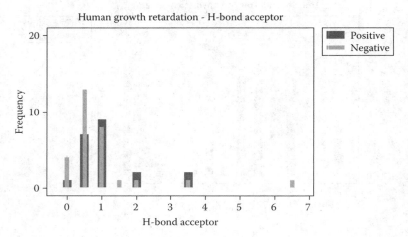

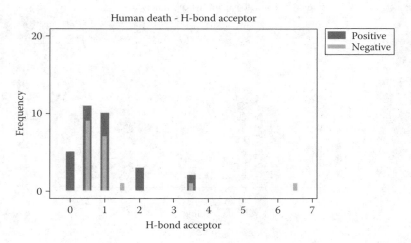

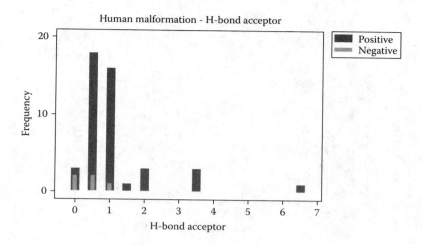

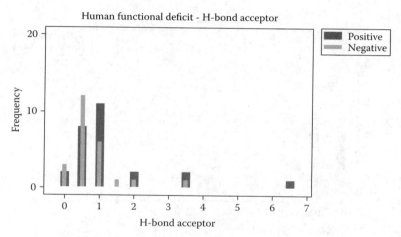

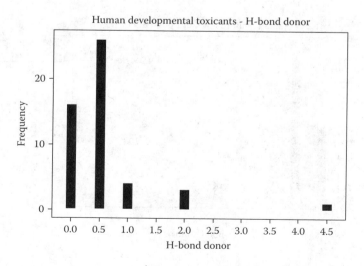

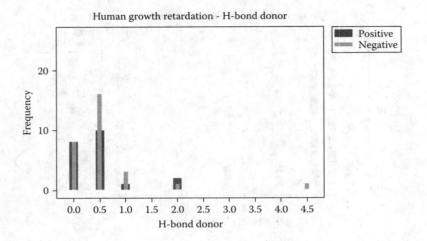

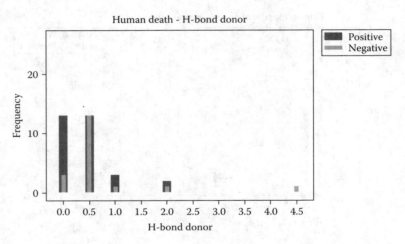

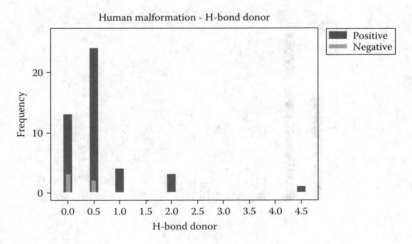

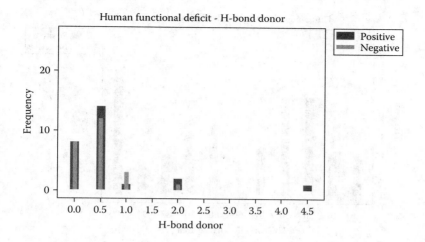

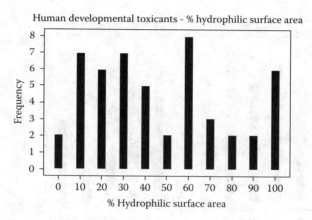

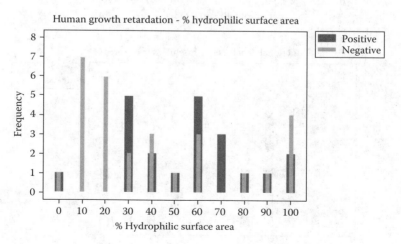

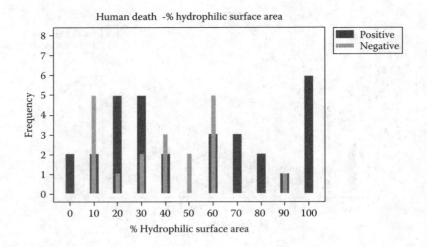

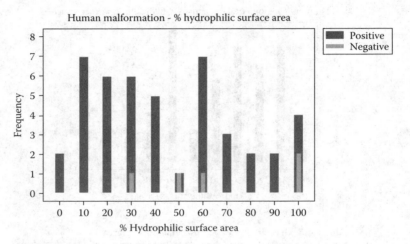

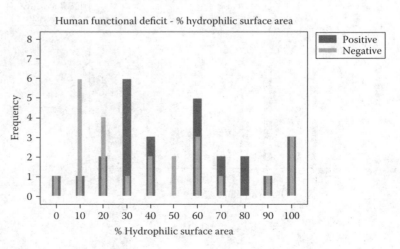

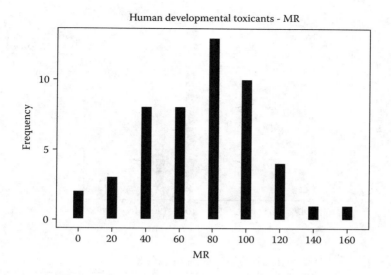

Human developmental toxicants - MR

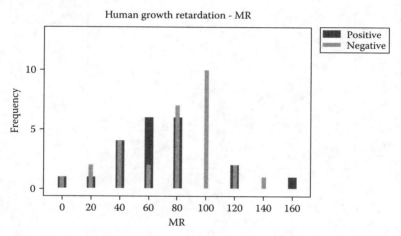

Human growth retardation - MR

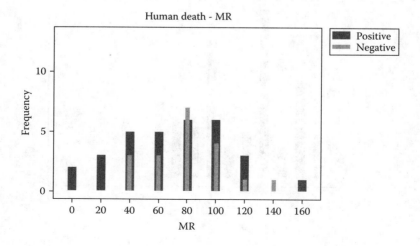

Human death - MR

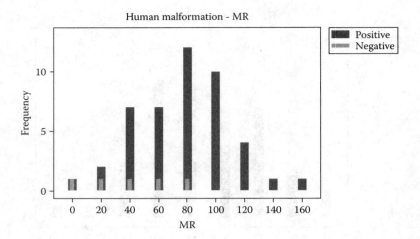

Human malformation - MR

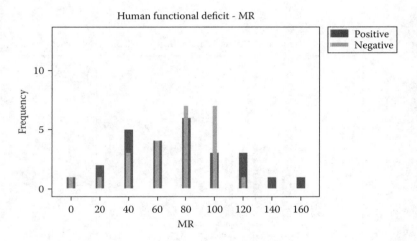

Human functional deficit - MR

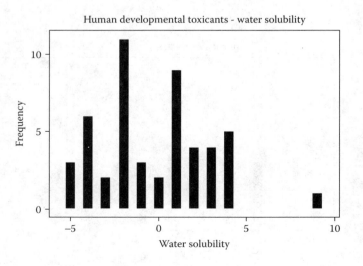

Human developmental toxicants - water solubility

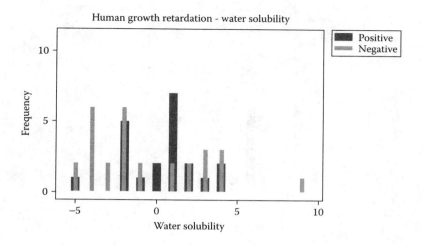

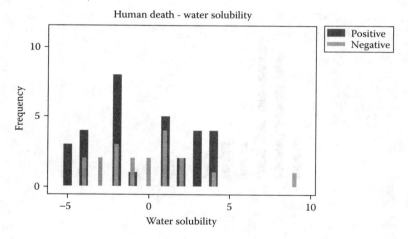

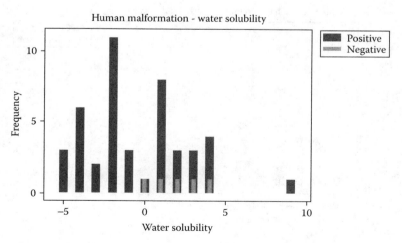

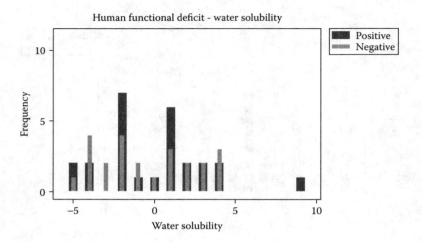

Human functional deficit - water solubility

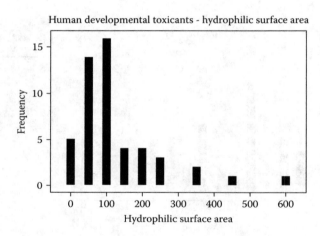

Human developmental toxicants - hydrophilic surface area

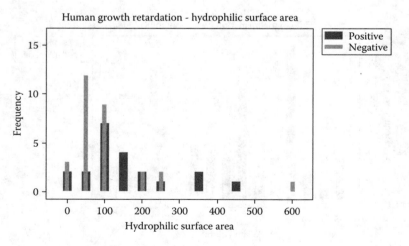

Human growth retardation - hydrophilic surface area

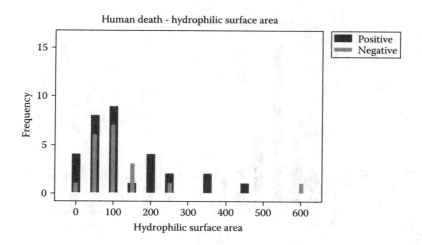

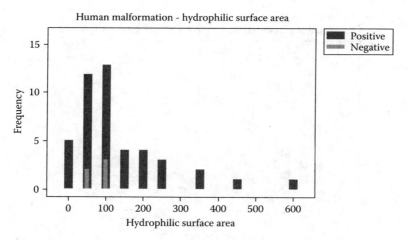

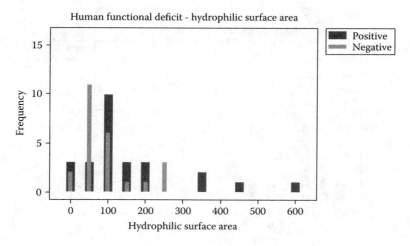

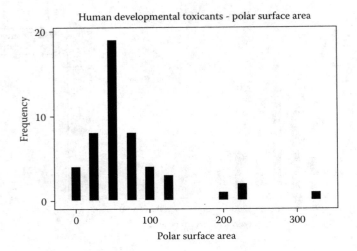

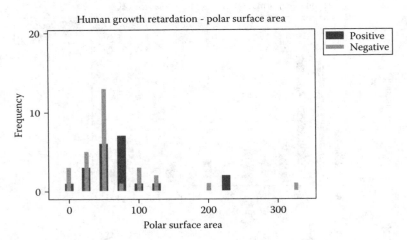

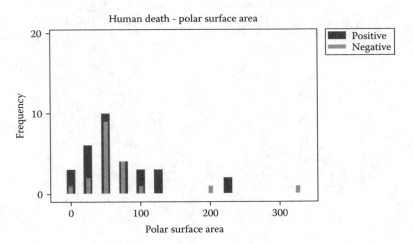

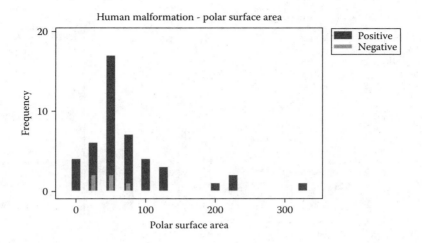

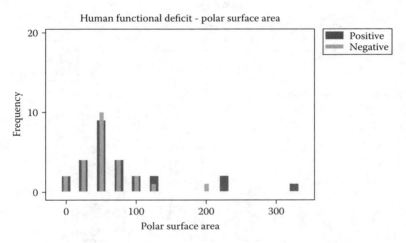

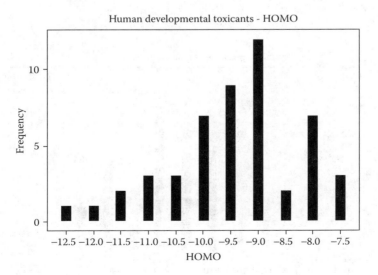

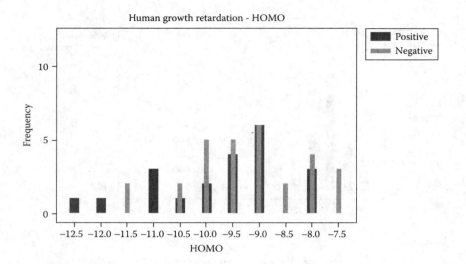

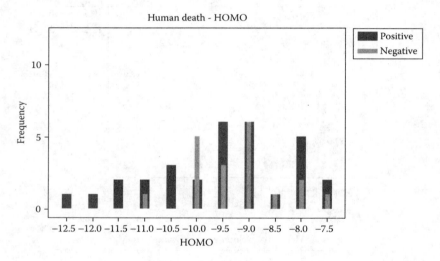

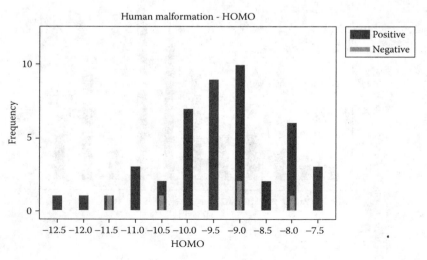

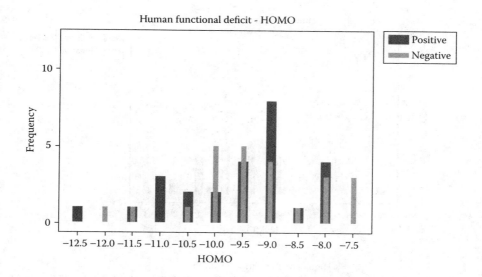

Human functional deficit - HOMO

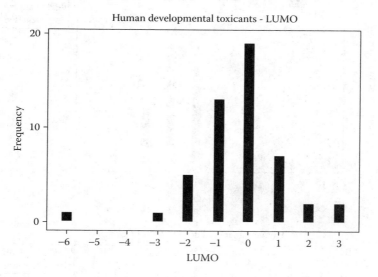

Human developmental toxicants - LUMO

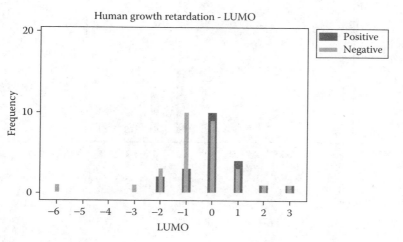

Human growth retardation - LUMO

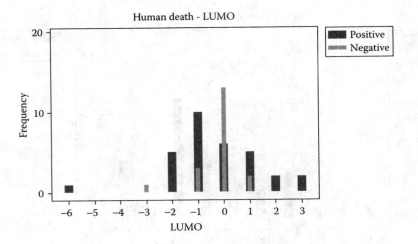

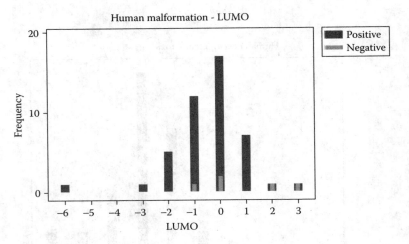

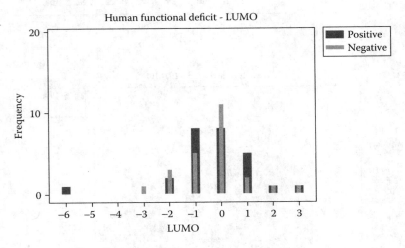

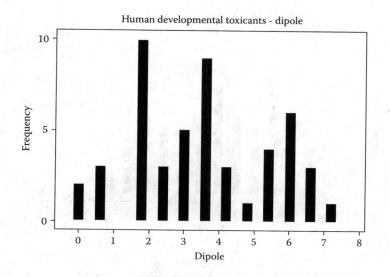

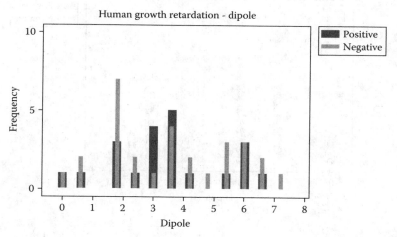

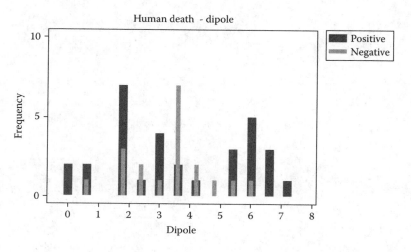

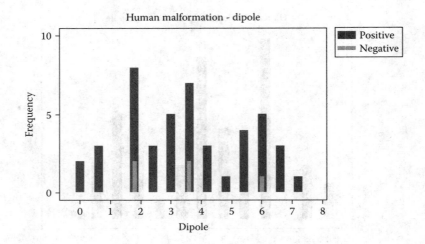

Human malformation - dipole

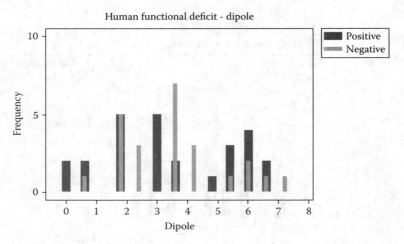

Human functional deficit - dipole

Appendix II
Topological Parameter Histograms

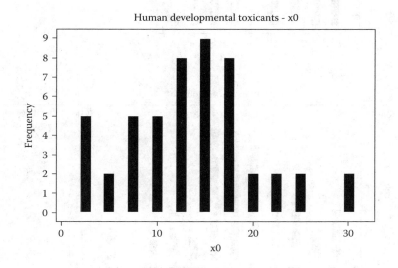

Human developmental toxicants - x0

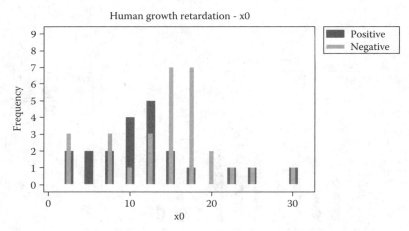

Human growth retardation - x0

Positive
Negative

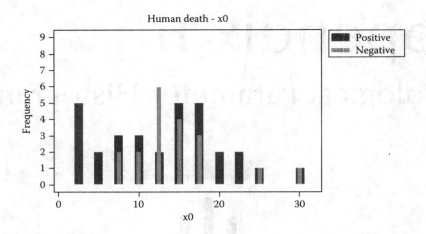

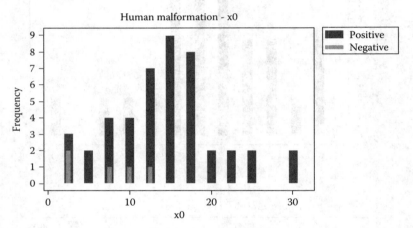

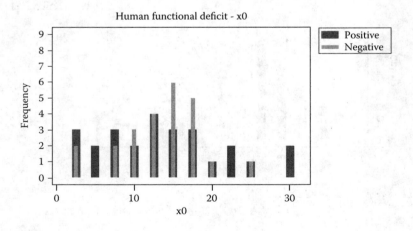

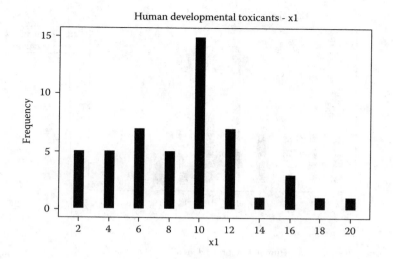

Human developmental toxicants - x1

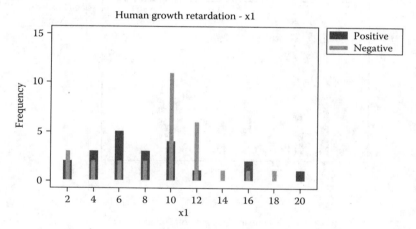

Human growth retardation - x1

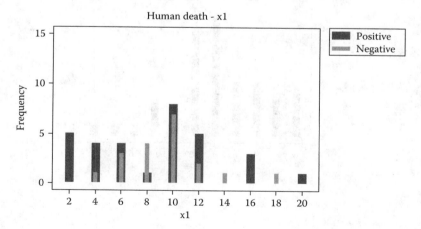

Human death - x1

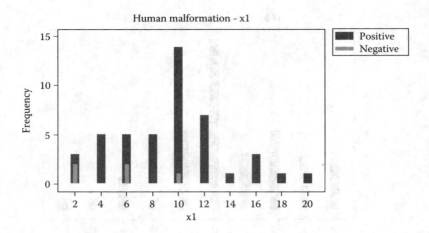

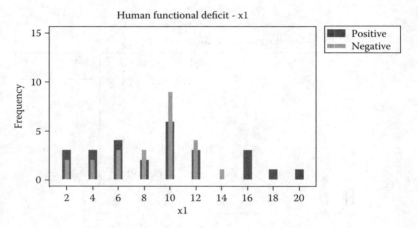

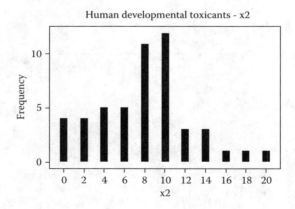

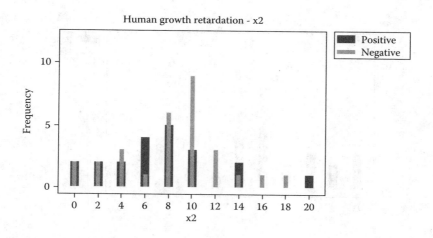

Human growth retardation - x2

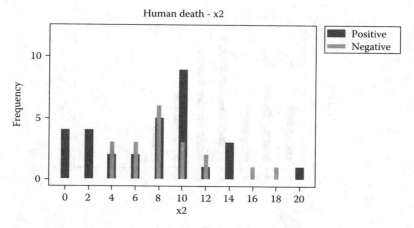

Human death - x2

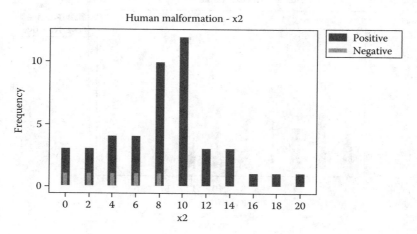

Human malformation - x2

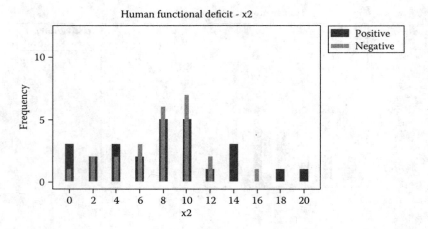

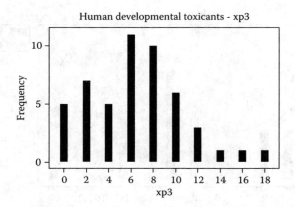

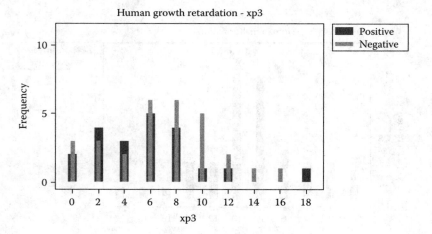

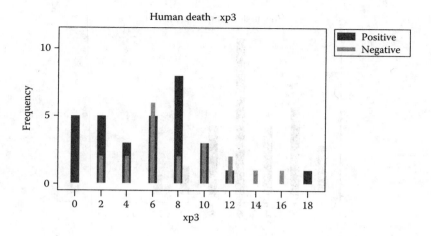

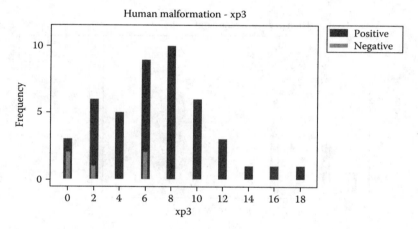

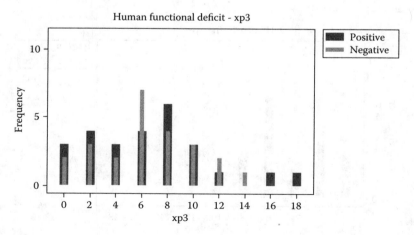

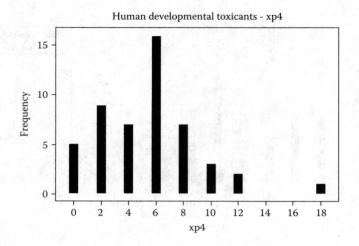

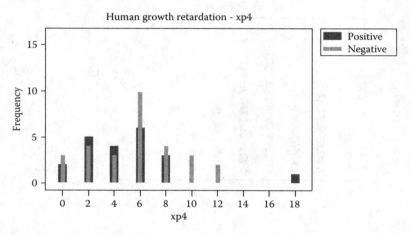

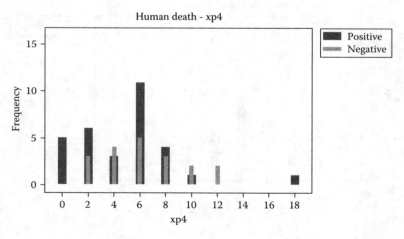

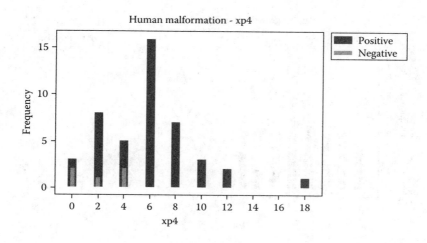

Human malformation - xp4

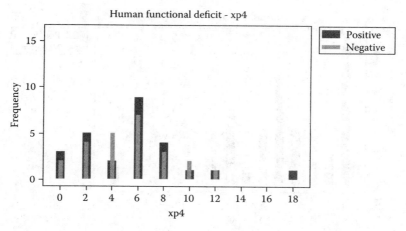

Human functional deficit - xp4

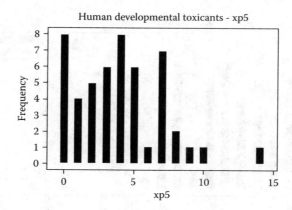

Human developmental toxicants - xp5

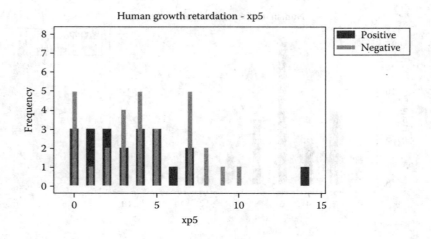

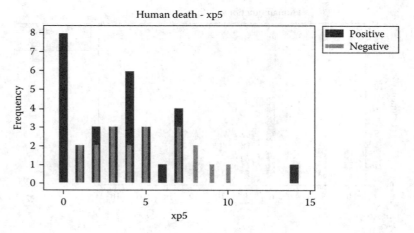

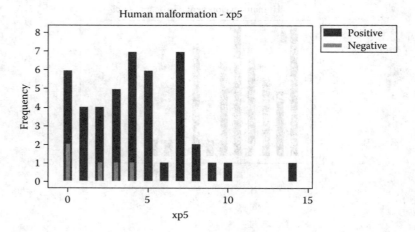

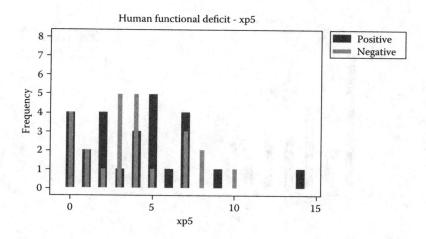

Human functional deficit - xp5

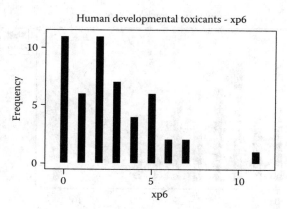

Human developmental toxicants - xp6

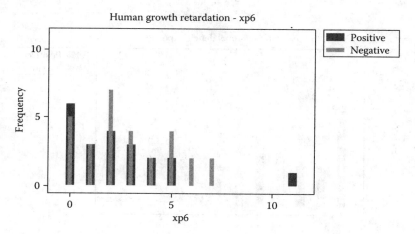

Human growth retardation - xp6

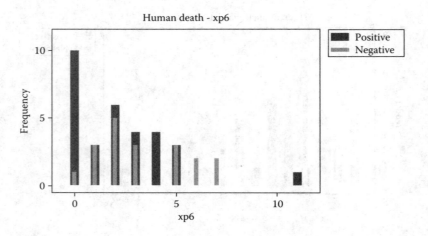

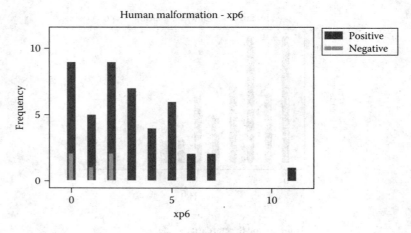

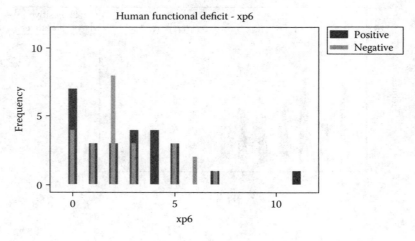

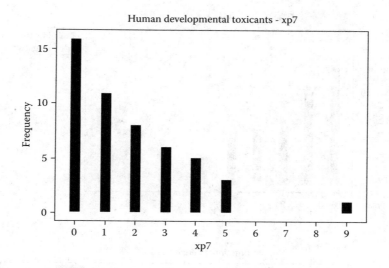

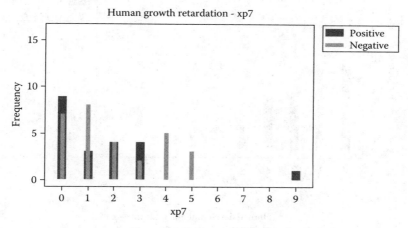

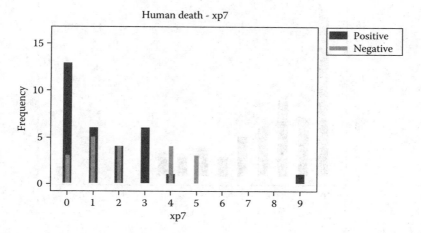

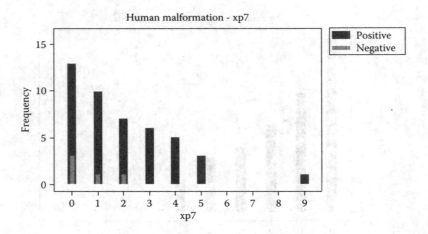

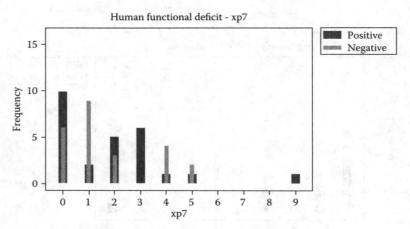

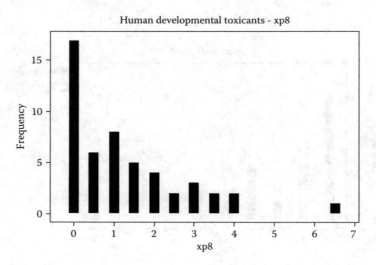

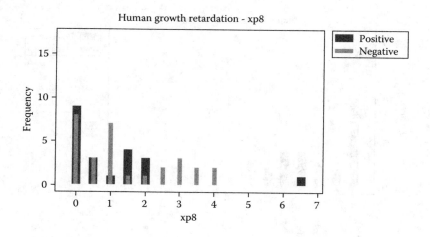

Human growth retardation - xp8

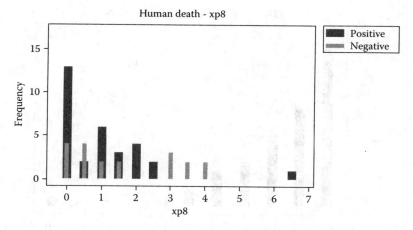

Human death - xp8

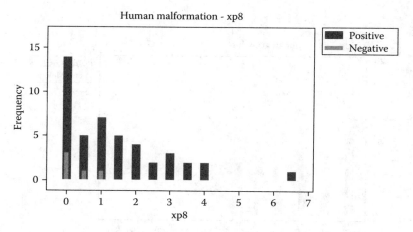

Human malformation - xp8

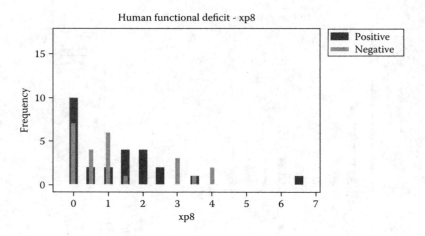

Human functional deficit - xp8

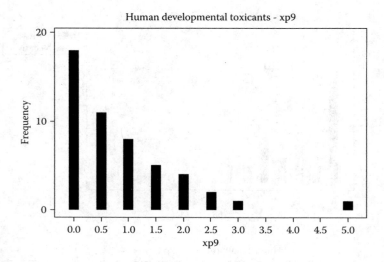

Human developmental toxicants - xp9

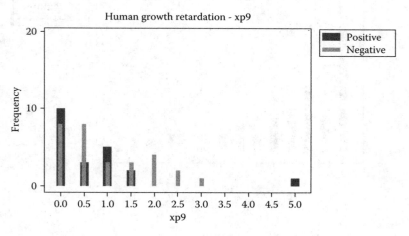

Human growth retardation - xp9

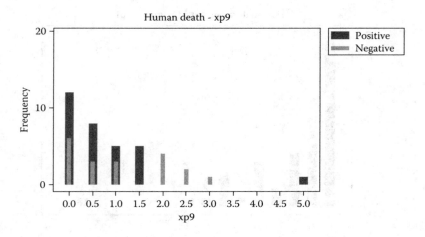

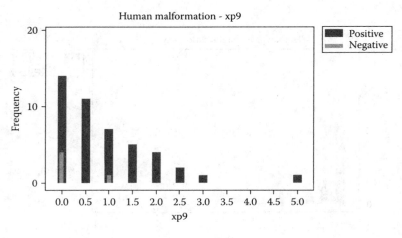

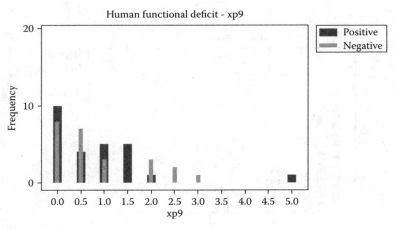

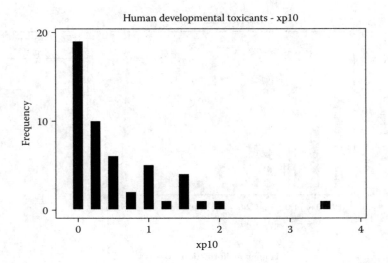

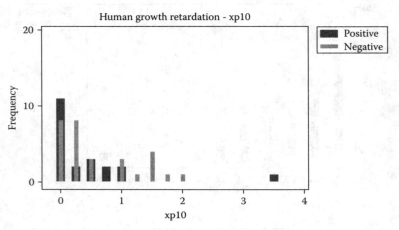

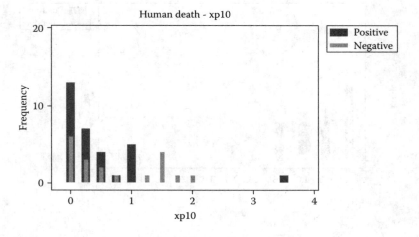

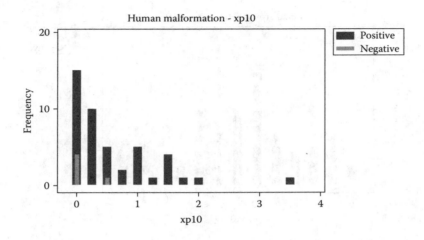

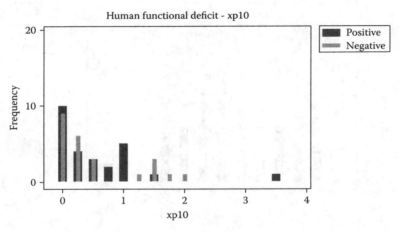

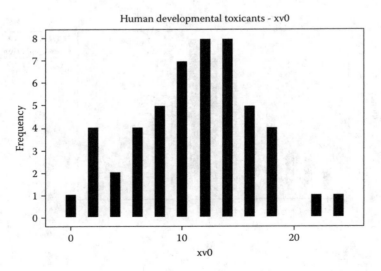

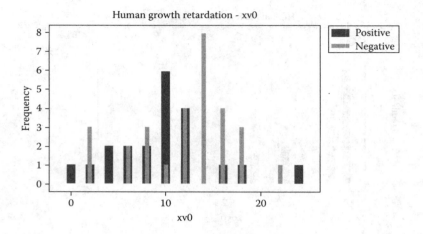

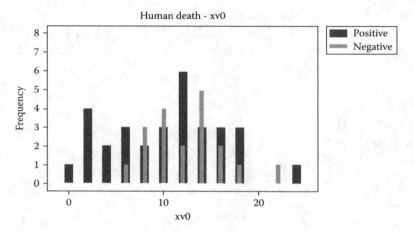

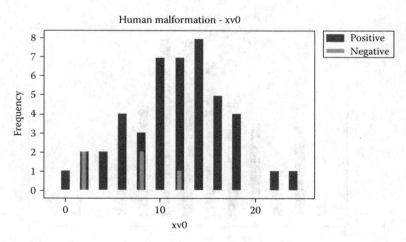

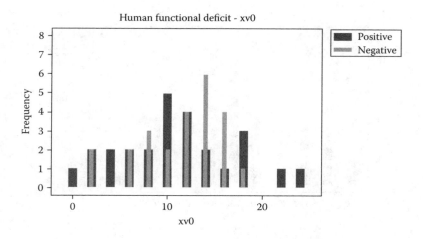

Human functional deficit - xv0

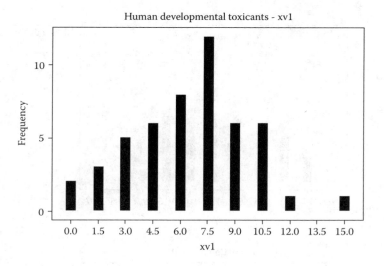

Human developmental toxicants - xv1

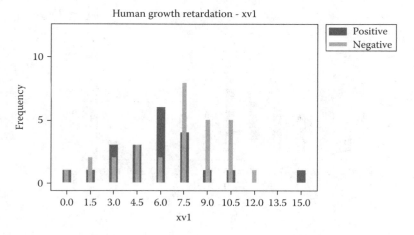

Human growth retardation - xv1

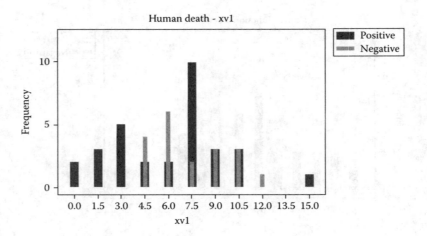

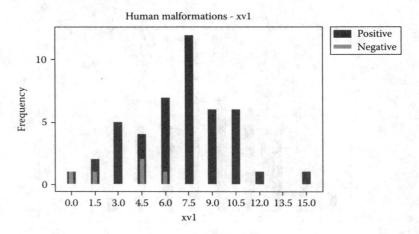

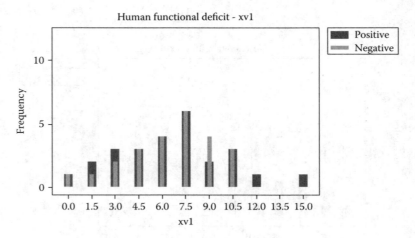

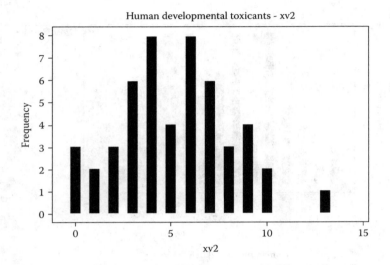

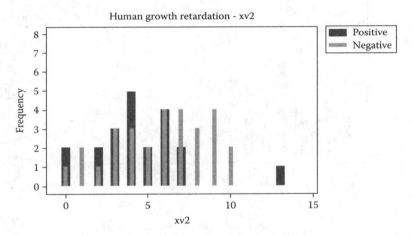

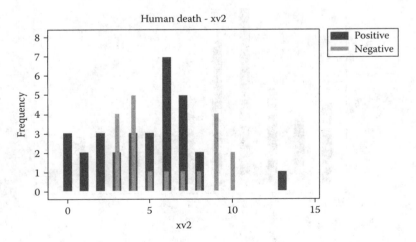

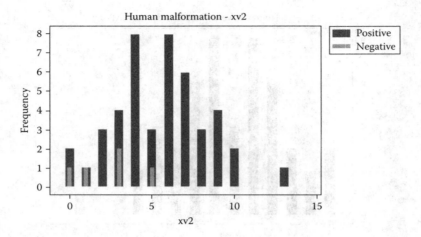

Human malformation - xv2

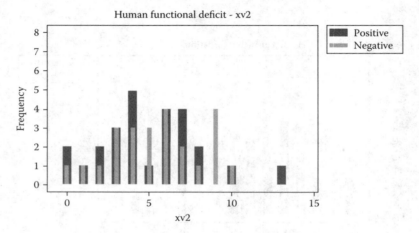

Human functional deficit - xv2

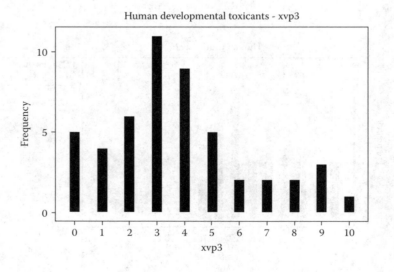

Human developmental toxicants - xvp3

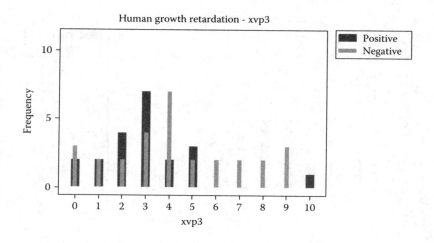

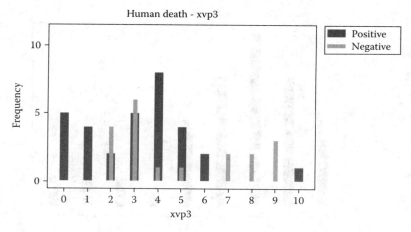

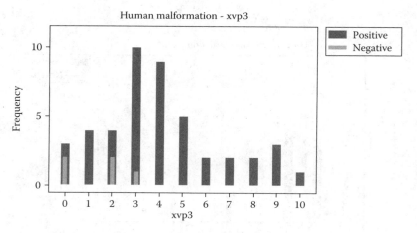

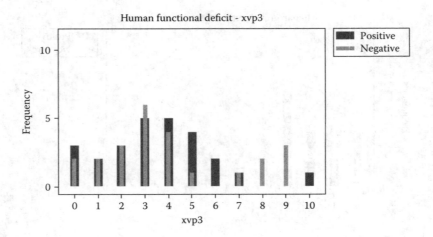

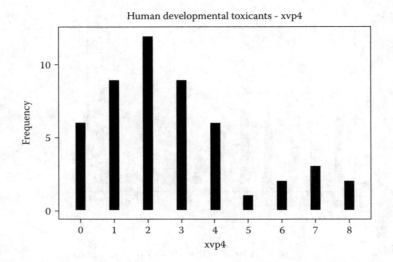

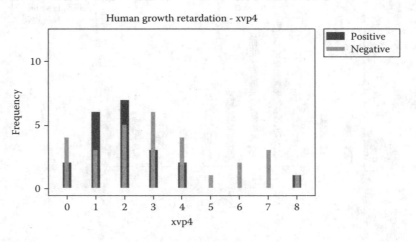

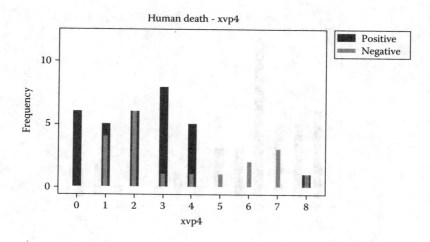

Human death - xvp4

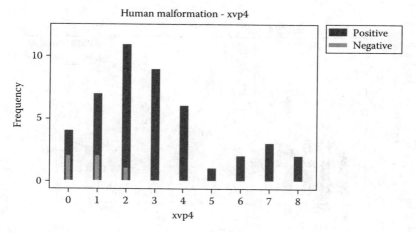

Human malformation - xvp4

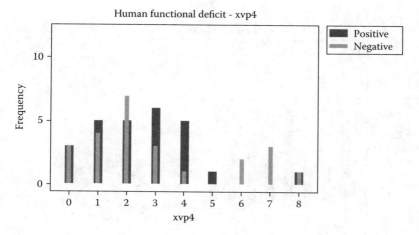

Human functional deficit - xvp4

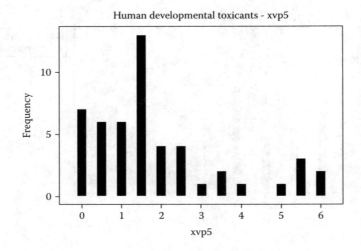

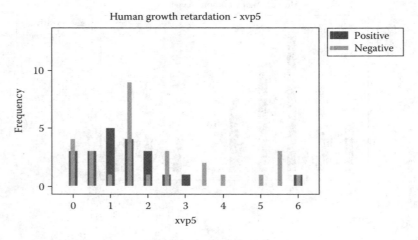

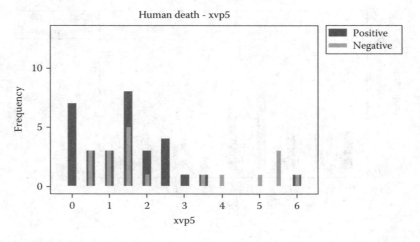

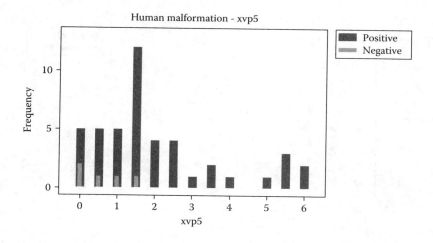

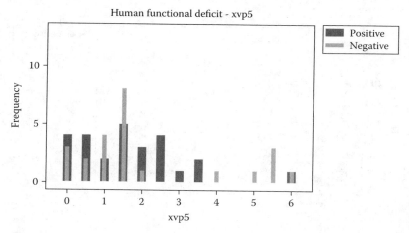

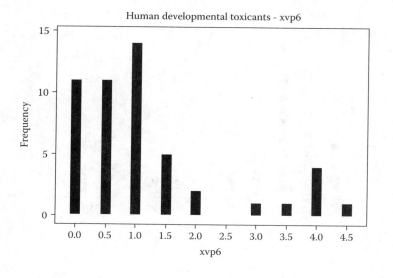

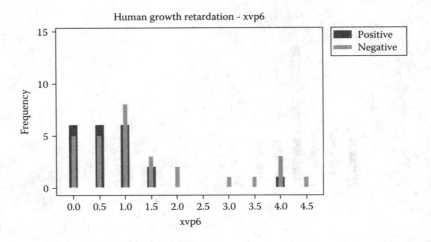

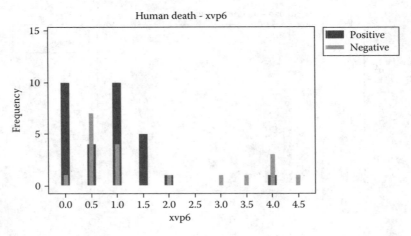

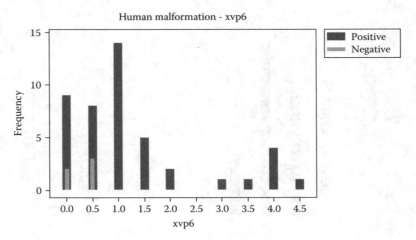

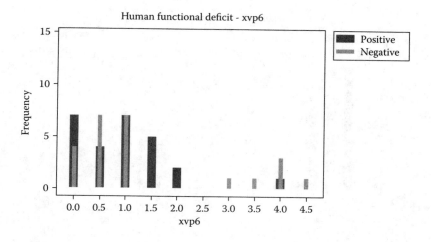

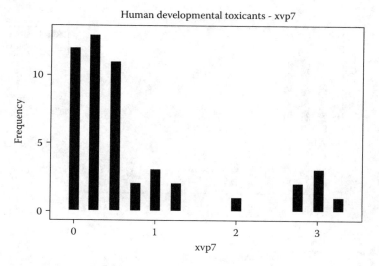

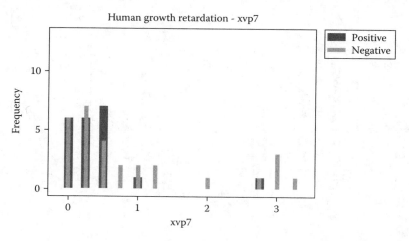

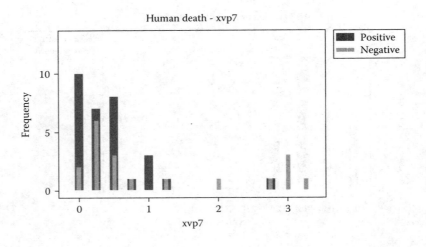

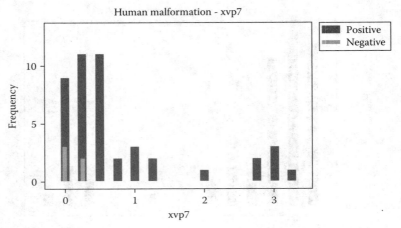

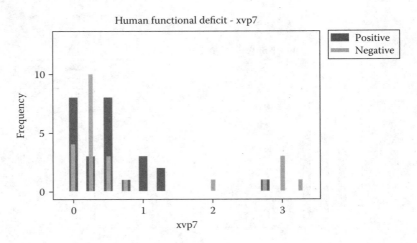

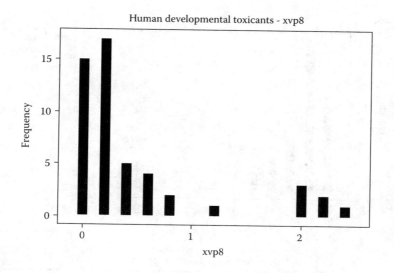

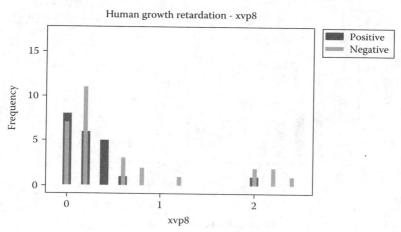

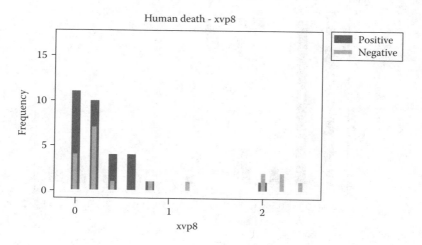

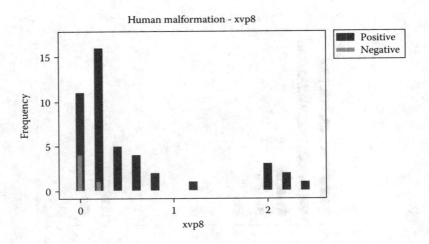

Human malformation - xvp8

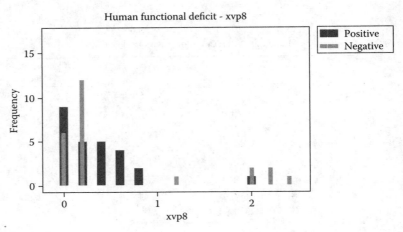

Human functional deficit - xvp8

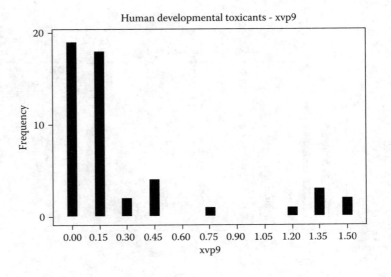

Human developmental toxicants - xvp9

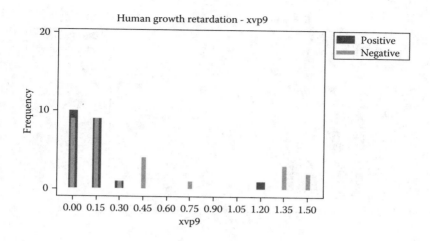

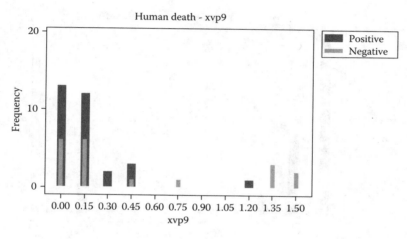

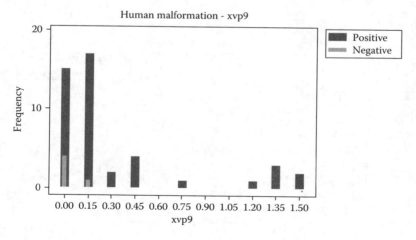

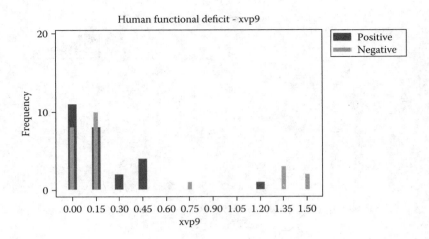

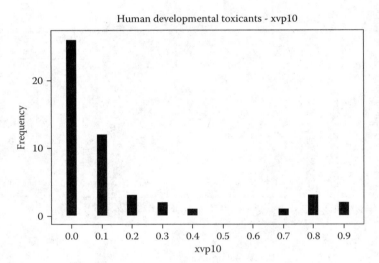

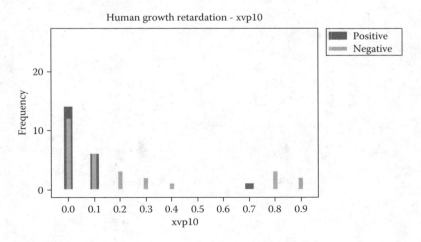

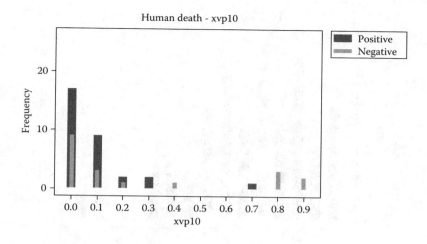

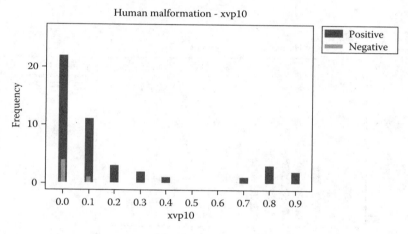

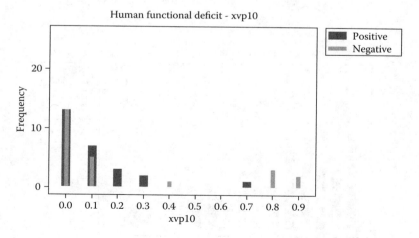

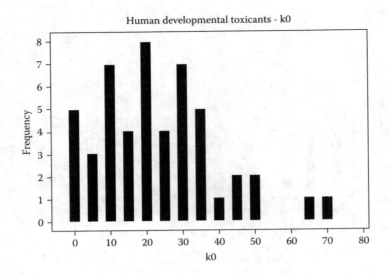

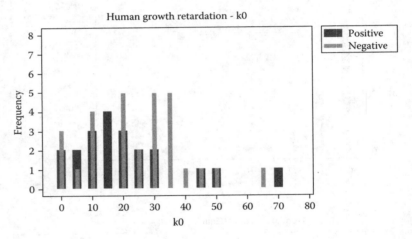

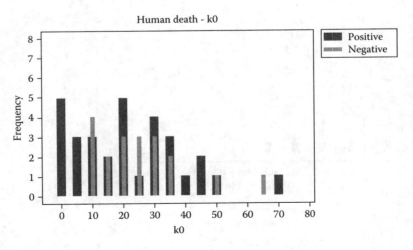

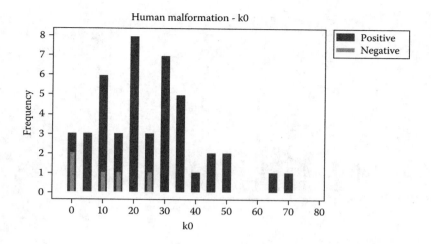

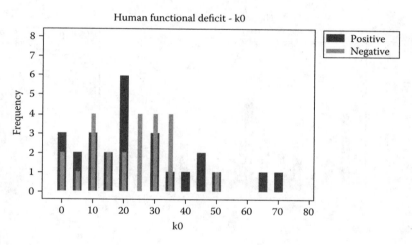

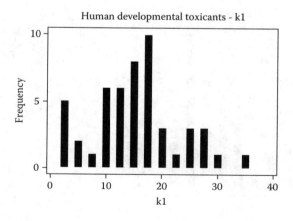

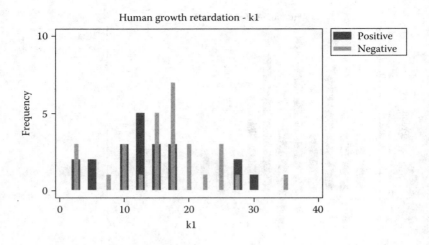

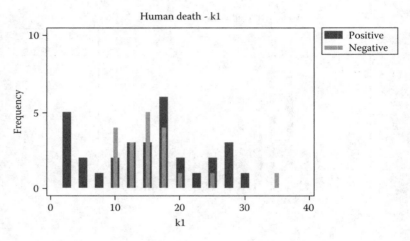

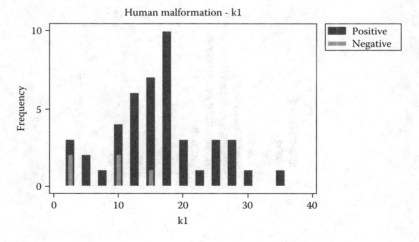

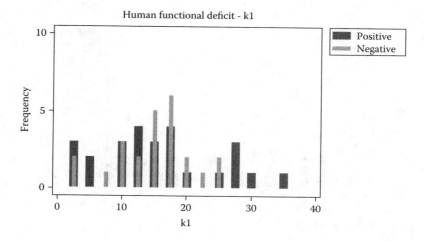

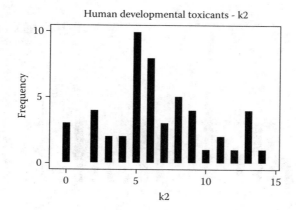

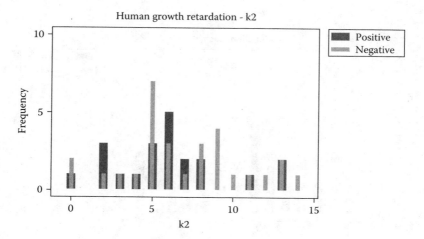

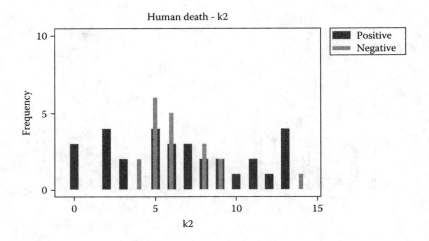

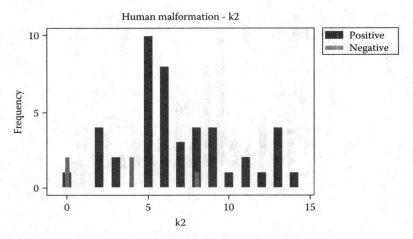

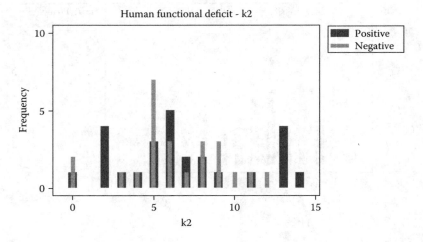

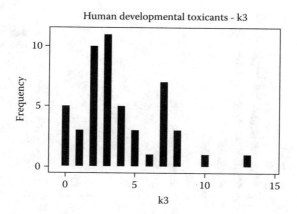

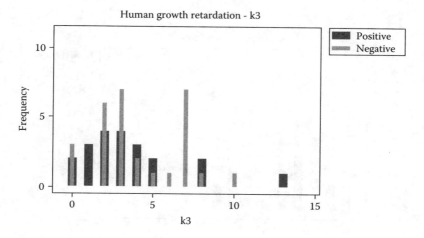

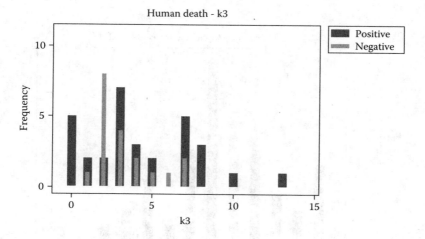

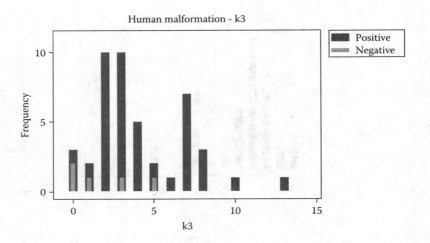

Human malformation - k3

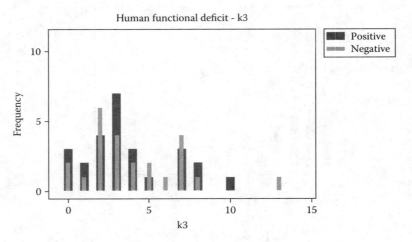

Human functional deficit - k3

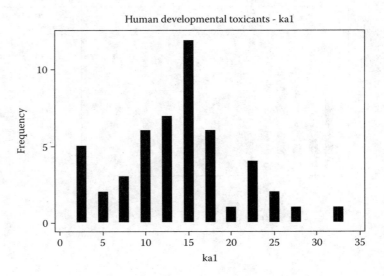

Human developmental toxicants - ka1

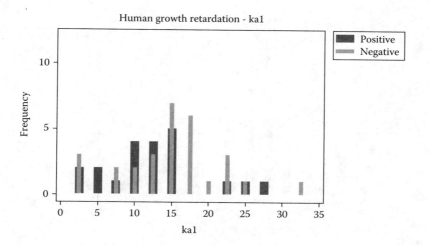

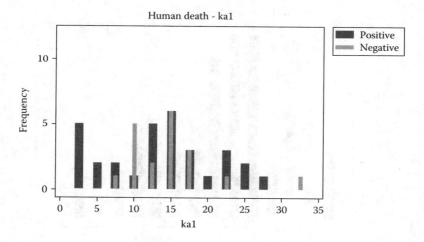

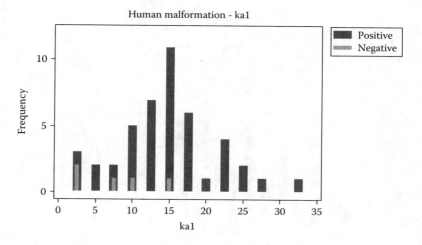

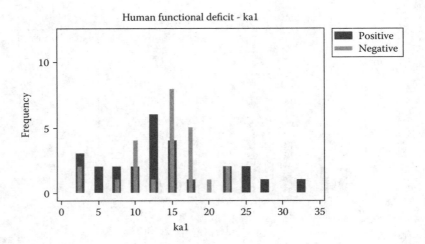

Human functional deficit - ka1

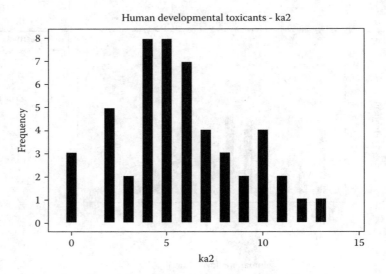

Human developmental toxicants - ka2

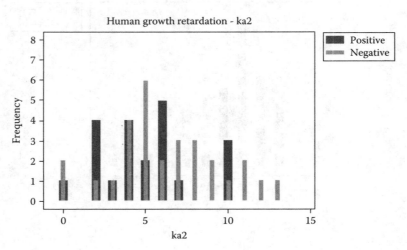

Human growth retardation - ka2

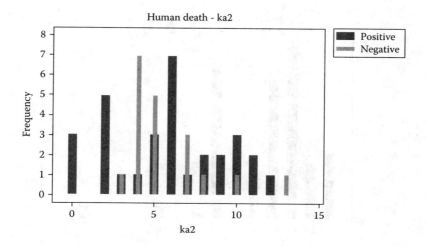

Human death - ka2

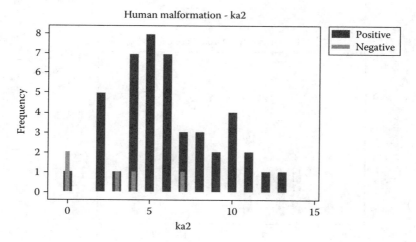

Human malformation - ka2

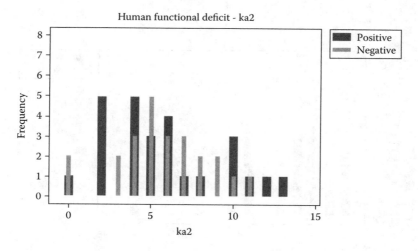

Human functional deficit - ka2

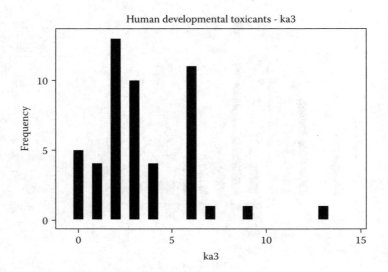

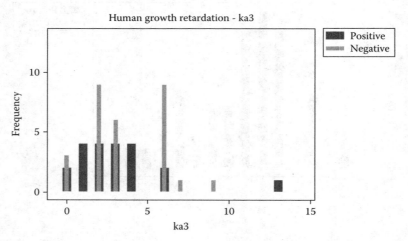

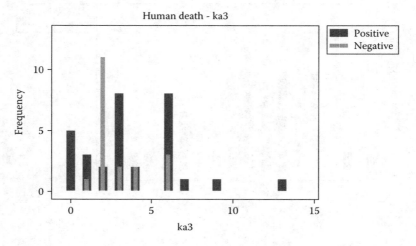

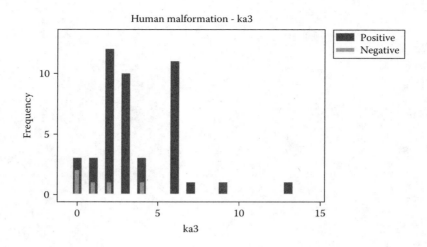

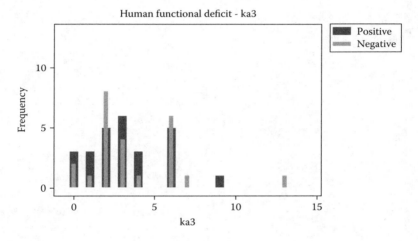

Appendix III
Description of Contents of Accompanying CD

Individual MDL MOL files of three-dimensional chemical structures
MDL SDF file with three-dimensional structures, biological activities (human/animal),
dosage/route/timing, calculated physicochemical/topological properties, SMILES
codes, CAS registry numbers
Excel XLS file identical to the SDF file without the three-dimensional chemical structures
MDL MOL/SDF files can be viewed by appropriate computational chemistry software (e.g.,
RasMol, Tripos, ChemOffice, etc.)

Index

A

ACE inhibitors, 87–91, 280
Acenocoumarol, 194
Acetaldehyde, 309
Acitretin, 273–277
    animal model, 274
    chemistry
        general properties, 273, 275
        physicochemical properties, 275
        topological properties, 276
    developmental toxicology
        in animals, 273, 274
        in humans, 274, 324, 325, 328, 329
        mechanism of toxicity, 275
    pregnancy category, 81, 273
    teratogenesis
        retinoic acid embryopathy, 274
        risk to, 275
    therapeutic use, 81, 273
Alcohol (*see* Ethanol)
Aminoglycosides, 99–103
Aminopterin, 1–5
    animal model, 3
    chemistry
        general properties, 1, 3
        physicochemical properties, 3
        topological properties, 4
    developmental toxicology
        in animals, 1
        in humans, 2,3, 323–325, 327, 329
            risk to teratogenesis, 3
    therapeutic use, 1, 17
Androgens, 63–66, 164, 213–216, 239, 270
Antibiotics, 99–103, 115–118, 209–212
Anticancer agents, 1–5, 7–10, 11–15, 17–21, 23–26, 27–30, 31–34
Anticonvulsants, 57–62, 67–70, 143–147, 203–208, 221–231, 243–254
Antithyroid agents, 105–110, 157–162

B

Barbiturates, 143–147, 203–209
Busulfan, 7–10
    chemistry
        general properties, 7, 8
        physicochemical properties, 8
        topological properties, 9
    developmental toxicology
        in animals, 7

        in humans, 7, 8, 324, 325, 327, 329
            risk to toxicity, 8
    pregnancy category, 7
    reproductive toxicology
        in animals, 7
        risk to toxicity in humans, 7
    therapeutic use, 7

C

Caffeine, 119–126
    chemistry
        general properties, 119, 123
        physicochemical properties, 123
        topological properties, 124
        uses, 119, 120, 153, 154
            content in products, 119, 120
    developmental toxicology
        in animals, 119, 120
        in humans, 121, 122, 323–325, 327, 329
            risk to toxicity, 121, 122
            risk to teratogenesis, 120–122
Candesartan, 280
Captopril, 87–91
    animal model, 88
    chemistry
        general properties, 87, 89
        physicochemical properties, 89
        topological properties, 89, 90
    developmental toxicology
        in animals, 87
        in humans, 88, 89, 323–325, 327, 329
            risk to teratogenesis, 89
            fetopathy, 88, 89
    pregnancy category, 87
    therapeutic use, 87
Carbamazepine, 57–62
    chemistry
        general properties, 57, 59
        physicochemical properties, 60
        topological properties, 60
    developmental toxicology
        in animals, 57
        in humans, 57–59, 324, 325, 327, 329
            risk to teratogenesis, 59, 144, 224
        mechanism of toxicity, 59
    pregnancy category, 57
    therapeutic use, 57
Carbimazole, 107
Carbon disulfide, 233–236
    chemistry
        general properties, 233, 234

W